TCP/IP Lean

Web Servers for Embedded Systems

Second Edition

Jeremy Bentham

CMP**Books**

San Francisco, CA • New York, NY • Lawrence, KS

CMP Books
CMP Media LLC
1601 West 23rd Street, Suite 200
Lawrence, Kansas 66046
USA
www.cmpbooks.com

Acquisitions Editor:	Robert Ward
Managing Editor:	Michelle O'Neal
Editor:	Rita Sooby
Layout production:	Kris Peaslee
Cover art:	Robert Ward
Cover design:	Damien Castaneda

Distributed in the U.S. and Canada by:
Publishers Group West
1700 Fourth Street
Berkeley, California 94710
1-800-788-3123
www.pgw.com

ISBN: 1-57820-108-X

Acc. No:	129662
Class:	004.62 BEN
Price:	€ 66.34

Printed in the United States of America
03 04 05 06 07   5 4 3 2

CMP Books

To Fred, Ilse, and Jane

Table of Contents

Chapter 11 PWEB: Miniature Web Server for the PICmicro® **331**

Chapter 12 ChipWeb — Miniature Ethernet Web Server. **369**

Chapter 13 Point-to-Point Protocol: PPP **411**

Preface

The Lean Plan

This is a hands-on book about TCP/IP (transmission control protocol/Internet protocol) networking. You can browse it to get an overview of the subject or study a particular section in detail, but, to get maximum benefit, I suggest you set up your own network and try out the software for real.

Not so long ago, I would have given you a detailed description of a computer network called the Internet and how it allowed academics to pass information between their computers using the TCP/IP protocol family. Now the Internet encroaches on all aspects of our lives, so an introduction to it seems totally unnecessary. Yet a hands-on introduction to TCP/IP seems highly necessary, because the very size of the Internet presents a massive barrier to those wishing to understand its inner workings.

My first attempt at implementing TCP was not a great success. I'd waded through the specifications and thought, "This isn't too bad," and waded through the few public domain sources I could find and thought, "This is horrendously complicated"; then I wrote my own implementation. When I came to test it, the problems started in earnest. I couldn't find a sensible set of software tools for testing; whenever I found a problem, I wasn't sure whether the fault lay with the test software, the software under test, or my understanding of the specification.

What I needed was:

- **an implementation I could understand** — not a heavyweight implementation for a large, multiuser operating system, but a lightweight one that clearly showed the underlying principles — and

- **software tools I could use**, that is, test utilities that allowed me to check my understanding and implementation of the protocols.

As time went by and my TCP/IP software matured, the Web became increasingly important. My industrial customers would browse the Web at home or work and could see the advantages of using a Web browser for remote control and to monitor their industrial equipment. TCP became just a vehicle for conveying Web pages. The focus shifted from "I want TCP/IP on my system" to "I want my system to produce Web pages," and these pages always included dynamic, real-time data.

History was repeating itself; the software to produce these dynamic Web pages was designed for large, multiuser systems, and I couldn't find small-scale implementations that were usable on simple, low-cost embedded systems hardware. I needed:

- **a description of the techniques** to insert live data into Web pages and
- **some simple platform-independent code** that I could adapt for specific projects.

Having implemented many small-scale Web servers of my own (generally an 80188 processor with 64Kb of ROM), I was delighted to hear of a 256-byte implementation on a microcontroller, although I was disappointed to discover that it could only produce fixed pages from its ROM, with no dynamic data. I wanted to know:

- **what compromises** were associated with implementing TCP and a Web server on a microcontroller and
- **what techniques** I could use to insert dynamic data into its Web pages.

Almost by chance, the first edition of this book included a miniature Web server running on a PICmicro®[1]. I wasn t the first to create such a server, but I was the first to publish a full description of the techniques used, including full source code. The success of the initial offering prompted me to update this book to broaden the range of networks and protocols supported on the PICmicro. Despite the Web servers in the title of this book, there are many ways to transfer data across a network, and I wanted to provide working examples of their use.

Hopefully, you'll find the answers you want in this book.

Embedded Systems

The term "embedded system" may be new to some of you and require some explanation, even though you use embedded systems every day of your life. Microwave ovens, TVs, cars, elevators, and aircraft are all controlled by computers, which don't necessarily have a screen, keyboard, and hard disk. A computer could be controlling your car without your knowledge: an engine-management system takes an input signal from the accelerator and provides outputs that control the engine.

These computers are embedded in a system, of which they may be only a small component. The embedded-system designer may have to work within tight constraints of size, weight, power consumption, vibration, humidity, electrical interference, and above all, cost and reliability. The PC architecture has been adapted for embedded-systems operation, and rugged, single-board computers (SBCs) are available from a wide variety of suppliers,

---

1.   PICmicro® is the registered trademark of Microchip Technology Inc.

together with the necessary add-on cards to process real-world signals. The ultimate in miniaturization is the microcontroller, which is a complete computer on a single chip, including all the necessary I/O interfaces.

Regardless of the user interface, most embedded systems have an external interface for status monitoring and system diagnosis. Traditionally, this has been in the form of a serial terminal, but industry is starting to see the advantages of remote diagnosis: because Web-browser usage is so widespread, it seems the logical choice for a user interface. The browser is technically a Web client, which implies that the embedded system must be a Web server; hence, the title of this book.

Whether you are an embedded-systems developer or not, I trust you will find plenty of interest in this book. I'll look at:

- what software components are needed,
- how these components work,
- clear, simple implementation, and
- effective test strategies.

The qualities of simplicity and clarity have much to recommend them. Modern programming toolkits are very useful because they can simplify a complex programming task so it becomes a join-the-dots exercise, but the resulting bloated code may require much more complex hardware than the slim-line code of your competitor; hence, the Lean Plan.

The Hardware

At the time of writing, the PC hardware platform, although distinctly showing its age, cannot be ignored. The second-hand market is awash with perfectly serviceable PCs that don't contain the latest and fastest technology but are more than adequate for your purposes. There are low-cost industrial SBCs that have a PC core, standard network interface, and the ability to accept interface cards for a wide variety of real-world signals.

My software will run on all these PC compatibles, and even on PC incompatibles (such as the 80188 CPU) with a very small amount of modification, because I have clearly isolated all hardware and operating-system dependencies.

In addition to the PC code, I have included a miniature TCP/IP stack and Web server for a Microchip PICmicro® microcontroller, using the Custom Computer Services PCM C compiler. A standard PICmicro evaluation board can be hand-modified to include the appropriate peripherals (a circuit diagram is given), or a complete, off-the-shelf board can be purchased instead. I won't pretend that it would be easy to adapt this software to another processor, but there is an in-depth analysis of the difficulties associated with microcontroller implementations, which would give you a very significant head-start if working with a different CPU.

The Network

Base-level Ethernet (10Mbit) is still widely available; complete kits, including interface cards and cabling, are available at low cost from computer retailers. My software directly supports two of the most popular Ethernet cards — Novell NE2000 compatibles and 3COM 3C509 — and can potentially (if using the Borland Compiler) support other cards through the packet driver interface, though the direct-hardware-interface approach is preferable because it makes experimentation and debugging much easier.

When developing network software, you are very strongly advised to use a separate scratch network, completely isolated from all other networks in the building. Not only does debugging become much easier, but you also avoid the possibility of disrupting other network traffic. It is remarkable how a minor change to the software can result in a massive increase in the network traffic and a significant disruption to other network users. You have been warned!

The software also supports serial links through SLIP (serial line Internet protocol), and a crossover serial cable between two PCs can, to a certain extent, be used as a substitute for a real network.

The Operating System

You may be surprised by the extent to which I ignore the operating system. In the embedded-systems market, there is always pressure to simplify the hardware and reduce the costs, and one way of achieving this is to use the simplest possible operating system, or none at all.

For those of you wedded to complex operating systems, and even more complex software-development environments, this will initially be an uncomfortable experience, because you are exposed to the harsh reality of real, bare-metal programming. However, I hope that you will soon come to appreciate the power, flexibility, and pure simplicity of this approach and gradually come to the realization that, for many common or garden-variety applications, an operating system (even a free operating system) is an expensive luxury. Luxury or not, I want to use my desktop PC for development, so the software is compatible with Windows 95 and 98, either in DOS, extended DOS, or Win32 console application mode.

My primary development system is a Windows 95 machine equipped with two network cards — only one of which is installed in the operating system. This is extremely useful because a single machine can simultaneously act as both network client (using a standard Web browser) and server (using my Web server), making experimentation much easier.

The final target machine can be a relatively humble SBC running DOS or a microcontroller compatible with PC code without an operating system, although the latter would entail some minor changes to the software provided.

The Development Environment

The following four PC compilers are supported:

Borland C++ v3.1. An excellent DOS-hosted compiler with an integrated development environment.

Borland (Inprise) C++ v4.52. Windows-hosted compiler, which seems to be the latest version that can generate executable files for DOS.

Microsoft Visual C++ v6. Windows-hosted compiler that can generate Win32 console applications.

DJGPP v2.02 with RHIDE v1.4. Part of the GNU project, this is a remarkably good clone of the Borland 3.1 development environment, which runs in a 32-bit extended DOS environment and can be downloaded free of charge.

The Borland compilers, though ostensibly obsolete, may be found on the CD-ROM of some C programming tutorial books or may be bundled with their 32-bit cousins. The high-level software can be compiled using all of these environments, but I have not been so fortunate with the low-level network interface code.

- The Borland compilers are the easiest to use because they allow the use of interrupts without the need for machine code inserts and so can support the full range of network interfaces.
- With the Microsoft compiler, the network card and SLIP interfaces are supported, but the packet driver interface is not.
- Only the direct-network-card interface is supported when using the DJGPP compiler.

Because the direct-network-card interface is the easiest to debug, and hence more suitable for experimentation, this restriction isn't as onerous as it might appear.

If your favorite compiler isn't on the list, I apologize for the omission, but I am very unlikely to add it. Each compiler represents a very significant amount of testing, and my preference is to reduce, rather than increase, the number of compilers supported. If your compiler is similar to the above (for example, an earlier version), then you should have little or no adaptation work to perform, though I can't comment on any compiler I haven't tried.

PICmicro Compilers.   The early software used the Custom Computer Services (CCS) PCM v2.693, but later developments are broadly compatible with the CCS and Hitech compilers for the PIC16xxx and PIC18xxx series microcontrollers. A detailed discussion of compatibility issues is beyond the scope of this chapter. See Appendix D and the software release notes on the CD-ROM for more information.

The Software

The enclosed CD-ROM contains complete source code to everything in this book so that you, as purchaser of the book, can experiment. However, the author retains full copyright to the software, and it may only be distributed in conjunction with the book; for example, you may not post any of the source code on the Internet or misrepresent its authorship by extracting fragments or altering the copyright notices.

If you want to sell anything that contains this software, a license is required for the "incorporation" of the software into each commercial product. This normally takes the form of a one-off payment that allows unlimited incorporation of any executable code derived from this source. There are no additional development fees (apart from purchase of the book), and license fees are kept low to encourage commercial usage. Full details and software updates are on the Iosoft Ltd. Web site at www.iosoft.co.uk.

Acknowledgments

The author owes a profound debt of gratitude to Berney Williams of CMP Books for being so keen on this project, Anthony Winter for his proofreading skills and advice, Glen Middleton of Arcom Control Systems Ltd. and Adrian Nicol of Io Ltd. for their help with the hardware, and, above all, to Jane McSweeney (now Jane Bentham) for her continued enthusiasm, support, and wonderful cakes.

Chapter 1

Introduction

The Lean Plan

This is a software book, so it contains a lot of code, most of which has been specially written (or specially adapted) for the book. The software isn't a museum piece, to be studied in a glass case, but rather a construction kit, to promote understanding through experimentation. The text is interspersed with source-code fragments that illustrate the points being discussed and provide working examples of theoretical concepts. All the source-code in the book and complete project configurations for various compilers are on the enclosed CD-ROM.

When I started writing this book, I intended to concentrate on the protocol aspects of embedded Web servers, but I came to realize that the techniques of providing dynamic content (on-the-fly Web page generation) and client/server data transfers were equally important, yet relatively unexplored. Here are some reasons for studying this book:

TCP/IP.   You want to understand the inner workings of TCP/IP and need some tools and utilities to experiment with.

Dynamic Web Content.   You have an embedded TCP/IP stack and need to insert dynamic data into the Web pages.

Miniaturization. You are interested in incorporating a miniature Web server in your system but need to understand what resources are required and what compromises will have to be made.

Prototyping. You want a prebuilt Web server that you can customize to evaluate the concept in a proposed application.

Data transfer. You need to transfer data across a network using standard protocols.

Client/server programming. You have to interface to standard TCP/IP applications, such as email servers.

Of course, these areas are not mutually exclusive, but I do understand that you may not want to read this book in a strict linear order. As far as possible, each chapter stands on its own and provides a stand-alone utility that allows you to experiment with the concepts discussed.

I won't assume any prior experience with network protocols, just a working knowledge of the C programming language. In the Preface, I detailed the hardware and software you would need to take full advantage of the source code in the book. You don't have to treat this book as a hands-on software-development exercise, but it would help your understanding if you did.

Getting Started

On the CD-ROM, you'll find the directory `tcplean` with several subdirectories.

`BC31` compiler-specific files for Borland C++ v3.1
`BC45` compiler-specific files for Borland C++ v4.52
`DJGPP` compiler-specific files for (GNU) DJGPP and RHIDE
`PCM` the PICmicro®-specific[1] files for Chapters 9–11
`ROMDOCS` sample documents for the PICmicro Web server
`SOURCE` all source code for PC systems
`VC6` compiler-specific files for Microsoft Visual C++ v6
`WEBDOCS` sample documents for the PC Web server

You ll also find the directory `chipweb` with two subdirectories containing the files for Chapters 12–16.

`ARCHIVE` zip files containing older versions of the ChipWeb source code
`P16WEB` latest ChipWeb source code

Executable copies of all the utilities, sample configuration files, and a `README` file with any late-breaking update information are in `tcplean`. Preferably, the complete directory tree `d:\tcplean` (where `d:` is the CD-ROM drive) should be copied to `c:\tcplean` on your hard disk,

---

1. PICmicro® is the registered trademark of Microchip Technology Inc.; PICDEM.net™ is the trademark of Microchip Technology Inc.

and d:\chipweb to c:\chipweb. If a different directory path is used, it will be necessary to edit the compiler project files.

The utilities read a configuration file to identify the network parameters and hardware configuration; the default is tcplean.cfg, read from the current working directory. It is unlikely that this will contain the correct hardware configuration for your system, so *it is important that you change the configuration file before running any of the utilities*. See Appendix A for details. If you attempt to use my default configuration without checking its suitability, it may conflict with your current operating system settings and cause a lockup.

It is possible to browse the source files on the CD-ROM and execute the utilities on it without loading them onto your hard disk, though you still need to adapt the configuration file and store it in the current working directory.

```
c:\>cd tcplean
c:\tcplean>d:\tcplean\ping 10.1.1.1
```

This would execute the utility on the CD-ROM using the configuration file.

```
c:\tcplean\tcplean.cfg
```

The default configuration file may be overridden using the -c command-line option.

```
c:\tcplean>ping -c slip 172.16.1.1
```

This uses the alternative configuration file slip.cfg, which makes it possible to experiment with multiple network configurations without having to rebuild the software each time.

If you are in any doubt about the command-line arguments for a utility, use the -? option.

```
c:\>cd tcplean
c:\tcplean>ping -?
```

Some of the utilities have the same name as their DOS counterparts (because they do the same job), so it is important to change to tcplean before attempting to run them.

A final word of warning: I strongly recommend that you create a new "scratch" network for your experimentation that is completely isolated from all other networks in the building. It is a very bad idea to experiment on a "live" network.

Network Configuration

The DOS software in this book supports the following network hardware.

Direct-drive network card   Novell NE2000-compatible or 3COM 3C509 Ethernet cards can be direct-driven by the software. This is the preferred option because of the ease of configuration and debugging.

Serial link   A serial line Internet protocol (SLIP) link between two PCs or a PC and the PIC-micro miniature Web server.

Packet driver   An otherwise unsupported network card may be used via a Crynwr packet driver supplied by the card manufacturer.

Some combinations of network hardware and compiler are not supported. Consult Appendix A and the README file for full information on the network-configuration options.

Compiler Configuration

Executable versions of all the DOS projects are included within the `tcplean` directory, so initial experimentation can take place without a compiler. The project files for each compiler reside in a separate directory, as described earlier, and all the compiler configuration information resides within the project files. All the source-code files reside in a single shared directory. There are a few instances where compiler-specific code (generally Win32-specific code) must be generated, in which case automatic conditional compilation is used.

Load specific projects for the following compilers:

Borland C++ v3.1   In a DOS box, change to the `BC31` directory and run BC using the project filename.

```
c:\>cd \tcplean\bc31
c:\tcplean\bc31>bc ping.prj
```

Borland C++ v4.52   Launch the Integrated Development Environment (IDE) and select Project–Open Project and the desired IDE file in the `BC45` directory.

DJGPP and RHIDE   Launch the RHIDE IDE and select Project–Open Project and the desired GPR file in the `DJGPP` directory.

Visual C++ v6   Launch the IDE and select File–Open Workspace and the desired DSW file in the `VC6` directory.

Custom Computer Services PCM   The PICmicro cross-compiler uses a completely different set of source code that resides in the PCM directory. Open a DOS box and change directory to `\tcplean\pcm`. Copy the necessary system files (`16C76.h` and `ctype.h`) into this directory from the standard PCM distribution. Run the PCM compiler, specifying `PWEB.C` on the command line.

```
c:\>cd \tcplean\pcm
c:\tcplean\pcm>copy \picc\examples\16c76.h
c:\tcplean\pcm>copy \picc\examples\ctype.h
c:\tcplean\pcm>\picc\pcm pweb.c
```

I run the PCM compiler from within the IDE of an emulator; see the emulator documentation for details on how to do this. When first using such a setup, make a minor but readily observable change, rebuild, and check that the new executable really has been downloaded into the emulator. It is all too easy to omit a vital step in the rebuild chain, such that the old file is still being executed.

Other PICmicro® Compilers

The software in Chapters 12–16 is broadly compatible with the later versions of the CCS and Hitech PICmicro compilers, for both the PIC16xxx and PIC18xxx series of devices. There are compatibility issues with some versions of these compilers; see Appendix D for guidance on compiler-specific issues, and always refer to the release notes (in file `readme.txt`) before using a specific ChipWeb release.

Software Introduction

For the rest of this chapter, I'll look at the low-level hardware and software functions needed to support software development.

- network hardware characteristics
- network device drivers
- process timing
- state machines
- buffering
- coding conventions

Even if you're keen to get on with the protocols, I suggest you at least skim this material, since it forms the groundwork for the later chapters.

Network Hardware

To help in setting up a serial or network link, I've included some sample configurations in Appendix A, together with the relevant software installations. Assuming one or both are installed, I will examine their characteristics with a view to producing the low-level hardware device drivers.

Figure 1.1   Serial link and network topologies.

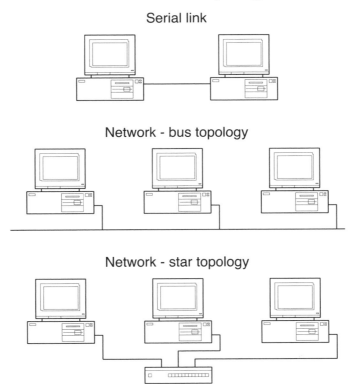

Figure 1.1 shows two types of networks (two "topologies"): the older-style bus network, where the computers are connected to a single common cable, and the newer star network, where the computers are individually connected to a common box (a hub), which electrically copies the network signals from one computer to all others. Fortunately, the operation of an Ethernet hub is completely transparent to the software, so you can still treat the network as if the computers were sharing a common cable.

Serial Hardware Characteristics

The simplest communication link between two PCs (A and B) consists of three wires: a ground connection, a wire from the A transmit to the B receive, and a wire from the B transmit to the A receive. A commercial serial crossover cable (often called a null modem or "Laplink" cable) generally has more wires connected so that the handshake signals are transferred, but you'll concentrate on the two data lines, which have the following characteristics:

Both computers have equal access to the serial link. The hardware simply acts as a "data pipe" between the two computers and does not prioritize one computer above another.

There are only two computers (nodes) on the network. Throughout this book, I'll use "node" as shorthand for "a computer on the network." Insofar as the simple serial link constitutes a network, it is clear that if one node transmits a message, it can only be received by the other node and no others.

A node can transmit data at any time. This is technically known as a full duplex system; both computers can transmit and receive simultaneously without any clash of data signals.

Message delivery is reliable. The assumption is that the two nodes are close to each other, with a short connecting cable, so there will be no corruption of data in transit. The predominant failure mode is a catastrophic link failure, such as a disconnection of the cable or a node powering down.

The serial data is a free-format stream of bytes, with little or no integrity checking. The serial hardware is only designed for short-distance interconnects, so it has a very simple error-checking scheme (parity bit), which is often disabled. To guarantee message integrity, error checking must be provided in software.

There is no limit on message size. Because the serial data is simply a stream of bytes with no predefined start or end, there is no physical restriction on its length.

There is no need for addressing. Because there is only one possible recipient for each message, there is no need to include an address identifying that recipient.

Network Hardware Characteristics

Whatever the actual topology, a base-level Ethernet network appears logically to be two or more computers transmitting and receiving on a single shared medium (cable).

All computers on the network have equal access to the network. This is called peer-to-peer networking, in which all nodes are equal. The alternative (master–slave networking) assumes that one or more special nodes control and regulate all network traffic; they are the masters, and their slaves only speak when spoken to. Master–slave operation is very useful for industrial data acquisition, where all data and control is to be funneled through a few large computer systems, but prohibits the kind of ad hoc communication that is required in an office or on the Internet.

All nodes have a 48-bit address that is unique on the network. Just as a postal address uniquely identifies a specific location in the world, so a node address (generally known as a media access and control, or MAC, address) must uniquely identify a node on the network. In fact, the standardization of Ethernet guarantees each node address to be also unique in the world; you can mix and match Ethernet adaptors from different manufacturers, secure in the knowledge that no two will have the same 48-bit address.

Any node may transmit on the network when it is idle. If a node is to communicate with another, it must wait for all others to be silent before it can transmit. Because all nodes are equal, they need not ask permission before transmitting on the network; they simply wait for a suitable gap in the network traffic.

Message delivery is unreliable. "Unreliable? Why don't you fix it?" Networks are, by their very nature, an unreliable way of sending data. The failure modes range from the catastrophic (the recipient's computer is powered down or physically disconnected from the network) to the intermittent (a packet has been corrupted by collision or electrical interference). The network hardware has the ability to detect and compensate for some intermittent faults (e.g., a retry in the event of a packet collision), but eventually an error will occur that has to be handled in software, so the software must assume the network is unreliable.

All data on the network is in blocks (frames) with a defined beginning and end and an integrity check. Nodes that are going to transmit when they want need a defined format for their transmissions so that others know when they are starting or finishing, assuming each transmission is a block with start and end markers and some form of checking (usually a CRC, or cyclic redundancy check) to ensure it hasn't been damaged in transit. The name given to this block differs according to the network used; Ethernet blocks are called frames.

The network can send a maximum of 1,500 bytes of data per frame. All networks have an upper limit on the size of data they can carry in one frame. This is called the maximum transfer unit, or MTU. Ethernet frames can contain up to 1.5KB, but TCP/IP software will work satisfactorily with a lot smaller MTU; the specifications state that an IP datagram size of at least 576 bytes must be supported.

All messages are equipped with a source and destination address. Frames are usually intended for a single recipient; this is known as unicast transmission. Occasionally, it may be necessary to send a frame to all nodes on the network, which is a broadcast transmission.

Device Drivers

It would be helpful if the driver software presented a common interface to the higher-level code, but it is clear from the preceding analysis that there are significant differences; these are summarized in Table 1.1.

Serial Driver Requirements

TCP/IP assumes the network data is sent in blocks, with a defined beginning and end, so the serial drivers must convert the free-format serial byte stream into well-defined blocks.

Table 1.1   RS232 serial versus Ethernet.

	RS232 Serial	**Ethernet**
Access	Equal	Equal
Address range	None	48-bit
Transmit	Any time	When network is idle
Delivery	Reliable	Unreliable
Format	None (data stream)	Frame
Data length	Unlimited	1.5Kb per frame
Addressing	None	Source, destination, broadcast

SLIP

Fortunately, one of the TCP/IP families of standards, SLIP, provides exactly this functionality. It uses simple escape codes inserted in the serial data stream to signal block boundaries as follows:

- The end of each block is signaled by a special End byte, with a value of C0h.
- If a data byte equals C0h, two bytes with the values DB, DC are sent instead.
- If a data byte equals DBh, two bytes with the values DB, DD are sent instead.

Additionally, most implementations send the End byte at the beginning of each block to clear out garbage characters prior to starting the new message (Figure 1.2).

Figure 1.2   SLIP frame.

END C0h	Data 1-1006 bytes	END C0h

There is effectively no limit to the size of the data block, but you have to decide on some value in order to dimension the data buffers. With old, slow serial links, a maximum size of 256 bytes was generally used, but you'll be using faster links, and a larger size is better for minimizing protocol overhead. By convention, 1,006 bytes is often used.

The encoding method can best be illustrated by an example (Figure 1.3). Assume a six-byte block of data with the hex values BF C0 C1 DB DC is sent; it is expanded to C0 BF DB DC C1 DB DD DC C0.

Figure 1.3   SLIP example.

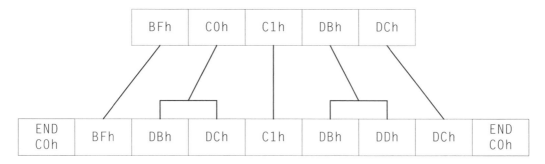

The original data has nearly doubled in size, due to my deliberately awkward choice of data values. In normal data streams, the overhead is much lower.

Modem Emulation

An additional problem with serial networking is that most PCs are configured to use a modem (Figure 1.4) to an Internet Service Provider (ISP).

Figure 1.4   Modem communication.

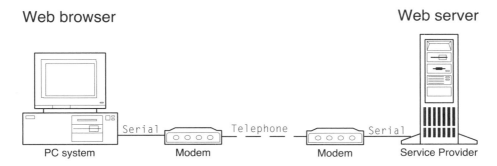

I'll create a Web server, but instead of two modems, I'll use a serial (null modem) cable to link it to the browser. The problem is that my Web server will then receive the browser's commands to its modem. If these go unanswered, the browser will assume its modem is faulty and report this to the user.

The easiest solution is to include a simple modem emulator in your serial driver so that the browser is fooled into thinking it is talking to a modem. Because modem commands are text based, you can easily distinguish between them and the SLIP message blocks prefixed by the delimiter character (C0h); when the latter appears, disengage the modem emulation.

Modem commands begin with the uppercase letters AT, followed by zero or more alphabetic command letters, with alphabetic or numeric arguments, terminated by an ASCII carriage return (<CR>) character. The usual reply is the uppercase letters OK, followed by a carriage return and line feed (<CR><LF>). Table 1.2 shows a few typical command–response sequences for a simple modem. This emulation would respond OK to all commands; this is normally sufficient.

Table 1.2 Modem command–response sequences.

Browser	Modem	Response
AT<CR>	OK<CR><LF>	Check modem present
ATZ<CR>	OK<CR><LF>	Reset modem
ATDT12345	OK<CR><LF> CONNECT 38400<CR><LF>	Dial phone number

Ethernet Driver Requirements

The Ethernet message (frame) is necessarily more complicated than the serial message (Figure 1.5). It contains the

- destination address,
- source address,
- type/length field,
- data, and
- cyclic redundancy check (CRC).

Figure 1.5 Ethernet frame.

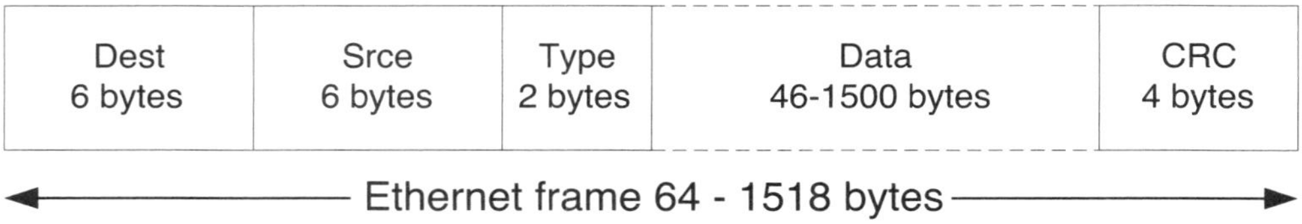

Dest 6 bytes	Srce 6 bytes	Type 2 bytes	Data 46-1500 bytes	CRC 4 bytes

◄─────────────── Ethernet frame 64 - 1518 bytes ───────────────►

It is traditional to include the CRC when quoting the Ethernet frame size (e.g. a maximum frame size of 1518 bytes), even though it is ignored by the software and is usually removed by the lower-level driver code.

```
#define MACLEN      6            /* Ethernet (MAC) address length */


/* Ehernet hardware Rx frame length includes the trailing CRC */
#define MAXFRAMEC   1518         /* Maximum frame size (incl CRC) */
#define MINFRAMEC   64           /* Minimum frame size (incl CRC) */


/* Higher-level drivers exclude the CRC from the frame length */
#define MAXFRAME    1514         /* Maximum frame size (excl CRC) */
#define MINFRAME    60           /* Minimum frame size (excl CRC) */
```

```
/* Ethernet (DIX) header */
typedef struct {
    BYTE dest[MACLEN];              /* Destination MAC address */
    BYTE srce[MACLEN];              /* Source MAC address */
    WORD ptype;                     /* Protocol type or length */
} ETHERHDR;


/* Ethernet (DIX) frame; data size is frame size minus header & CRC */
#define ETHERMTU (MAXFRAME-sizeof(ETHERHDR))
typedef struct {
    ETHERHDR h;                     /* Header */
    BYTE data[ETHERMTU];            /* Data */
    LWORD crc;                      /* CRC */
} ETHERFRAME;
```

This is the basic Ethernet frame, also known as Ethernet 2 (Ethernet 1 is obsolete), or DIX Ethernet (after its creators, DEC, Intel, and Xerox).

Destination and Source Addresses

These six-byte values identify the recipient and sender of the frame and are generally known as media access and control (MAC) addresses. They are standardized by the IEEE; the first three bytes identify the network hardware vendor, and the next three are used by that vendor to guarantee the address is unique, so they are different for every network adaptor that the manufacturer has ever produced.

Each adaptor has its six-byte address burned into a memory device at manufacture, but it is normally the responsibility of the networking software to copy this value into the appropriate field of the network packet. A destination address of all ones indicates a broadcast address.

Type/Length Field

Unfortunately, there are several Ethernet standards, and they make different use of this two-byte field. One standard uses it as a length, giving the total count of bytes in the data field. Others use it as a protocol type, indicating the protocol that is being used in the data field. Mercifully, there are simple ways of detecting and handling these standards, which are discussed in Chapter 3.

Data

This area contains user data in any format; the only restrictions are that its minimum size is 46 bytes and its maximum is 1,500 bytes. The minimum is necessary to ensure that the overall frame is at least 64 bytes. If it were smaller, there would be a danger that frame collisions wouldn't be detected on large networks.

Cyclic Redundancy Check

This is a check value that allows the network controller to discard corrupted frames. It is automatically appended by the Ethernet controller on transmit and checked on receive. The bit-by-bit algorithm is particularly suited to hardware implementation. The following code fragment is equivalent but operates on byte values:

```
#define ETHERPOLY 0xedb88320L


/* Update CRC for next input byte */
unsigned long crc32(unsigned long crc, unsigned char b)
{
    int i;

    for (i=0; i<8; i++)
    {
        if ((crc ^ b) & 1)
            crc = (crc >> 1) ^ ETHERPOLY;
        else
            crc >>= 1;
        b >>= 1;
    }
    return(crc);
}
```

A starting CRC value of FFFFFFFFh is sent to this function, together with the first byte value. A new CRC value is returned, which is sent to this function together with the next byte value, and so on. When all bytes have been processed, the final CRC value is inverted (one's complement) to produce the four-byte Ethernet CRC, which would be transmitted least-significant byte first.

Generic Driver Functions

You need some generic network driver functions that are usable for a variety of network types and hardware configurations. This node-specific information will be in a configuration file and read from disk at boot time. The following code fragments show what a line in this file might look like:

```
net ether ne 0x280
```

This specifies an Ethernet interface using an NE2000-compatible card at I/O address 280h. See Appendix A for details on the cards and networks supported.

This string passed to a network initialization function, to open the required interface.

```
WORD open_net(char *cfgstr);
```

This function opens up the network driver, given a string specifying the type of driver and configuration parameters, and returns a driver type, which must be used in all subsequent accesses, or a 0 on error (e.g., when the hardware is in use by other software).

```
void close_net(WORD dtype);
```

This function shuts down the network driver. The returned value for the driver type serves two purposes: it provides a unique handle for the interface, and its flags inform you of the type of interface in use. This allows you to create software that can handle multiple network interfaces, each with different hardware characteristics.

You need a generic frame that can accommodate any one of the different frame types. Its header includes the driver type.

```
/* General-purpose frame header, and frame including header */
typedef struct {
    WORD len;                    /* Length of data in genframe buffer */
    WORD dtype;                  /* Driver type */
    WORD fragoff;                /* Offset of fragment within buffer */
} GENHDR;


typedef struct {
    GENHDR g;                    /* General-pupose frame header */
    BYTE buff[MAXGEN];           /* Frame itself (2 frames if fragmented) */
} GENFRAME;
```

The header also has a length word to assist in low-level buffering (e.g., polygonal buffering, described later) and support for fragmentation. This is where a frame that exceeds the MTU size is broken up, sent as two smaller frames, and reassembled at the far end. This will be discussed further in Chapter 3; for now, you need to be aware that the maximum frame size (MAXGEN in the above definitions) need not be constrained to the maximum Ethernet frame size. You'll use a MAXGEN of slightly more than 3Kb, so two complete Ethernet frames can be stored in the one GENFRAME.

Having standardized on a generic frame, you can create the driver functions to read and write these frames.

WORD get_net(GENFRAME *gfp); Checks for an incoming frame. If present, it copies it into the given buffer and returns the data length. If there is no frame, it returns 0.

WORD put_net(GENFRAME *gfp, WORD len); Sends a frame, given its length, and returns the total transmitted length or 0 if error.

You don't need to specify which network interface is used, because the function can examine the driver-type field to determine this. Sample device drivers have been included on the CD-ROM, but they will not be discussed here because they are highly specific to the hardware (and operating system).

Configuration File Format

As part of the experimentation in this book, you'll frequently need to change the software parameters at run time. Because it is tedious to type these in every time the program runs, they'll be incorporated into a configuration file called `tcplean.cfg`. By default, utilities will read this file from the default file path, although an alternative configuration filename can be specified on the command line.

The file consists of ASCII text lines, each line referring to one configuration item.

```
# TCP/IP Lean configuration file

net     ether ne 0x280
id      node1
ip      10.1.1.1
gate    10.1.1.111


# EOF
```

Blank lines, or lines beginning with #, are treated as comments. At the start of each line is a single lowercase configuration parameter name delimited by white space and followed by a string giving the required parameter value(s).

The content of the file is specific to the software being run; if any configuration parameter is unrecognized, it is ignored. In the above example, the `net` entry defines the network driver to be used and its base I/O address. The node name is identified as `node1`, with IP address `10.1.1.1` and gateway address `10.1.1.111` given. Appendix A gives guidance on how to customize the configuration file for the network hardware you are using.

Process Timer

When implementing a protocol, an event for a future time is often scheduled. Whenever you send a packet on the network, you must assume that it, or the response to it, might go astray. After a suitable time has elapsed, you may want to attempt a retry or alert the user.

Most modern operating systems have a built-in provision for scheduling such events, but I am very keen to keep the code Operating System (OS) independent and to be able to run it on the bare metal of small-embedded systems. To this end, my software includes a minimal event scheduler of its own, which requires a minimum of OS support and can be adapted to use the specific features of your favorite OS.

The simplest scheduling algorithm is to delay between one event and another.

```c
putpacket(...);              /* Packet Tx */
delay(2000);                 /* Wait 2 seconds */
if (getpacket(...))          /* Check for packet Rx */
{
    /* Handle response packet */
}
else
{
    /* Handle error condition */
}
```

The dead time between transmission and reception is highly inefficient. If the response arrives within 100 milliseconds (ms), the system would wait a further 900ms before processing it. With a multitasking OS, you could use `sleep` instead of `delay`, which would wake up on time-out or when the packet arrived (a method called blocking, since it blocks execution until an event occurs). An alternative pseudo-multitasking method is to use timer interrupts to keep track of elapsed time and to initiate corrective action as necessary, but this approach would be highly specific to the OS.

A simple compromise, not entirely unfamiliar to old-style Windows programmers, is to have the software check for its own events and handle them appropriately.

```
putpacket(...);          /* Packet Tx */
timeout(&txtimer, 0);    /* Start timer */
while (1)
{
    /* Check for packet Rx */
    if (getpacket(...))
    {
        /* Handle response packet */
    }
    /* Check for timeout on response */
    else if (timeout(&txtimer, 2))
    {
        /* Handle error condition */
    }
    /* Check for other events */
    else if ...

}
```

The `timeout()` function takes two arguments: the first is a pointer to a variable that will hold the starting time (tick count), and the second is the required time-out in seconds. When the time-out is exceeded, the function triggers an event by reloading the starting time with the current time and returning a non-zero value. For example, the following code fragment prints a seconds count every second.

```
WORD sectimer, secs=0;

timeout(&sectimer, 0);
while (1)
{
    if (timeout(&sectimer, 1))
        printf("%u sec\n", ++secs);
}
```

Before a timer is used, a `timeout()` call must be made using time value 0. This forces an immediate time-out, which loads the current (starting) time into the timer variable. The `timeout()` function is easy to implement, providing you take care with the data types.

```
/* Check for timeout on a given tick counter, return non-zero if true */
int timeout(WORD *timep, int sec)
{

    WORD tim, diff;
    int tout=0;

    tim = (WORD)time(0);
    diff = tim - *timep;
    if (sec==0 || diff>=sec)
    {
        *timep = tim;
        tout = 1;
    }
    return(tout);
}
```

If the use of unsigned arithmetic appears counterintuitive, consider the following code:

```
WORD a, b, diff;
a = <any starting value>;
b = a + 10;
diff = b - a;
```

What is the value of `diff`? It must be 10, whatever the starting value.

There is a hidden trap that is due to timer granularity. The `if` statement in the code

```
timeout(&sectimer, 0);
if (timeout(&sectimer, 1))
    ...
```

will sometimes return `TRUE`, even though much less than a second has elapsed. This is because the two statements happen to bracket a timer tick, so it appears that one second has elapsed when it has not.

A cure for this problem is to change the unit of measurement to milliseconds, although the nonstandard millisecond timer, `mstime()`, must be coded for each operating system.

```
/* Check for timeout on a given msec counter, return non-zero if true */
int mstimeout(LWORD *timep, int msec)
{
```

```
    LWORD tim;
    long diff;
    int tout=0;

    tim = mstime();
    diff = tim - *timep;
    if (msec==0 || diff>=msec)
    {
        *timep = tim;
        tout = 1;
    }
    return(tout);
}
```

Alternatively, you can just document this feature by saying that there is a tolerance of −1/+0 seconds on the time measurement. Given this timing tolerance, you might be surprised that my trivial example of printing seconds works as suggested.

```
WORD sectimer, secs=0;

timeout(&sectimer, 0);
while (1)
{
    if (timeout(&sectimer, 1))
        printf("%u sec\n", ++secs);
}
```

It works because the state changes in the main loop are locked to the timer tick changes. The whole operation has become synchronous with the timer, so after a random delay of up to one second, the one-second ticks are displayed correctly.

When working with protocols, you will frequently see software processes synchronizing with external events, such as the arrival of data frames, to form a pseudo-synchronous system. When testing your software, you must be sure that this rhythm is regularly disrupted (e.g., by interleaving accesses to another system) to ensure adequate test coverage.

State Machines

When learning to program, I always avoided state machines and skipped the examples (which always seemed to be based on traffic lights) because I couldn't see the point. Why go to all the effort of drawing those awkward diagrams when a simple bit of procedural code would do the job very effectively?

Tackling network protocols finally convinced me of the error of my ways. You may think a network transaction is a tightly specified sequence of events that can be handled by simple procedural code, but that is to deny the unpredictability (or unreliability, as I discussed earlier) of any network. In the middle of an orderly transaction, your software might see some strangely inconsistent data, perhaps caused by a bug in the someone else's software or your own. Either way, your software must make a sensible response to this situation, and it can't do that if you didn't plan for this possibility. True, you can't foresee every problem that may occur, but with proper analysis you can foresee every *type* of problem and write in a strategy to handle it.

Only the simplest of network transactions are stateless; that is, neither side needs to keep any state information about the other. Usually, each side keeps track of the other and uses the network to

- signal a change of state,
- signal the other machine to change its state, or
- check whether the other machine has signaled a change of state.

The key word is *signal*. Signals are sent and received over the network to ensure that two machines remain in sync; that is, they track each other's state changes. The signals may be explicit (an indicator variable set to a specific value) or implicit (a quantity exceeding a given threshold). Either way, the signals must be detected and tracked by the recipient.

Any error in this tracking will usually lead to a rapid breakdown in communications. When such problems occur, inexperienced network programmers tend to concentrate on the data, rather than the states. If a file transfer fails, they might seek deep meaning in the actual number of bytes transferred, whereas an older hand would try to establish whether a state change had occurred and what caused it at the moment of failure. This process is made much easier if the protocol software has specifically defined states and has the ability to display or log the state information while it is running.

At the risk of creating a chapter that you will skip, I'd like to present a simple, worked example of state machine design, showing the relationship between state diagram, state table, and software for a simple communications device, the telephone.

Telephone State Machine

If you ignore outgoing calls, what states can a telephone be in?

Idle   on-hook, unused

Ringing   on-hook, bell ringing

Connected   off-hook, connected to another phone

Sending   sending speech to other phone

Receiving   receiving speech from other phone

The last two states are debatable, since a telephone can send and receive simultaneously. However, most human beings possess a half-duplex audio system (they seemingly can't speak and listen at the same time), so the separation into transmission and reception is logical.

A telephone changes state by a combination of electrical messages down the phone cable and by user actions. From the point of view of a hypothetical microcontroller in the telephone, these might all be considered *signals*.

Line ring   ring signal from another phone

Line idle   no signal on phone line

Pick up   user picks up handset

Mic. speech   user speaks into microphone

Line speech   speech signal from other phone

Hang up   user replaces handset

It is now necessary to define which signals cause *transitions* between states; for example, to change state from *idle* to *ringing*, a *ring signal* is required.

It is traditional to document these state changes using a *state diagram* such as Figure 1.6, which is a form of flowchart with special symbols. Each circle represents a defined state, and the arrows between circles are the state transitions, labeled with the signal that causes the transition. So *line speech* causes a transition from the *connected* state to the *receiving* state, and *line idle* causes the transition back to *connected*.

Because of the inherent limitations of the drawing method, these diagrams tend to over-simplify the state transitions; for example, Figure 1.6 doesn't show a state change if the user hangs up while receiving.

A more rigorous approach is to list all the states as rows of a table and all the signals as columns (Table 1.3). The table entries give a new state or are blank if there is no change of state.

Figure 1.6   Telephone state diagram.

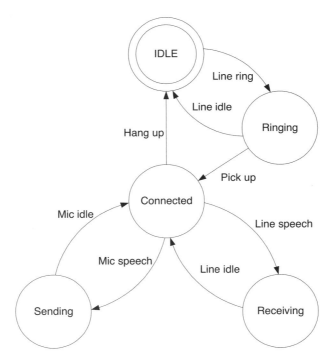

Table 1.3 Telephone state table.

	Line Ring	Line idle	Pick up	Mic. speech	Mic. idle	Line speech	Hang up
Idle	Ringing						
Ringing		Idle	Connected				Idle
Connected				Sending		Receiving	Idle
Sending					Connected		Idle
Receiving		Connected					Idle

Once the table has been created, it isn't difficult to generate the corresponding code. You could use a two-dimensional lookup table, although a series of conditional statements are generally more appropriate.

```
switch(state)
{
    case STATE_IDLE:
        if (signal == SIG_LINE_RING)
            newstate(STATE_RINGING);
        break;
    case STATE_RINGING:
        if (signal == SIG_PICKUP)
            newstate(STATE_CONNECTED);
        else if (signal == SIG_LINE_IDLE)
            newstate(STATE_IDLE);
        break;

    case STATE_CONNECTED:
        // ..and so on
}
```

I have created an *explicit state machine* where the states, signals, and relationship between them are clearly and explicitly identified. Contrast this with an *implicit* state machine, where the current state is buried in function calls.

```c
void idle(void)
{
    while (1)
    {
        if (signal == SIG_LINE_RING)
            ringing();
    }
}
void ringing(void)
{
    while (signal != SIG_HANGUP)
    {
        if (signal == SIG_PICKUP)
            connected();
    }
}
void connected(void)
{
    // ... and so on
```

Here, the current state is indicated *implicitly* by the current position in the code, and it is far harder to keep control of all the possible state transitions, particularly under error conditions. The stack-based call return mechanism imposes a hierarchical structure that is ill suited to the arbitrary state transitions required. It is important that the state machine is explicitly created, rather than being an accidental by-product of the way the software has been structured. The requirements of the state machine must dictate the software structure, not (as is often the case) the other way around.

Buffering

To support the protocols, three special buffer types will be used. The first is a modified version of the standard first in, first out (FIFO) to accommodate an extra trial pointer; the second is a fixed-data-length variant of this, and the third is a FIFO specifically designed for bit-wide, rather than byte-wide, transfers.

FITO Buffer

The FITO (first in, trial out) is a variant of the standard FIFO, or circular buffer (Figure 1.7). A normal FIFO has one input and one output pointer; data is added to the buffer using the input pointer and removed using the output pointer. For example, assume that a 10-character FIFO has the letters "ABCDEFG" added, then "ABCDE" removed, then "HIJKL" added.

Figure 1.7   FIFO example.

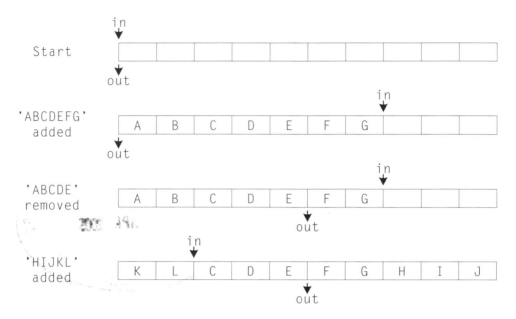

The circularity of the buffer is demonstrated in Figure 1.7 by the second addition; instead of running off the end, the input pointer wraps around to the start, providing there is sufficient space (i.e., the pointers do not collide). Note that, after removal, the characters "ABCDE" are shown as still present in the buffer; only the output pointer has changed position. This reflects standard practice, in that there is little point in clearing out unused locations, so the old characters remain until overwritten.

Now imagine this FIFO is being used in a Web server; the input text is a Web page stored on disk, and the output is being transmitted on the network. Due to network unreliability, you don't actually know whether the transmitted data has been received or has been lost in transit. If the latter, then the data will have to be retransmitted, but it is no longer in the FIFO, so it must be refetched from disk.

It would be better if the FIFO had the ability to retain transmitted data until an acknowledgment was received; that is, it keeps a marker for output data that may still be needed, which I will call *trial* data, in contrast to *untried* data, which is data in the buffer that hasn't been transmitted yet; hence, the FITO buffer has one input and two output pointers, as shown in Figure 1.8.

Having loaded "ABCDEFG" in the buffer, data fragments "ABC" and "DE" are sent out on the network, and the trial pointer is moved up to mark the end of the trial data. "ABC" is then acknowledged, so the output pointer can be moved up, but the rest of the data is not, so the unacknowledged data between the output and trial pointers is retransmitted on the network, followed by the remaining untried data. Finally, that is all acknowledged, so the output pointer can be moved up to join the input pointer.

Figure 1.8   FITO example.

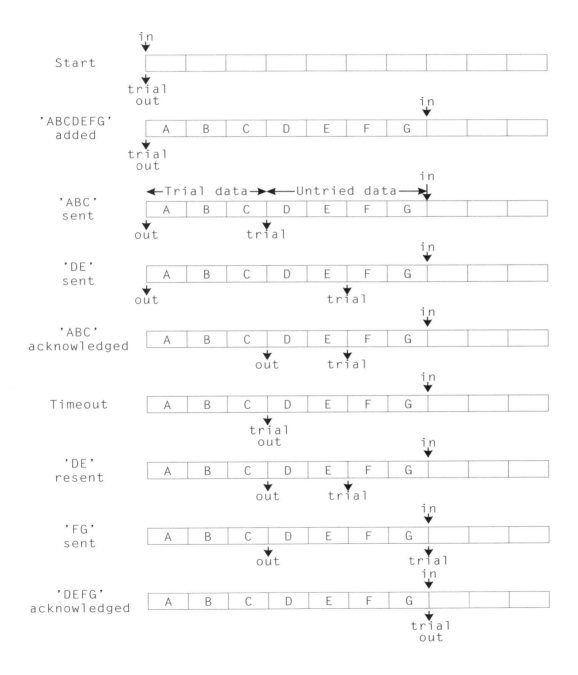

A structure stores the data and its pointers (as index values into the data array). The first word indicates the buffer length, which allows for a variety of buffer sizes. For speed, the buffer size is constrained to be a power of two.

```
#ifndef _CBUFFLEN_
#define _CBUFFLEN_ 0x800
#endif
/* Circular buffer structure */
typedef struct
{
    WORD len;                    /* Length of data (must be first) */
    LWORD in;                    /* Incoming data */
    LWORD out;                   /* Outgoing data */
    LWORD trial;                 /* Outgoing data 'on trial' */
    BYTE data[_CBUFFLEN_];       /* Buffer */
} CBUFF;
```

A default buffer size of 2Kb is provided, which may be overridden if required. This permits a buffer to be declared as a simple static structure.

```
#include "netutil.h"
CBUFF rxpkts = {_CBUFFLEN_};
```

Or, consider the code when using dynamically allocated memory:

```
#define BUFFLEN 0x2000
CBUFF *rxp;
if ((rxp = (CBUFF *)malloc(BUFFLEN+16))!=0)
{
    rxp.len = BUFFLEN;
    ...
}
```

In both cases, the length value is set when the buffer is created; this is very important if strange bugs are to be avoided.

The use of LWORD (unsigned 32-bit) buffer pointers with WORD (unsigned 16-bit) data length may seem strange. The former is part of a Cunning Plan to map the TCP 32-bit sequencing values directly onto these pointers, whereas the latter permits the code to be compiled into a 16-bit memory space (e.g., small model), if necessary. All should become clear in subsequent chapters.

In creating the buffer-handling software, it is important to retain a clear idea of what is meant by *untried* data (not yet sent), and *trial* data (sent but not acknowledged).

```c
/* Return total length of data in buffer */
WORD buff_dlen(CBUFF *bp)
{
    return((WORD)((bp->in - bp->out) & (bp->len - 1)));
}
/* Return length of untried (i.e. unsent) data in buffer */
WORD buff_untriedlen(CBUFF *bp)
{
    return((WORD)((bp->in - bp->trial) & (bp->len - 1)));
}
/* Return length of trial data in buffer (i.e. data sent but unacked) */
WORD buff_trylen(CBUFF *bp)
{
    return((WORD)((bp->trial - bp->out) & (bp->len - 1)));
}
/* Return length of free space in buffer */
WORD buff_freelen(CBUFF *bp)
{
    return(bp->len ? bp->len - 1 - buff_dlen(bp) : 0);
}

/* Set all the buffer pointers to a starting value */
void buff_setall(CBUFF *bp, LWORD start)
{
    bp->out = bp->in = bp->trial = start;
}
```

When loading data into the buffer, the simple but slow method is to copy it byte by byte. Instead, I'll use either one or two calls to a fast block-copy function, depending on whether the new data wraps around the end of the buffer. If the data is too big for the buffer, it is truncated, because I'm assuming the programmer has checked the free space before calling this function. The free space is always reported as one byte less than the actual space, so there is no danger of the input pointer catching up with the output pointer.

```c
/* Load data into buffer, return num of bytes that could be accepted
** If data pointer is null, adjust pointers but don't transfer data */
WORD buff_in(CBUFF *bp, BYTE *data, WORD len)
{
    WORD in, n, n1, n2;
```

```
    in = (WORD)bp->in & (bp->len-1);          /* Mask I/P ptr to buffer area */
    n = minw(len, buff_freelen(bp));          /* Get max allowable length */
    n1 = minw(n, (WORD)(bp->len - in));       /* Length up to end of buff */
    n2 = n - n1;                              /* Length from start of buff */
    if (n1 && data)                           /* If anything to copy.. */
        memcpy(&bp->data[in], data, n1);      /* ..copy up to end of buffer.. */
    if (n2 && data)                           /* ..and maybe also.. */
        memcpy(bp->data, &data[n1], n2);      /* ..copy into start of buffer */
    bp->in += n;                              /* Bump I/P pointer */
    return(n);
}


/* Load string into buffer, return num of chars that could be accepted */
WORD buff_instr(CBUFF *bp, char *str)
{
    return(buff_in(bp, (BYTE *)str, (WORD)strlen(str)));
}
```

Removal of untried data from the buffer, so that it becomes trial data, is essentially the inverse of the above.

```
/* Remove waiting data from buffer, return number of bytes */
** If data pointer is null, adjust pointers but don't transfer data */
WORD buff_try(CBUFF *bp, BYTE *data, WORD maxlen)
{
    WORD trial, n, n1, n2;

    trial = (WORD)bp->trial & (bp->len-1);  /* Mask trial ptr to buffer area */
    n = minw(maxlen, buff_untriedlen(bp));  /* Get max allowable length */
    n1 = minw(n, (WORD)(bp->len - trial));  /* Length up to end of buff */
    n2 = n - n1;                            /* Length from start of buff */
    if (n1 && data)                         /* If anything to copy.. */
        memcpy(data, &bp->data[trial], n1); /* ..copy up to end of buffer.. */
    if (n2 && data)                         /* ..and maybe also.. */
        memcpy(&data[n1], bp->data, n2);    /* ..copy from start of buffer */
    bp->trial += n;                         /* Bump trial pointer */
    return(n);
}
```

Functions to remove data from the buffer are required to complete the set and to wind back the trial pointer so that the data is waiting for retransmission.

```
/* Remove data from buffer, return number of bytes
** If data pointer is null, adjust pointers but don't transfer data */
WORD buff_out(CBUFF *bp, BYTE *data, WORD maxlen)
{
    WORD out, n, n1, n2;
    out = (WORD)bp->out & (bp->len-1);         /* Mask O/P ptr to buffer area */
    n = minw(maxlen, buff_dlen(bp));           /* Get max allowable length */
    n1 = minw(n, (WORD)(bp->len - out));       /* Length up to end of buff */
    n2 = n - n1;                               /* Length from start of buff */
    if (n1 && data)                            /* If anything to copy.. */
        memcpy(data, &bp->data[out], n1);      /* ..copy up to end of buffer.. */
    if (n2 && data)                            /* ..and maybe also.. */
        memcpy(&data[n1], bp->data, n2);       /* ..copy from start of buffer */
    bp->out += n;                              /* Bump O/P pointer */
    if (buff_untriedlen(bp) > buff_dlen(bp))/* ..and maybe trial pointer */
        bp->trial = bp->out;
    return(n);
}
/* Rewind the trial pointer by the given byte count, return actual count */
WORD buff_retry(CBUFF *bp, WORD len)
{
    len = minw(len, buff_trylen(bp));
    bp->trial -= len;
    return(len);
}
```

As a useful extra feature, a null data pointer can be given to the function, in which case it goes through the same motions, but without copying any actual data. This is handy for discarding unwanted data (e.g., trial data that has been acknowledged).

I've made extensive use of minw(), which returns the lower of two word values and so is similar to the standard function min().

```
WORD minw(WORD a, WORD b)
{
    return(a<b ? a : b);
}
```

Why define my own? Because min() is usually implemented as a macro,

```
#define min(a, b) (a<b ? a : b)
```

and any function arguments may be executed twice, which is a major problem in interrupt-driven (reentrant) code. Take a line from buff_out().

```
n = minw(maxlen, buff_dlen(bp));          /* Get max allowable length */
```

The macro expands this to the following:

```
n = maxlen < buff_dlen(bp) ? maxlen : buff_dlen(bp);
```

Imagine that the first time buff_dlen() is executed, the source buffer is almost empty, so all its data can be transferred into the destination. However, before the function is executed a second time, an interrupt occurs that fills the buffer with data, so the actual data length copied exceeds the maximum the destination can accept, with disastrous results. The easiest way to avoid this problem is to buffer the comparison values in a function's local variables; hence, the usage of minw().

Polygonal Buffer

A circular buffer is useful for handling unterminated streams of data, but sometimes you'll need to store blocks of known length. The classic case is a packet buffer, in which you can queue packets prior to transmission or on reception. The standard technique is to have a *buffer pool*, from which the storage for individual packets can be allocated. A simpler technique is to use a circular buffer as before but to prefix each addition to it with a length word, to show how much data is being added.

```
if (len>0 && buff_freelen(&rxpkts) >= len+2)/* If space in circ buffer..*/
{
    buff_in(&rxpkts, (BYTE *)&len, 2);       /* Store data len.. */
    buff_in(&rxpkts, buff, len);             /* ..and data */
}
```

The smooth circle of data has been replaced by indivisible straight-line segments; when recovering the data, check that the whole block is available (if there is a risk that part of the block may be in transit). The trial system comes in handy because you can retry (i.e., push back) the length if the entire data block isn't available yet.

```
if ((dlen=buff_dlen(&txpkts)) >= 2)
{
    buff_try(&txpkts, (BYTE *)&len, 2); /* Get length */
    if (dlen >= len+2)                  /* If all there.. */
    {
        buff_out(&txpkts, 0, 2);        /* ..remove len */
        buff_out(&txpkts, buff, len);   /* ..and data */
    }
    else
        buff_retry(&txpkts, 2);         /* Else push back len */
}
```

This explains the length parameter on the front of the generic frame. It allows you to store and retrieve GENFRAME structures from circular buffers without having to understand the contents of the frame.

Coding Conventions

It isn't essential that you use the same coding conventions (source code formatting) as I do, though it may help if I describe the rules I've used, so you can choose whether to follow them or not.

Data Types

When defining protocol structures, it is really important to use the correct data width. You may be used to assuming that an int is 16 bits wide, but that isn't true in a 32-bit system. I've made the following assumptions for all the DOS compilers:

- char is an 8-bit signed value
- short is 16 bits
- long is 32 bits

From these, I have derived the following width-specific definitions.

```
#define BYTE unsigned char
#define WORD unsigned short
#define LWORD unsigned long
```

I have used #define in preference to typedef because compilers use better optimization strategies for their native data types. A notable omission is a Boolean (TRUE/FALSE) data type; I use an integer value and assume TRUE is any non-zero value.

Keeping compatibility with both 16- and 32-bit compilers also necessitates the addition of some redundant-looking typecasts.

```
WORD a, b;
a = (WORD)(b + 1);
```

If the typecast is omitted, the Visual C++ compiler issues a warning message because b is promoted to a 32-bit value for the addition, which must be truncated when assigned to a.

Another tendency of 32-bit compilers is, by default, to pad data elements out to four- or eight-byte boundaries, which blows gaping holes in the structures.

```
typedef struct {
    BYTE a;
    BYTE b;
    LWORD c;
    WORD d;
} MYSTRUCT;
```

If the mix of bytes, words, and long words is to be transmitted on the network, it is vital that the compiler is set so that it does not introduce any padding between these structure elements; that is, the structure-member alignment is one byte, not four or eight bytes.

All the necessary configuration data is included in the appropriate compiler-specific project file, so should be loaded automatically. The PICmicro cross-compiler is very different from all the others; its peculiarities are explored in Chapter 9.

Source Code Format

I have attempted to illustrate all the techniques in this book by embedding source-code fragments in the text. Because the full source code is included on the accompanying CD and you won't be retyping the code from the book, I have used some artistic license in the preparation of these fragments by stripping out all nonessential stuff so the essence of the code is clear to see.

You will see frequent ellipses (...) indicating deletion of the lines I think are irrelevant to the topic under discussion. If the absence of these items hinders your comprehension of the text, I suggest you print out a copy of the latest source code and use that in conjunction with this book.

Also, the version printed in the book is not necessarily the latest version. All released versions of the software have revision headers, containing entries as follows:

```
/*
** v0.01 JPB 1/1/00  First version
** v0.02 JPB 2/1/00  Added widget-handling option
*/
```

The version number of the complete utility is taken from the version number of the file bearing its name. For example, if test.exe is built from utils.c v0.12, test.c v1.23, and driver.c v2.34, then MYTEST has the overall version number v1.23. To confirm the build status, each project has a CRC file containing the file names, version numbers, and a 32-bit CRC for each file.

```
Project PING v0.17 archived 24-Mar-00 11:49

ether3c.c    6376B05D v0.05
etherne.c    9759AE0A v0.24
ip.c         C3BB52D4 v0.10
net.c        1EB2A188 v0.07
netutil.c    73320977 v0.11
ping.c       B66F030C v0.17
pktd.c       183DCB73 v0.09
serpc.c      DD6477B8 v1.08

ether.h      4FF45517
ip.h         C739761D
net.h        C19A2BBD
netutil.h    B8B96B7F
pktd.h       0ACBA7F6
serpc.h      089DEAF1
```

The Iosoft Ltd. Web site (www.iosoft.co.uk) has up-to-date versions of the software in this book and useful extra information, such as application notes.

Chapter 2

Introduction to Protocols: SCRATCHP

Overview

In this chapter, I start by looking at what a protocol is; then I show how it can be implemented in software. I'll examine

- the definition of a protocol,
- the standard way of describing a protocol,
- the client–server model,
- modal and modeless clients, and
- logical connections — open, close, and data transfer

Because this is a hands-on book, I'll illustrate these points by creating a protocol from scratch (called SCRATCHP) and writing a utility that allows you to exercise the protocol. While implementing the protocol, you'll have an opportunity to explore the following areas:

- Storage of Ethernet and SLIP frames
- Ethernet addressing
- Protocol identification
- Byte swapping
- Low-level packet transmission and reception

You'll end up with a stand-alone utility that can be used to exercise the protocol that's been created, or it can be used as a base for implementing another protocol. The foundations will have been laid for the TCP/IP protocols to come.

Protocol

For two computers to communicate, they must speak the same language. A communication language framework is generally called a *protocol*. The name is derived from the framework employed by diplomats when attempting to communicate across cultural boundaries. Two computers may employ different processors, languages, and operating systems, but if they both use a common protocol, then they can communicate.

Protocols don't just enable communications, they also restrict them. Neither party may stray outside the bounds of protocol without facing incomprehension or rejection. So a protocol doesn't just define how communication may occur but also provides a framework for the information that is communicated. But how can any one protocol encompass all the variety of present-day computer communications? It can't, so you need a family of protocols, each of which is designed for a specific task. As with a software project, you need a tree structure, with the simpler, network-oriented tasks at the bottom and the higher, user-oriented tasks at the top. Such a structure is often called a protocol *stack*, though this leads to confusion with the last in, first out (LIFO, push/pop, or call/return) stack data storage mechanism also used by programmers.

Here, the term stack refers to the way protocol components are stacked on top of each other to give the desired functionality. If I want to transfer a file, I might take a standard file transfer protocol and stack it on top of a communications protocol. The communications protocol wouldn't understand about files — it simply moves bocks of data around. Conversely, the file transfer protocol wouldn't understand about networks — it simply converts files into blocks of data. Combine the two, and you have network file transfer capability.

The separation into protocol layers doesn't necessarily make for easier reading. Older protocol specifications used to simply record the pattern of "bytes on the wire" for achieving a given result (and also, if you're lucky, a smattering of timing information). In a layered world, a protocol specification must tie down the upper and lower application programming interfaces (APIs) and the operations to be performed on them. Any relation to bytes on the wire (i.e., actual, visible work) is purely coincidental. There is the danger that the APIs may become operating system specific, so vendor-independent standardization is very important.

Standardization

The international community, in its wisdom, decided to standardize on the number of protocol layers in a stack, and the International Standards Organization (ISO) Open Systems Interconnection (OSI) model was born (Figure 2.1). Their layers are listed below from top down.

7. ApplicationUser interface

6. PresentationData formatting

5. SessionLogical connections

4. TransportError-free communication

3. NetworkNetwork addressing

2. Data linkTransmission and reception

1. PhysicalNetwork hardware

For local area networks (LANs), the data link layer is further subdivided into a component called medium access control (MAC), which resides within the network hardware, and the software-based logical link control (LLC), which provides a uniform software interface (packet driver interface) to the higher levels.

Figure 2.1   OSI seven-layer model.

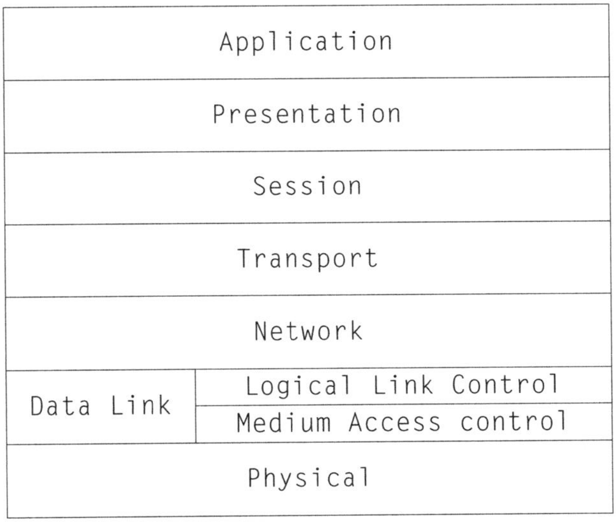

| Application |
| Presentation |
| Session |
| Transport |
| Network |
| Data Link | Logical Link Control / Medium Access control |
| Physical |

When two applications are communicating over a network, it can be useful to think in terms of the data entering at the top of one protocol stack, then traveling downward on that machine, across to the other machine at the physical layer, and back up the second stack (Figure 2.2).

Of course, not all data will originate at application level: resolving addresses requires communication between network layers, and maintenance of a connection requires session-to-session communications. The user is generally unaware of these until he or she happens to see a diagnostic log of all packet transfers; then the reaction is one of amazement that a simple transfer between applications can generate so much traffic. As with a duck crossing a pond, the smooth visible motion belies the furious paddling underneath.

The TCP/IP family of protocols predates the ISO standardization effort, so it does not fit comfortably within the model. Also, the higher layers are remarkably difficult to standardize because they must encompass the totality of network applications. Confronted by the remarkable growth of the Internet, this overall ISO standardization effort has been completely sidelined, though the seven-layer terminology and the lower-level standards are still in widespread use. Although I'll implement SCRATCHP as a single protocol, the seven-layer model does provide important pointers on how your software might be structured.

Figure 2.2   **Application-to-application transfer.**

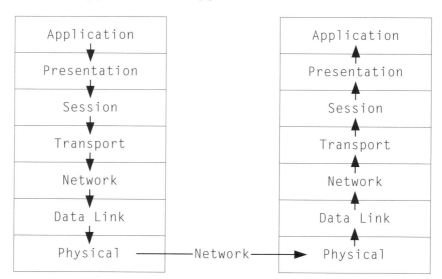

SCRATCHP Services

Just as an operating system offers the user a range of commands, so SCRATCHP will offer the network user a range of services (e.g., remotely accessible functions). The usual TCP/IP approach is to create a separate specification for each service, but to save time, I'll combine several services into one protocol. I'll start with a minimum of these, but the protocol must permit the addition of services at a later date. A preliminary list of services is

1. IDENT (ID resolution)
2. ECHO (connection diagnostic)
3. DIR (file directory)
4. GET (file transfer: read)
5. PUT (file transfer: write)

The ident service is used for converting computer IDs into addresses and is explained later. The echo service allows simple diagnostic tests to be performed. It duplicates incoming data and returns it to the sender. In this way, you can check response times and error rates. File transfer is a fundamental requirement of computer networking and is useful for examining bulk data transfer techniques. I have provided simple dir, get, and put functions.

Client–Server Model

A useful piece of terminology would be to refer to one machine (the requester of the service) as a *client* and the other (the provider of the service) as the *server*. In reality, you might as well write the software so that every machine has the potential to become a client or a server and use keyboard or network commands to determine which mode should be activated at any given time.

The `ident` service is used to identify potential servers, so it must be as simple as possible — one packet transmitted and one packet received. The command is sent as a string, followed by an optional argument string. If a server responds, it returns a copy of the command string to confirm which command it is responding to.

A potential problem is that the response is indistinguishable from the command, so it may be interpreted as another command, generating another response, and so on ad infinitum. There are two approaches to solving this problem: one modal, the other modeless.

Modal Client

Every time a client issues a command, it could go into some sort of command mode, so it knows the next communication it receives is going to be a response to that command. This mode would typically be stored as a state variable. The transaction would be:

1. client goes into command mode and
2. client sends command to server; then,

either

3. client receives response and goes back into normal mode

or

3. client receives no response, times out, and goes back into normal mode.

There are two risks with this approach.

1. The client time-out occurs while the response is still in transit, so it is no longer in command mode when it arrives.
2. While in command mode, the client receives an unexpected packet from another node, which it can't handle because it is in the wrong mode.

Modal techniques are frequently used in simple point-to-point serial links, but they must be used with care in networking, where it is impossible to anticipate what will happen next.

Modeless Client

If you want to keep your client as modeless (i.e., stateless) as possible, you must include more information in the data packet that is transmitted. Instead of sending pure data and storing the command mode *internally*, the transmitted packet must contain an indication that says, "I am a command packet." The server's reply packet must then have a different indication that says, "I am a response packet." By expressing the information *externally*, the client doesn't have to store it internally, and debugging is made easier because you can determine the client's intentions by examining the packets it has sent, rather than having to pry on its internal data.

It is interesting to note that one of the key factors in the success of the World Wide Web has been that the upper protocol layers are stateless. At any one time, a Web server may be handling hundreds of clients, and, in the course of a day, it may handle millions. If it had to keep detailed information on each, there would be a major storage problem. A simple Web server stores no information about any user. Contrast this with a typical multiuser system, where a large number of settings and preferences are stored in an individual user's account.

So, keep the `ident` command stateless for simplicity, but what about the file transfer commands? If you're going to handle bulk data transfers, it is hard to keep the machines completely stateless. If nothing else, they have to remember which files they have opened and why. Ideally, the network would be treated as a simple *pipe*, through which data would flow (or stream).

For this, you really need to establish a *logical connection*, or bidirectional *data pipe* between the client and server: anything fed in one end of the pipe will emerge unaltered at the other end. This is an important concept, which is much used in networking.

Logical Connections

From the earliest days, networks have usually been used for the purpose of establishing logical connections between two computers. When you use a browser to contact a Web site, you are setting up one or more logical connections between your client and their server. The Web pages and graphics are then fed down these connections, like water down a pipe, until the client has all the necessary data to display the page.

Logical connections are reliable. To maintain the connection, the protocol software has to keep track of all packets sent and received and have a retry strategy to cover any packets that go astray. Unfortunately, this reliability comes at a price: writing the protocol software for opening, maintaining, and closing logical connections is a nontrivial task, involving the creation of state machines in both client and server and an exchange of signals between them to ensure the state machines remain in sync.

You may spot an apparent contradiction with my previous assertion that Web client–server communications are stateless. Clearly, they must keep state information about each other for the duration of a transfer. That's why I was careful to say their *applications* are stateless; the lower levels are continuously making and breaking connections, with all the state tracking that entails.

Opening and Closing a Connection

In a simplified protocol, clients have to initiate all actions, so they will request a service that requires the establishment of a connection. The host can then agree to the establishment of a connection by acknowledging the request, ignoring the request if it disagrees, or setting an error flag (it may have insufficient resources to support another connection).

Closure of the connection may be initiated by either party. In a file transfer, it will normally be the sender of the data who closes the connection after the data is transferred, although the recipient may also do this if it can't handle the data any more (e.g., its disk is full).

Data Flow in a Connection

For the duration of the connection, data may flow bidirectionally between the two parties. Both sides need to keep track of the amount of data sent and received to ensure no data has been skipped or duplicated. Commonly used techniques to do this are listed below:

Lock-step   One packet is sent, and the sender waits until an acknowledgment is received before sending another.

Block sequencing   Each packet contains one data block, with a sequential number (sequence number). The recipient may acknowledge receipt of each block, using its sequence number, or wait until a few have been received and then acknowledge them all, using the sequence number of the latest block.

Byte sequencing This is similar to block sequencing, but the sequence number reflects the byte count, rather than the block count.

Figure 2.3   Sequencing methods.

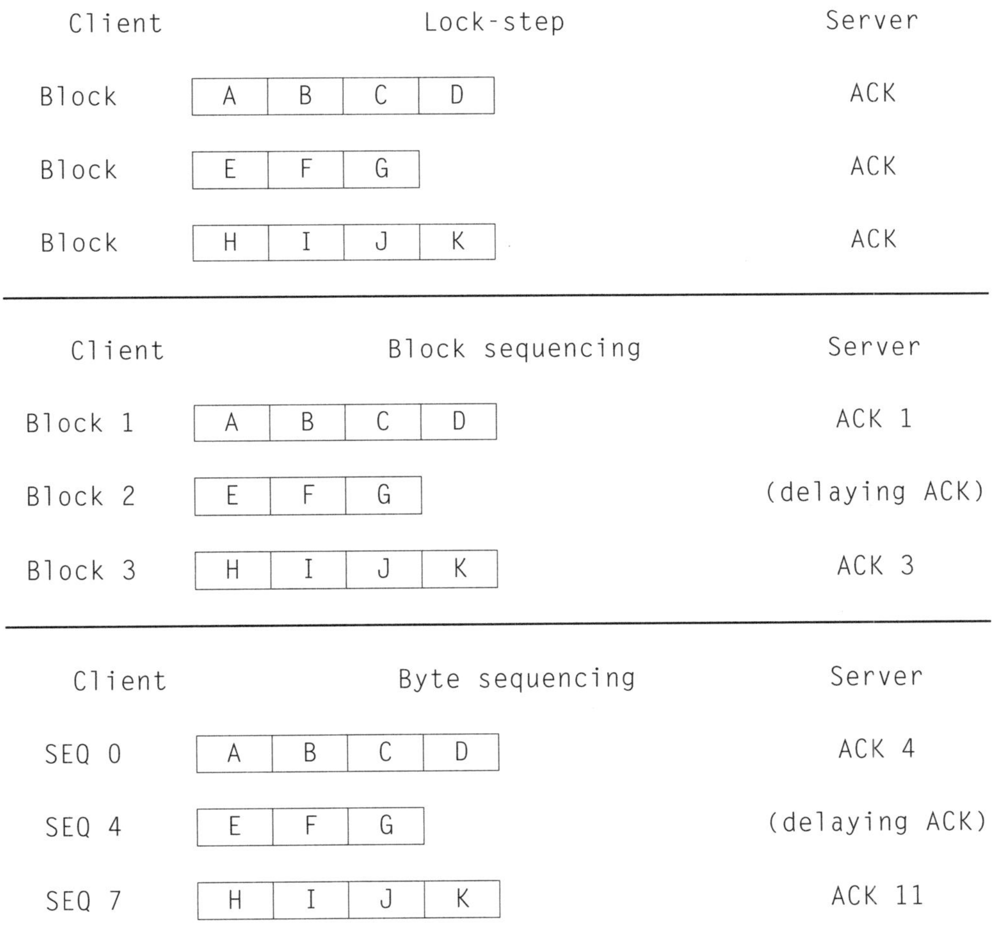

```
      Client                  Lock-step                Server

     Block         | A  |  B  |  C  |  D  |             ACK

     Block         | E  |  F  |  G  |                   ACK

     Block         | H  |  I  |  J  |  K  |             ACK
     _____

      Client              Block sequencing             Server

    Block 1        | A  |  B  |  C  |  D  |             ACK 1

    Block 2        | E  |  F  |  G  |              (delaying ACK)

    Block 3        | H  |  I  |  J  |  K  |             ACK 3
     _____

      Client               Byte sequencing             Server

    SEQ 0          | A  |  B  |  C  |  D  |             ACK 4

    SEQ 4          | E  |  F  |  G  |              (delaying ACK)

    SEQ 7          | H  |  I  |  J  |  K  |             ACK 11
```

Figure 2.3 shows the client–server interactions, assuming the client is sending 11 bytes in three blocks to the server. The lock-step method doesn't need to identify each block individually, since only one can be in transit at any one time (in railway parlance: one engine in steam). The sequencing methods differ in that they identify a block using either an incrementing block number or the total number of bytes sent prior to the current block. The acknowledgment reflects either the latest block received or the latest byte received (i.e., the sequence number plus the byte count).

For simplicity, I have shown the first block as having a byte count of zero. In reality, it is better to start with a pseudorandom base value, which is negotiated at the start of the transaction and is subsequently increased to reflect the actual byte count transferred. The value is typically stored as a 32-bit `LWORD` and is allowed to wrap around past zero when it gets too large, on the assumption that there won't be several gigabytes of data in transit for any one transaction at any one time.

Note that the lock-step method has built-in flow control: the sender cannot out-pace the receiver because the receiver will only acknowledge if it has spare buffer space for the next data block. Flow control can be added to the other techniques by the simple expedient of placing a limit on the maximum number of blocks (or bytes of data) that can be in transit and unacknowledged. When the sender exceeds this "window," it must stop transmitting data until it receives an acknowledgment.

There is little to choose between the two sequencing methods. Block acknowledgment is used in the ISO link layer LLC and is slightly easier to implement than byte sequencing, provided a fixed block size is used. TCP has a variable block size (a "sliding window"), so it employs a byte-sequencing method. This is what I'll use for SCRATCHP.

Packet Format

Having decided on the basic structure of transactions, I can define a packet format to suit (Figure 2.4). Because of my minimalist approach, there is relatively little in it.

- protocol version (one byte)
- flags (one bit each)
  - command
  - response
  - start connection
  - connected
  - stop connection
  - error
- sequence and acknowledgment numbers (four bytes each)
- data length (two bytes)

Figure 2.4   SCRATCHP packet format.

Version 1 byte	Flags 1 byte	Seq 4 bytes	Ack 4 bytes	Datalen 2 bytes	Data 0 - 1488 bytes

←————————————————— 12 - 1500 bytes —————————————————→

The *protocol version* is a useful way of retaining compatibility as SCRATCHP evolves. It can be checked by any recipient to ensure that it is equipped to decode this version of the protocol and to give the user sensible error messages if there is a problem (e.g., "This utility does not support SCRATCHP version 5").

The *data length* field may seem redundant, since the underlying network protocol should provide an overall length value from which the data length could be derived. Unfortunately, the Ethernet frame length will include any padding applied to undersized frames, so it won't always give the correct answer.

```c
/* Flag values */
#define FLAG_CMD        0x01    /* Data area contains command */
#define FLAG_RESP       0x02    /* Data area contains response */
#define FLAG_START      0x04    /* Request to start connection */
#define FLAG_CONN       0x08    /* Connected; sequenced transfer in use */
#define FLAG_STOP       0x10    /* Stop connection */
#define FLAG_ERR        0x20    /* Error; abandon connection */


/* SCRATCHP packet header */
typedef struct {
    BYTE ver;                   /* Protocol version number */
    BYTE flags;                 /* Flag bits */
    LWORD seq;                  /* Sequence value */
    LWORD ack;                  /* acknowledgment value */
    WORD dlen;                  /* Length of following data */
} SCRATCHPHDR;


/* SCRATCHP packet */
#define SCRATCHPDLEN 994
typedef struct {
    SCRATCHPHDR h;              /* Header */
    BYTE data[SCRATCHPDLEN];    /* Data (or null-terminated cmd/resp string) */
} SCRATCHPKT;
```

In common with most other Ethernet protocols, the integer values will be sent with the most significant byte first. The SCRATCHP data array is dimensioned at 994 bytes, which allows it to fit within the 1,500-byte Ethernet or 1,006-byte SLIP data area.

Internal Storage

Having fixed the external appearance of the SCRATCHP packet, I need to decide the internal storage format. You will recall that my network drivers work on a generic frame format, which has a two-byte frame type (which will identify whether it is an Ethernet or SLIP packet and maybe provide a system-specific handle for the network adaptor), followed by a block of data up to the maximum Ethernet frame size.

```
typedef struct {
    GENHDR g;                   /* General-pupose frame header */
    BYTE buff[MAXGEN];          /* Frame itself (2 frames if fragmented) */
} GENFRAME;
```

The SCRATCHP packet will be contained within the data area of an Ethernet or SLIP packet (Figure 2.5).

Figure 2.5   Ethernet and SLIP packets.

Ethernet

Dest	Srce	Pcol	Ver	Flag	Seq	Ack	Dlen	Data

SLIP

Ver	Flag	Seq	Ack	Dlen	Data

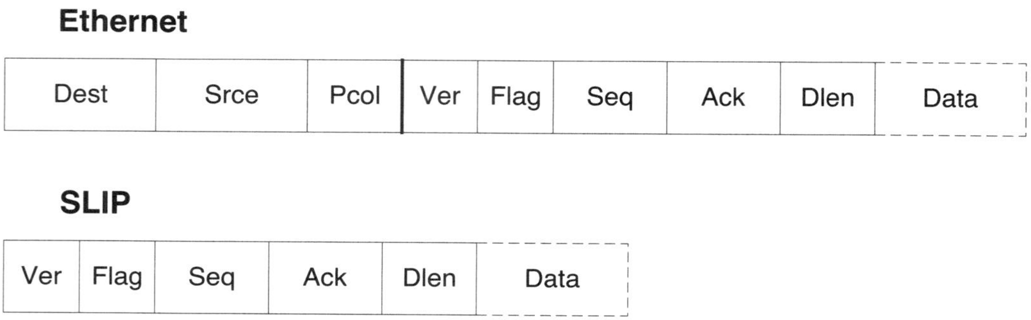

Because Ethernet and SLIP packets have different header lengths (14 bytes and zero bytes), you need a standard way of determining where the network header ends and the SCRATCHP packet starts. A function can do this by checking the packet type and indexing into the packet data area accordingly.

```
/* Get pointer to the data area of the given frame */
void *getframe_datap(GENFRAME *gfp)
{
    return(&gfp->buff[dtype_hdrlen(gfp->g.dtype)]);
}
/* Return frame header length, given driver type */
WORD dtype_hdrlen(WORD dtype)
{
    return(dtype&DTYPE_ETHER ? sizeof(ETHERHDR) : 0);
}
```

Note that a pointer to the frame *data* area also points to the SCRATCHP *header*, and a pointer to the SCRATCHP *data* area may also point to a command *header*. In this nested world, one packet's data is generally another packet's header, so the term "data" must always be qualified by the context in which it appears.

There are other awkward differences between Ethernet and SLIP: the former has a source address, which will be useful when sending a reply, and a protocol-type identifier, which is discussed later. Any functions attempting to access these features need to check the packet type first.

```c
/* Get pointer to the source address of the given frame, 0 if none */
BYTE *getframe_srcep(GENFRAME *gfp)
{
    ETHERHDR *ehp;
    BYTE *srce=0;

    if (gfp->g.dtype & DTYPE_ETHER)            /* Only Ethernet has address */
    {
        ehp = (ETHERHDR *)gfp->buff;
        srce = ehp->srce;
    }
    return(srce);
}
/* Copy the source MAC addr of the given frame; use broadcast if no addr */
BYTE *getframe_srce(GENFRAME *gfp, BYTE *buff)
{
    BYTE *p;

    p = getframe_srcep(gfp);
    if (p)
        memcpy(buff, p, MACLEN);
    else
        memcpy(buff, bcast, MACLEN);
    return(p);
}
/* Get pointer to the destination address of the given frame, 0 if none */
BYTE *getframe_destp(GENFRAME *gfp)
{
    ETHERHDR *ehp;
    BYTE *dest=0;

    if (gfp->g.dtype & DTYPE_ETHER)            /* Only Ethernet has address */
    {
        ehp = (ETHERHDR *)gfp->buff;
        dest = ehp->dest;
    }
    return(dest);
}
```

```
/* Copy destination MAC addr of the given frame; use broadcast if no addr */
BYTE *getframe_dest(GENFRAME *gfp, BYTE *buff)
{
    BYTE *p;

    p = getframe_destp(gfp);
    if (p)
        memcpy(buff, p, MACLEN);
    else
        memcpy(buff, bcast, MACLEN);
    return(p);
}
/* Get the protocol for the given frame; if unknown , return 0 */
WORD getframe_pcol(GENFRAME *gfp)
{
    ETHERHDR *ehp;
    WORD pcol=0;

    if (gfp->g.dtype & DTYPE_ETHER)           /* Only Ethernet has protocol */
    {
        ehp = (ETHERHDR *)gfp->buff;
        pcol = ehp->ptype;
    }
    return(pcol);
}
```

Using these functions, you can safely access the address and protocol fields on all packets, even though SLIP frames don't possess them. This avoids the necessity for frame-specific features in the SCRATCHP code layer, since all frames can be treated equally.

Addressing

I have already talked about the client contacting the server, but I have given no indication as to how this is achieved. How are the client and server identified so that they can contact each other? Of course, the server can simply respond to the address of any client that contacts it, but there is still the burden on the client to make the initial contact, and, to do that, it needs some way of addressing the host, since there might be multiple hosts on the network.

Each Ethernet card has a unique six-byte *physical address*, so the client could use that. But imagine the complaints from the users if they have to type a 12-digit hexadecimal number every time they want to contact a new host. Also, the number would be highly specific to that item of hardware. If the network card failed and had to be replaced, the number would change, even though the computer still seemed to be the same from the user's point of view.

It is far better to assign each computer on the network a *logical address* and then invent some scheme to map the logical address onto the physical address of the Ethernet card. For convenience, I will refer to the logical address as the Ident (ID) of the computer and the physical address as the *address*. The logical-to-physical mapping process is called *address resolution*.

What is an ID, and where does it come from? An ID can be numeric (123) or a null-terminated string (fileserver). I'll use the latter format for maximum flexibility. It must either be permanently burned into the software when it is created (a nuisance, since all nodes on the network would have to run different copies of the software) or read when the software is loaded — either from the command line or from a configuration file. Either way, it is essential that each computer on the network acquire a unique ID.

To resolve an ID into an address, the client must broadcast the ID on the network as an invitation for the designated server to respond. The server responds, giving its physical address, which the client stores and uses for all subsequent communications.

Figure 2.6   Sample ident **transactions.**

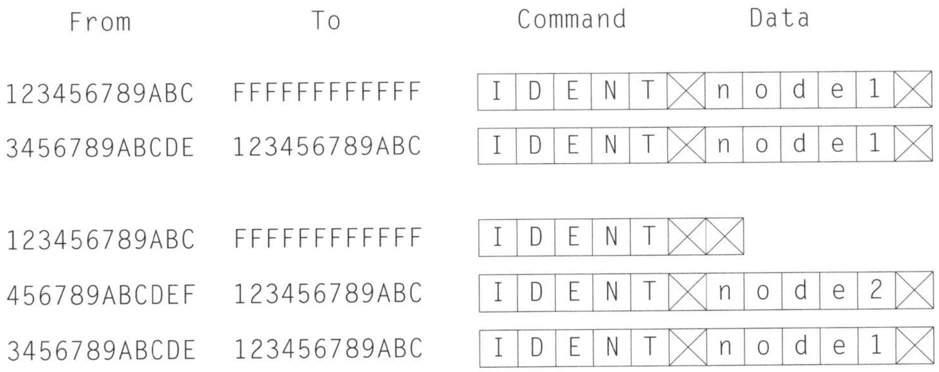

From	To	Command	Data
123456789ABC	FFFFFFFFFFFF	I D E N T ⊠	n o d e 1 ⊠
3456789ABCDE	123456789ABC	I D E N T ⊠	n o d e 1 ⊠
123456789ABC	FFFFFFFFFFFF	I D E N T ⊠	
456789ABCDEF	123456789ABC	I D E N T ⊠	n o d e 2 ⊠
3456789ABCDE	123456789ABC	I D E N T ⊠	n o d e 1 ⊠

Figure 2.6 shows a client broadcasting an identification request for the machine node1, using two null-terminated strings in the SCRATCHP data area — the null character is indicated by a strikethrough of the box. The client receives a reply containing a duplicate of the request, with the all-important node address, which will be used for subsequent communications. The second transaction illustrates the use of a null ident string to identify all nodes on the (hopefully very small) network. Two responses are obtained in a pseudorandom order. There is no knowing which node will answer first.

Protocol Identification

Ethernet is capable of carrying several protocols at the same time without the risk of confusion over which data belongs to which protocol. It achieves that by tagging each frame with a 16-bit protocol type, which uniquely identifies that protocol; for example, Internet Protocol (IP) has a hexadecimal value of 800h. If SCRATCHP was intended to coexist with other protocols, you would need to obtain an official protocol identifier from the Institution of Electrical and Electronic Engineers (IEEE). At the time of writing, this cost $5,000; however,

SCRATCHP should only be run on a "scratch" network, so you can use any identifier you like. Prudence dictates you should pick a high number that is out of the range of currently assigned protocols, so the hexadecimal value FEEBh is used.

Multiplexing and Buffering

The software that gathers transmit packets from a variety of senders is a *multiplexer* (mux, for short), and the corresponding software that accepts received packets and dispatches them to the appropriate recipient is called a *demultiplexer* (demux, for short).

Figure 2.7   Data flow between nodes.

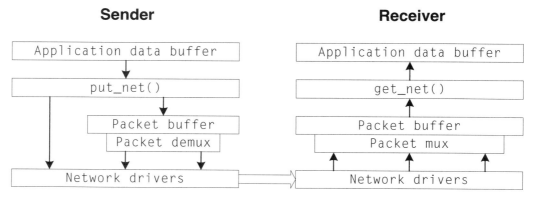

The mux/demux operation (Figure 2.7) is automatically performed by the network driver layer. Submitting a packet to put_net() automatically routes it to the appropriate network driver, possibly via a (polygonal, as described in the previous chapter) packet buffer, if the interface doesn't have its own Transmit buffer. All received packets are stored in a similar polygonal incoming packet buffer.

Control flow, as shown in Figure 2.8, is more convoluted since there must be some provision for polling the network interfaces, as they may be interrupt driven.

The receive_ether() and receive_slip() functions take the place of Ethernet and serial interrupt handlers, in that they are called from get_net(), call get_ether() or get_slip() for each packet received, and then do an up-call to save the packet, which in turn uses the standard circular buffer input routine (Figure 2.8). Having done that, get_net() calls the buffer output routine to fetch any stored packets.

If interrupts are available, the two receive_ functions are redundant, and the interrupt handlers call the get_ functions directly.

Figure 2.8   Control flow for packet reception.

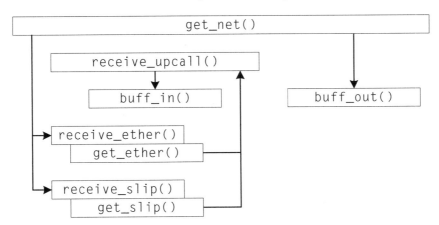

Byte Swapping

SCRATCHP will normally run on a little-endian (least significant byte first) PC architecture, so it was tempting to use this storage method for the two-byte values in the SCRATCHP packets. However, most Ethernet protocols use big-endian (most significant byte first) storage, and I wanted to explore byte-swapping issues, so I made SCRATCHP little-endian. If you happen to run my software on a little-endian machine, then the byte-swapping stage must be skipped (preferably using conditional compilation), but the underlying software structure remains the same.

I have seen protocol software that is liberally sprinkled with byte-swap functions, which is a nightmare to debug because you're never quite sure whether a value is in its swapped or unswapped state. To avoid this, you have to have a byte-swapping philosophy and stick rigidly to it. My philosophy is that byte swapping is the *last* action to be performed when *sending* a packet and the *first* action to be performed when *receiving* a packet.

This means that a transmit packet that has been byte swapped is fit only for transmission: it may not be used for other purposes such as diagnostic printouts because the printout function won't display the swapped values correctly. After transmission, a transmit packet must be discarded because it is useless; on the relatively rare occasions a retransmission is required, the packet can easily be rebuilt from the original data. This approach also helps to minimize the storage requirements and forces you to think clearly about a retry strategy, rather than relying on resending old packets that happen to be around. This rigorous approach is perhaps slightly too dogmatic and inflexible for a simple protocol such as SCRATCHP, but it prepares the ground for the more complex protocols to come.

Reception and Transmission

When a packet is received, do the necessary testing and byte swapping; then forward it to do_scratchp() for action.

```
/* Demultiplex incoming packets */
int get_pkts(GENFRAME *nfp)
{
    int rxlen, txlen=0;

    if ((rxlen=get_frame(nfp)) > 0)              /* If any packet received.. */
    {
        if (is_scratchp(nfp, rxlen))             /* If SCRATCHP.. */
        {
            swap_scratchp(nfp);                      /* ..do byte-swaps.. */
            txlen = do_scratchp(nfp, rxlen, 0);  /* ..action it.. */
        }
    }                                            /* ..and maybe return a response */
    return(txlen);                               /* (using the same pkt buffer) */
}
```

To economize on storage, do_scratchp() reuses the Receive buffer as a Transmit buffer to hold any response it wants to make and simply returns a transmit length value, or 0 if no response has been generated.

```
/* Check Ethernet frame, given frame pointer & length, return non-0 if OK */
int is_ether(GENFRAME *gfp, int len)
{
    int dlen=0;

    if (gfp && (gfp->g.dtype & DTYPE_ETHER) && len>=sizeof(ETHERHDR))
    {
        dlen = len - sizeof(ETHERHDR);
        swap_ether(gfp);
    }
    return(dlen);
}
/* Make a frame, given data length. Return length of complete frame
** If Ethernet, set dest addr & protocol type; if SLIP, ignore these */
int make_frame(GENFRAME *gfp, BYTE dest[], WORD pcol, WORD dlen)
{
    ETHERHDR *ehp;
```

```
    if (gfp->g.dtype & DTYPE_ETHER)
    {
        ehp = (ETHERHDR *)gfp->buff;
        ehp->ptype = pcol;
        memcpy(ehp->dest, dest, MACLEN);
        swap_ether(gfp);
        dlen += sizeof(ETHERHDR);
    }
    return(dlen);
}
/* Byte-swap an Ethernet frame, return header length */
void swap_ether(GENFRAME *gfp)
{
    ETHERFRAME *efp;

    efp = (ETHERFRAME *)gfp->buff;
    efp->h.ptype = swapw(efp->h.ptype);
}
/* Check SLIP frame, return non-zero if OK */
int is_slip(GENFRAME *gfp, int len)
{
    return((gfp->g.dtype & DTYPE_SLIP) && len>0);
}
/* Check for SCRATCHP, given frame pointer & length */
int is_scratchp(GENFRAME *nfp, int len)
{
    WORD pcol;
                                        /* SLIP has no protocol field.. */
    pcol = getframe_pcol(nfp);          /* ..so assume 0 value is correct */
    return((pcol==0 || pcol==PCOL_SCRATCHP) && len>=sizeof(SCRATCHPHDR));
}


/* Byte-swap a SCRATCHP packet, return header length */
int swap_scratchp(GENFRAME *nfp)
{
    SCRATCHPKT *sp;

    sp = getframe_datap(nfp);
    sp->h.dlen = swapw(sp->h.dlen);
```

```
    sp->h.seq = swapl(sp->h.seq);
    sp->h.ack = swapl(sp->h.ack);
    return(sizeof(SCRATCHPHDR));
}
```

Transmission is a fill-in-the-blanks exercise, followed by the necessary byte swaps.

```
/* Make a SCRATCHP packet given command, flags and string data */
int make_scratchpds(GENFRAME *nfp, BYTE *dest, char *cmd,
                    BYTE flags, char *str)
{
    return(make_scratchp(nfp, dest, cmd, flags, str, strlen(str)+1));
}


/* Make a SCRATCHP packet given command, flags and data */
int make_scratchp(GENFRAME *nfp, BYTE *dest, char *cmd, BYTE flags,
                    void *data, int dlen)
{
    SCRATCHPKT *sp;
    ETHERHDR *ehp;
    int cmdlen=0;

    sp = (SCRATCHPKT *)getframe_datap(&genframe);
    sp->h.ver = SCRATCHPVER;                    /* Fill in the blanks.. */
    sp->h.flags = flags;
    sp->h.seq = txbuff.trial;                   /* Direct seq/ack mapping.. */
    sp->h.ack = rxbuff.in;                       /* ..to my circ buffer pointers! */
    if (cmd)
    {
        strcpy((char *)sp->data, cmd);          /* Copy command string */
        cmdlen = strlen(cmd) + 1;
    }
    sp->h.dlen = cmdlen + dlen;                  /* Add command to data length */
    if (dlen && data)                            /* Copy data */
        memcpy(&sp->data[cmdlen], data, dlen);
    if (nfp->g.dtype & DTYPE_ETHER)
    {
        ehp = (ETHERHDR *)nfp->buff;
        ehp->ptype = PCOL_SCRATCHP;              /* Fill in more blanks */
        memcpy(ehp->dest, dest, MACLEN);
    }
```

```
    diaghdrs[diagidx] = sp->h;                /* Copy hdr into diagnostic log */
    diaghdrs[diagidx].ver = DIAG_TX;
    diagidx = (diagidx + 1) % NDIAGS;
    return(sp->h.dlen+sizeof(SCRATCHPHDR)); /* Return length incl header */
}


/* Transmit a SCRATCHP packet. given length incl. SCRATCHP header */
int put_scratchp(GENFRAME *nfp, WORD txlen)
{
    int len=0;

    if (txlen >= sizeof(SCRATCHPHDR))         /* Check for min length */
    {
        if (pktdebug)
        {
            printf ("Tx ");
            disp_scratchp(nfp);
            printf("   ");
        }
        swap_scratchp(nfp);                   /* Byte-swap SCRATCHP header */
        if (is_ether(nfp, txlen+sizeof(ETHERHDR)))
            txlen += sizeof(ETHERHDR);
        txcount++;
        len = put_net(nfp, txlen);            /* Transmit packet */
    }
    return(len);
}
```

Implementation

If you have read the first chapter, you'll not be surprised that I'm about to embark on a states-and-signals exercise. The software receives the following signals:

- User (keystrokes)
- Network (packets)
- Timer (time-outs)
- Null (idle)

When it receives one of these, it may take any or none of the following actions:

- Change state
- Send a packet
- Update user display

I'll start with the simplest command, ident, which is completely stateless.

ident **Command**

When the user presses the I key, a broadcast Ident packet is emitted. If any responses are received, the software displays them as part of its normal idle-state network polling.

First, I have a main loop that translates the key press into a signal.

```
GENFRAME *nfp;
WORD txlen;
...
nfp = &genframe;                          /* Open net driver.. */
nfp->ftype = frametype = open_net(netcfg);  /* ..get frame type */
...
int i, keysig, connsig, sstep=0;

while (cmdkey != 'Q')                /* Main command loop.. */
{
    txlen = keysig = connsig = 0;
    if (sstep || kbhit())            /* If single-step or keypress..*/
    {
        k = getch();                 /* ..get key */
        if (sstep)
            timeout(&errtimer, 0);   /* If single-step, refresh timer */
        cmdkey = toupper(k);         /* Decode keystrokes.. */
        switch (cmdkey)              /* ..and generate signals */
        {
        case 'I':                    /* 'I': broadcast ident */
            if (connstate != STATE_CONNECTED)
            printf("Broadcast ident request\n");
            keysig = SIG_USER_IDENT;
            break;
        }
    }
}
connsig = do_apps(&rxbuff, &txbuff, keysig);
txlen = do_scratchp(nfp, 0, connsig);
put_scratchp(nfp, txlen);          /* Transmit packet (if any) */
```

The user key press is translated into a key signal, SIG_USER_IDENT. This signal is bounced straight through the application code, do_apps(), without change (more on this function later). It is then sent to the main SCRATCHP state machine, do_scratchp(), to be translated into a network packet.

```
int do_scratchp(GENFRAME *nfp, int rxlen, int sig)
{
    ...
    if (connstate == STATE_IDLE)               /* If idle state.. */
    {
        timeout(&errtimer, 0);                 /* Refresh timer */
        switch (sig)                           /* Check signals */
        {
        case SIG_USER_IDENT:                       /* User IDENT request? */
            txlen = make_scratchpds(nfp, bcast, CMD_IDENT, FLAG_CMD, "");
            break;
        ...
        }
    }
    ...
}
```

The packet (in the buffer indicated by network frame pointer `nfp`) is then transmitted by `put_scratchp()`.

```
txlen = make_scratchpds(nfp, bcast, CMD_IDENT, FLAG_CMD, "");
put_scratchp(nfp, txlen);
...
```

So what happens when you press the I key? With a bit of luck, your first packet is sent on the network. If you're fortunate enough to possess a protocol analyzer (which captures and displays all network traffic), you might see a display similar to this:

```
Packet #1
  Packet Length:64
  Ethernet Header
  Destination:  FF:FF:FF:FF:FF:FF  Ethernet Broadcast
  Source:       00:C0:26:B0:0A:93  Rack2
  Protocol Type:0xFEEB
  Packet Data:
...........iden  01 01 00 00 00 00 00 00 00 00 00 07 69 64 65 6E
t...............  74 00 00 00 00 00 00 00 00 00 00 00 00 00 00 00
............      00 00 00 00 00 00 00 00 00 00 00 00 00 00 00
```

This is the actual byte stream on the network. The analyzer doesn't understand the packet contents, so some manual decoding is necessary. The six-byte Ethernet addresses are unique to each adaptor, so yours should not be the same as mine! The analyzer has identified the node name as `Rack2`, which, not coincidentally, is the same ID name as in the SCRATCHP configuration file.

The actual data is below the 64-byte minimum frame size, so there is a significant amount of padding. You can see the protocol version number (01) followed by the command flag (01). Skipping the four-byte sequence and acknowledgment numbers, there is a length value of seven (most significant byte first). The `ident` string is only six bytes, including a null terminator, so one extra null character is significant, indicating that this is a wildcard search for all nodes.

Such a broadcast would be inadvisable on a network of any size, since I'd get a flood of responses, but I'll assume I have only two other nodes on the network, named `vale` and `sun`, to get the responses.

```
Packet #2
  Packet Length:64
  Ethernet Header
  Destination:  00:C0:26:B0:0A:93  Rack2
  Source:       00:20:18:3A:ED:64  Sun  Protocol Type:0xFEEB
  Packet Data:
...........iden   01 02 00 00 00 00 00 00 00 00 00 0A 69 64 65 6E
t.sun..........   74 00 73 75 6E 00 00 00 00 00 00 00 00 00 00 00
..............    00 00 00 00 00 00 00 00 00 00 00 00 00 00 00

Packet #3
  Packet Length:64
  Ethernet Header
  Destination:  00:C0:26:B0:0A:93  Rack2
  Source:       00:50:04:F7:7C:CA  Vale
  Protocol Type:0xFEEB
  Packet Data:
...........iden   01 02 00 00 00 00 00 00 00 00 00 0B 69 64 65 6E
t.vale.........   74 00 76 61 6C 65 00 00 00 00 00 00 00 00 00 00
..............    00 00 00 00 00 00 00 00 00 00 00 00 00 00 00
```

The order in which these responses arrive is not significant, since both are transmitting at more or less the same time.

The responses are received and decoded and displayed by the SCRATCHP application.

```
ident 'sun' address 00:20:18:3a:ed:64
ident 'vale' address 00:50:04:f7:7c:ca
```

In a real application, the Ident-to-address mapping would be stored (cached) for reuse later. I will just display the addresses and discard them.

```c
int do_scratchp(GENFRAME *nfp, int rxlen, int sig)
{
    if (rxlen)                          /* If packet received.. */
    {
        rxflags = sp->h.flags;          /* Decode command & data areas */
        if (rxflags&FLAG_CMD || rxflags&FLAG_RESP)
            crlen = strlen((char *)sp->data) + 1;
        dlen = sp->h.dlen - crlen;      /* Actual data is after command */
        if (rxflags & FLAG_ERR)         /* Convert flags into signals */
            sig = SIG_ERR;
        ...
        else if (rxflags & FLAG_CMD)
```

```
            sig = SIG_CMD;
        else if (rxflags & FLAG_RESP)
            sig = SIG_RESP;
        ...
    }
    if (connstate == STATE_IDLE)                /* If idle state.. */
    {
        timeout(&errtimer, 0);                  /* Refresh timer */
        switch (sig)                            /* Check signals */
        {
        case SIG_CMD:                           /* Command signal? */
            if (!strcmp((char *)sp->data, CMD_IDENT))
            {                                   /* IDENT cmd with my ID or null? */
                if (dlen<2 || !strncmp((char *)&sp->data[crlen], locid, dlen))
                {                               /* Respond to sender */
                    txlen = make_scratchp(nfp, getframe_srcep(nfp), CMD_IDENT,
                                          FLAG_RESP, locid, strlen(locid)+1);
                }
            }
            break;

        case SIG_RESP:                          /* Response signal? */
            if (!strcmp((char *)sp->data, CMD_IDENT))
            {                                   /* IDENT response? */
                printf("Ident '%s'", (char *)&sp->data[crlen]);
                if ((p=getframe_srcep(nfp)) !=0 )
                {
                    printf(" address ");
                    pr6byt(p);
                }
                printf("\n");
            }
            break;
        ...
        }
    }
}
```

If a packet is received (`rxlen` is non-zero), then a signal is raised. Because I am the command originator, I'm interested in the response signal, `SIG_RESP`, which simply prints the `ident` name and address.

Note that the same function also handles the case where I have *received* a command, that is, I am the host being queried. In this case, a `SIG_CMD` is raised, and I must respond by putting my (local) `ident` string, `locid`, in the response.

It may seem strange placing the client and server code side-by-side in the same function, and this can make the code slightly more difficult to read, since these are two mutually exclusive execution strands. However, the commonality of the support code (e.g., packet composition and decomposition) and the vital necessity of keeping any modifications to the client and server in sync do favor this approach, even at the expense of some confusion over identity ("… so, is this a client, or server, or what?").

Connection

The bulk of the services require a logical connection between the two machines. I can put off the creation of state and signal tables no longer (Table 2.1).

Table 2.1   State and signal table.

Signals	States				
	IDLE	IDENT	OPEN	CONNECTED	CLOSE
CMD	Send RESP			Send to app.	
RESP	Check RESP	Send START <OPEN>		Send to app.	
START	Send CONN <CONNECTED>		Send CONN <CONNECTED>	Send CONN	
CONN	Send ERR		Send CONN <CONNECTED>	Send to app.	
STOP	Send ERR		Send STOP <IDLE>	Send STOP <IDLE>	<IDLE>
ERR			<IDLE>	Send STOP <IDLE>	Send STOP <IDLE>
timeout		Resend IDENT	Resend START	Resend data	Resend STOP
fail		<IDLE>	Send ERR <IDLE>	Send ERR <IDLE>	<IDLE>
open	Send IDENT <IDENT>				
close			Send END <IDLE>	Send STOP <CLOSE>	

Connection State Machine

The state changes have been marked with angle brackets, so `<IDENT>` indicates a change to the `IDENT` state. Network signals (from received packets) are in uppercase, whereas user and system signals (key presses and time-outs) are in lowercase. The `fail` signal is raised after several successive time-outs (i.e., the retry count has been exceeded).

Opening and Closing a Connection

To assist you in reading these tables, here's a sample connection sequence for a client. That is, the node requests the connection starting from the IDLE state.

1. receive open signal from user; send ident command; go to IDENT state
2. receive ident response; send start; go to OPEN state
3. receive conn; send conn; go to CONNECTED state

The client also shoulders the burden of handling connection errors. Each step is retried on time-out.

1. no ident response; resend ident command
2. no conn response; resend start command

The sequence for the server, starting from IDLE, is much simpler.

1. receive ident command; send response; no state change
2. receive start; send conn; go to CONNECTED state

However, you must exercise a small amount of caution when assuming that the client is responsible for all error handling. Imagine that the server's conn response is corrupted; the server then thinks it is connected, but the client doesn't realize this, so it resends a start signal. Although already connected, the server must accept this error condition (the duplicate start packet) and resend the conn.

There are two closure sequences: abrupt, in the event of an error, or slightly more graceful under normal conditions.

The graceful closure involves the exchange of stop signals, whereas the abrupt closure is the unilateral sending of an error packet. A potential problem with the latter is that the error packet may go astray; then one side would think the connection was still open, while the other thought it was closed. The only remedy for this situation is that, sooner or later, the open side would send a data packet to the closed side and receive an error packet in response, thus closing the connection.

The state machine software is simply a large set of nested conditionals, with entries for each state–signal combination that requires an action.

```c
int do_scratchp(GENFRAME *nfp, int rxlen, int sig)
{
    ...
    if (connstate == STATE_IDLE)            /* If idle state.. */
    {
        timeout(&errtimer, 0);              /* Refresh timer */
        switch (sig)                        /* Check signals */
        {
        case SIG_USER_IDENT:                /* User IDENT request? */
            txlen = make_scratchpds(nfp, bcast, CMD_IDENT, FLAG_CMD, "");
            break;
```

```
       case SIG_USER_OPEN:                    /* User OPEN request? */
           txlen = make_scratchpds(nfp, bcast, CMD_IDENT, FLAG_CMD, remid);
           buff_setall(&txbuff, 1);           /* My distinctive SEQ value */
           newconnstate(STATE_IDENT);         /* Start ident cycle */
           break;

       case SIG_CMD:                          /* Command signal? */
           ...
           break;

       case SIG_RESP:                         /* Response signal? */
           ...
           break;

       case SIG_START:                        /* START signal? */
           getframe_srce(nfp, remaddr);
           buff_setall(&txbuff, 0x8001);      /* My distinctive SEQ value */
           txack = sp->h.seq;                 /* My ack is his SEQ */
           buff_setall(&rxbuff, txack);
           *remid = 0;                        /* Clear remote ID */
           txlen = make_scratchp(nfp, remaddr, 0, FLAG_CONN, 0, 0);
           newconnstate(STATE_CONNECTED);     /* Go connected */
           break;

       case SIG_CONN:                         /* CONNECTED or STOP signal? */
       case SIG_STOP:
           txlen = make_scratchp(nfp, getframe_srcep(nfp), 0, FLAG_ERR, 0, 0);
           break;                             /* Send error */
       }
   }
   else if (connstate == STATE_IDENT)         /* If in identification cycle.. */
   {
       switch (sig)                           /* Check signals */
       {
       case SIG_RESP:                         /* Got IDENT response? */
           if (!strcmp((char *)sp->data, CMD_IDENT) && dlen<=IDLEN)
           {
               if (!remid[0] || !strcmp((char *)&sp->data[crlen], remid))
               {                              /* Get remote addr and ID */
                   getframe_srce(nfp, remaddr);
```

```
                        strcpy(remid, (char *)&sp->data[crlen]);
                    txlen = make_scratchp(nfp, remaddr, 0, FLAG_START, 0, 0);
                    newconnstate(STATE_OPEN);
                }                                 /* Open up the connection */
            }
            break;

        case SIG_ERR:                             /* Error response? */
            newconnstate(STATE_IDLE);             /* Go idle */
            break;

        case SIG_TIMEOUT:                         /* Timeout on response? */
            n = strlen(remid) + 1;                /* Resend IDENT command */
            txlen = make_scratchp(nfp, bcast, CMD_IDENT, FLAG_CMD, remid, n);
            break;

        case SIG_FAIL:                            /* Failed? */
            newconnstate(STATE_IDLE);             /* Go idle */
            break;
        }
    }
    else if (connstate == STATE_OPEN)             /* If I requested a connection.. */
    {
        switch (sig)                              /* Check signals */
        {
        case SIG_START:
        case SIG_CONN:                            /*  Response OK? */
            buff_setall(&rxbuff, sp->h.seq);
            txlen = make_scratchp(nfp, remaddr, 0, FLAG_CONN, 0, 0);
            newconnstate(STATE_CONNECTED);        /* Send connect, go connected */
            break;

        case SIG_STOP:                            /* Stop already? */
            txlen = make_scratchp(nfp, remaddr, 0, FLAG_STOP, 0, 0);
            newconnstate(STATE_IDLE);             /* Send stop, go idle */
            break;

        case SIG_ERR:                             /* Error response? */
            newconnstate(STATE_IDLE);
            break;                                /* Go idle */
```

```
        case SIG_TIMEOUT:                       /* Timeout on response? */
            txlen = make_scratchp(nfp, remaddr, 0, FLAG_START, 0, 0);
            break;                              /* Resend request */


        case SIG_FAIL:                          /* Failed? */
            newconnstate(STATE_IDLE);           /* Go idle */
            break;



    }
}
else if (connstate == STATE_CONNECTED)  /* If connected.. */
{
    switch (sig)                        /* Check signals */
    {
    case SIG_START:                         /* Duplicate START? */
        txlen = make_scratchp(nfp, remaddr, 0, FLAG_CONN, 0, 0);
        break;                              /* Still connected */


    case SIG_TIMEOUT:                       /* Timeout on acknowledge? */
        buff_retry(&txbuff, buff_trylen(&txbuff));
                                            /* Rewind data O/P buffer */
        /* Fall through to normal connect.. */
    case SIG_CONN:                          /* If newly connected.. */
    case SIG_NULL:                          /* ..or still connected.. */
        ...
        break;


    case SIG_USER_CLOSE:                    /* User closing connection? */
        txlen = make_scratchp(nfp, remaddr, 0, FLAG_STOP, 0, 0);
        newconnstate(STATE_CLOSE);          /* Send stop command, go close */
        break;


    case SIG_STOP:                          /* STOP command? */
        txlen = make_scratchp(nfp, remaddr, 0, FLAG_STOP, 0, 0);
        newconnstate(STATE_IDLE);           /* Send ack, go idle */
        break;
```

```
        case SIG_ERR:                         /* Error command? */
            newconnstate(STATE_IDLE);         /* Go idle */
            break;

        case SIG_FAIL:                        /* Application failed? */
            txlen = make_scratchp(nfp, remaddr, 0, FLAG_ERR, 0, 0);
            newconnstate(STATE_IDLE);         /* Send stop command, go idle */
            break;

        }
    }
    else if (connstate == STATE_CLOSE)        /* If I'm closing connection.. */
    {
        switch (sig)                          /* Check signals */
        {
        case SIG_STOP:                        /* Stop or error command? */
        case SIG_ERR:
            newconnstate(STATE_IDLE);         /* Go idle */
            break;

        case SIG_TIMEOUT:                     /* Timeout on response? */
            txlen = make_scratchp(nfp, remaddr, 0, FLAG_STOP, 0, 0);
            break;                            /* Resend stop command */
        }
    }
    return(txlen);
}
```

The state changes are handled by newconnstate(), which allows a simple diagnostic print-out if the appropriate debug option is enabled. It also refreshes the time-out timer, on the assumption that no time-out is required if the system is constantly changing state (or re-entering the same state).

```
/* Do a connection state transition, refresh timer, do diagnostic printout */
void newconnstate(int state)
{
    if (state!=connstate)
    {
        if (statedebug)
            printf("connstate %s\n", connstates[state]);
```

```
        if (state != STATE_CONNECTED)
            newappstate(APP_IDLE);              /* If not connected, stop app. */
    }
    connstate = state;
    errcount = 0;
    timeout(&errtimer, 0);                      /* Refresh timeout timer */
}
```

Maintaining a Connection

A connection supports the transfer of data between the two systems. The software must

- send and receive data, keeping in sync with the other node,
- reject duplicate data,
- resend lost data,
- avoid sending too much data to the other node, and
- avoid sending too little data in each packet.

To address the first point, imagine that both nodes have circular buffers of data, and you are simply trying to keep the circular buffer pointers in sync. The circular buffer pointers have 32-bit values (even though the buffer size doesn't warrant it) to allow a simple mapping onto the sequence and acknowledgment values. What is this mapping? Imagine a data block is traveling from one application into the transmit circular buffer, across the network, into the receive circular buffer, and into another application.

Figure 2.9 shows the data ABCDE in transit, on the assumption that it had to be transmitted over the network in two blocks, and a single acknowledgment was generated for both blocks.

It can be seen that the sequence pointer for the transfer is equivalent to the sender's trial pointer, whereas the acknowledgment value is equivalent to the sender's in pointer. This accounts for the following code in the routine used to create SCRATCHP packets.

```
/* Make a SCRATCHP packet given command, flags and data */
int make_scratchp(GENFRAME *nfp, BYTE *dest, char *cmd, BYTE flags,
                void *data, int dlen)
{
    SCRATCHPKT *sp;
    ...
    sp = (SCRATCHPKT *)getframe_datap(&genframe);
    ...
    sp->h.seq = txbuff.trial;           /* Direct seq/ack mapping.. */
    sp->h.ack = rxbuff.in;              /* ..to my circ buffer pointers! */
    ...
}
```

Figure 2.9   Data flow through a connection.

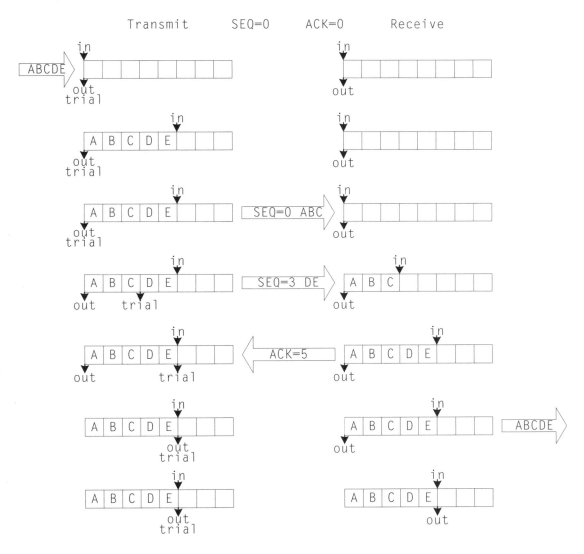

There are many ways to structure the connection code. The hardest job is to keep a clear indication of how it reaches its decisions as to whether to accept incoming packet data and whether to send data, acknowledgments, or both. First, I present the code for the receive decisions.

```
int do_scratchp(GENFRAME *nfp, int rxlen, int sig)
{
    ...
    LWORD oldrx, rxw, acked=0;
```

```
    ...
            /* Check received packet */
            if (rxlen > 0)                      /* Received packet? */
            {
                newconnstate(connstate);    /* Refresh timeout timer */

                /* Rx seq shows how much of his data he thinks I've received */
                oldrx = rxbuff.in - sp->h.seq; /* Check for his repeat data */
                if (oldrx == 0)                     /* Accept up-to-date data */
                    buff_in(&rxbuff, &sp->data[crlen], dlen);
                else if (oldrx <= WINDOWSIZE)  /* Respond to repeat data.. */
                    tx = 1;                         /* ..with forced (repeat) ack */
                else                            /* Reject out-of-window data */
                    errstr = "invalid SEQ";

                /* Rx ack shows how much of my data he's actually received */
                acked = sp->h.ack - txbuff.out; /* Check amount acked */
                if (acked <= buff_trylen(&txbuff))
                    buff_out(&txbuff, 0, (WORD)acked);  /* My Tx data acked */
                else if (acked > WINDOWSIZE)
                    errstr = "invalid ACK";
                rxw = rxbuff.in - txack;        /* Check Rx window.. */
                if (rxw >= WINDOWSIZE/2)         /* ..force Tx ack if 1/2 full */
                    tx = 1;
                if (errstr)                     /* If error, close connection */
                {
                    printf("Protocol error: %s\n", errstr);
                    txlen = make_scratchp(nfp, remaddr, 0, FLAG_ERR, 0, 0);
                    newconnstate(STATE_IDLE);
                }
            }
        ...
}
```

Usually, the incoming sequence value will equal the Receive buffer in value, so the incoming data block can be accepted. If it is not, but it is still within the data window size, then the block is probably a duplicate of a previous one and may be ignored (although the most likely reason for the duplicate is that the latest acknowledgment has gone astray, so it's best to retransmit it). If the incoming data block is outside the data window, then it can't be a duplicate, so an error is flagged.

A similar test is applied to the incoming acknowledgment value. This must be within the data window to be meaningful. If it is outside, it is an error condition.

The decision to transmit is contingent on having data to transmit or a pressing need to send an acknowledgment. It is tempting to generate an acknowledgment for every incoming packet, but this would significantly increase network traffic and the workload of the sender and receiver. Instead, wait until the data window is half full, the sender has duplicated a packet, or you have data to send (don't forget that every one of the data transmissions always has an acknowledgment field). This is hardly an optimal strategy, but it serves reasonably well.

```
/* Check whether a transmission is needed */
txw = WINDOWSIZE - buff_trylen(&txbuff);/* Check Tx window space */
trylen = minw(buff_untriedlen(&txbuff), /* ..size of data avail */
              minw(SCRATCHPDLEN, txw)); /* ..and max packet len */
if (trylen>0 || sig==SIG_TIMEOUT || tx) /* If >0, or timeout.. */
{                                        /* ..or forced Tx.. */
    txlen = make_scratchp(nfp, remaddr, 0, FLAG_CONN, 0, trylen);
    buff_try(&txbuff, sp->data, trylen);/* ..do a transmission */
    txack = rxbuff.in;
}
if (buff_trylen(&txbuff) == 0)  /* If all data acked.. */
    newconnstate(connstate);    /* refresh timer (so no timeout) */

break;
```

The Applications

Now that all the hard work of creating, maintaining, and destroying connections is done, there is the relatively simple job of creating application code for

- ECHO (connection diagnostic),
- DIRectory of files,
- GET (file transfer: read), and
- PUT (file transfer: write).

To isolate them from the vagaries of the network, these applications preside over two circular buffers: a Receive buffer that is automatically filled by incoming network data and a Transmit buffer that is automatically emptied into outgoing network packets. As far as the applications are concerned, data transfers are *reliable*. The only error they may see is a catastrophic failure of the connection. All other errors are handled by the lower levels.

To also isolate the applications from the vagaries of the user, they receive predigested user actions in the form of signals. They can also emit signals to the lower layers — for example, to close a connection if the user requests it (Figure 2.10).

Figure 2.10   Application and connection signals.

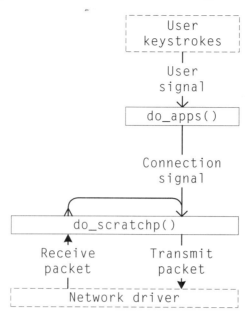

There is an inherent symmetry between the sending and receiving of files over the connection; to exploit this, I have a *sender* state and a *receiver* state, where a put command makes the client a sender and the server a receiver, and the get command does the converse.

```c
/* Do application-specific tasks, given I/P and O/P buffers, and user signal
** Return a connection signal value, 0 if no signal */
int do_apps(CBUFF *rxb, CBUFF *txb, int usersig)
{
    WORD len;
    BYTE lenb;
    int connsig=0;
    char cmd[CMDLEN+1];

    if (sigdebug && usersig && usersig>=USER_SIGS)
        printf("Signal %s ", signames[usersig]);
    connsig = usersig;                      /* Send signal to connection */
    if (connstate != STATE_CONNECTED)       /* If not connected.. */
        ;                                   /* Do nothing! */
    else if (appstate == APP_IDLE)          /* If application is idle.. */
    {
        if (usersig == SIG_USER_DIR)        /* User requested directory? */
        {                                   /* Send command */
```

```
                buff_in(txb, (BYTE *)CMD_DIR, sizeof(CMD_DIR));
    }
    else if (usersig == SIG_USER_GET)   /* User 'GET' command? */
    {
        filelen = 0;                        /* Open file */
        if ((fhandle = fopen(filename, "wb"))==0)
            printf("Can't open file\n");
        else
        {                                   /* Send command & name to remote */
            buff_instr(txb, CMD_GET " ");
            buff_in(txb, (BYTE *)filename, (WORD)(strlen(filename)+1));
            newappstate(APP_FILE_RECEIVER); /* Become receiver */
        }
    }
    else if (usersig == SIG_USER_PUT)   /* User 'PUT' command? */
    {
        filelen = 0;                        /* Open file */
        if ((fhandle = fopen(filename, "rb"))==0)
            printf("Can't open file\n");
        else
        {                                   /* Send command & name to remote */
            buff_instr(txb, CMD_PUT " ");
            buff_in(txb, (BYTE *)filename, (WORD)(strlen(filename)+1));
            newappstate(APP_FILE_SENDER);   /* Become sender */
        }
    }
    else if (usersig == SIG_USER_ECHO)  /* User equested echo? */
    {
        buff_in(txb, (BYTE *)CMD_ECHO, sizeof(CMD_ECHO));
        txoff = rxoff = 0;                  /* Send echo command */
        newappstate(APP_ECHO_CLIENT);      /* Become echo client */
    }
    else if ((len=buff_strlen(rxb))>0 && len<=CMDLEN)
    {
        len++;                             /* Possible command string? */
        buff_out(rxb, (BYTE *)cmd, len);
        if (!strcmp(cmd, CMD_ECHO))        /* Echo command? */
            newappstate(APP_ECHO_SERVER);  /* Become echo server */
        else if (!strcmp(cmd, CMD_DIR))    /* DIR command? */
            do_dir(txb);                   /* Send DIR O/P to buffer */
```

```
        else if (!strncmp(cmd, CMD_GET, 3)) /* GET command? */
        {                                   /* Try to open file */
            filelen = 0;
            strcpy(filename, &cmd[4]);
            if ((fhandle = fopen(filename, "rb"))!=0)
                newappstate(APP_FILE_SENDER);   /* If OK, become sender */
            else                            /* If not, respond with null */
                buff_in(txb, (BYTE *)"\0", 1);
        }
        else if (!strncmp(cmd, CMD_PUT, 3)) /* PUT command? */
        {
            filelen = 0;
            strcpy(filename, &cmd[4]);       /* Try to open file */
            fhandle = fopen(filename, "wb");
            newappstate(APP_FILE_RECEIVER); /* Become receiver */
        }
    }
    else                                    /* Default: show data from remote */
    {
        len = buff_out(rxb, apptemp, TESTLEN);
        apptemp[len] = 0;
        printf("%s", apptemp);
    }
}
else if (appstate == APP_ECHO_CLIENT)   /* If I'm an echo client.. */
{
    if (usersig==SIG_USER_CLOSE)            /* User closing connection? */
        newappstate(APP_IDLE);
    else
    {                                       /* Generate echo data.. */
        if ((len = minw(buff_freelen(txb), TESTLEN)) > TESTLEN/2)
        {
            len = rand() % len;             /* ..random data length */
            buff_in(&txbuff, &testdata[txoff], len);
            txoff = (txoff + len) % TESTLEN;/*..move & wrap data pointer*/
        }
        if ((len = buff_out(rxb, apptemp, TESTLEN)) > 0)
        {                                   /* Check response data */
            if (!memcmp(apptemp, &testdata[rxoff], len))
```

```
                    {                                    /* ..match with data buffer */
                rxoff = (rxoff + len) % TESTLEN;/*..move & wrap data ptr*/
                testlen += len;
                printf("%lu bytes OK       \r", testlen);
                }
            else
                {
                printf("\nEcho response incorrect!\n");
                connsig = SIG_STOP;        /* If error, close connection */
                }
            }
        }
    }
else if (appstate == APP_ECHO_SERVER)   /* If I'm an echo server.. */
    {
    if (usersig == SIG_USER_CLOSE)          /* User closing connection? */
        newappstate(APP_IDLE);
    else if ((len = minw(buff_freelen(txb), TESTLEN))>0 &&
            (len = buff_out(rxb, apptemp, len)) > 0)
        buff_in(txb, apptemp, len);        /* Else copy I/P data to O/P */
    }
else if (appstate == APP_FILE_RECEIVER) /* If I'm receiving a file.. */
    {
    while (buff_try(rxb, &lenb, 1))        /* Get length byte */
        {                                  /* If rest of block absent.. */
        if (buff_untriedlen(rxb) < lenb)
            {
            buff_retry(rxb, 1);            /* .. push length byte back */
            break;
            }
        else
            {
            filelen += lenb;
            buff_out(rxb, 0, 1);           /* Check length */
            if (lenb == 0)                 /* If null, end of file */
                {
                if (!fhandle || ferror(fhandle))
                    printf("ERROR writing file\n");
                fclose(fhandle);
```

```
                            fhandle = 0;
                            newappstate(APP_IDLE);
                    }
                    else                        /* If not null, get block */
                    {
                            buff_out(rxb, apptemp, (WORD)lenb);
                            if (fhandle)
                                    fwrite(apptemp, 1, lenb, fhandle);
                    }
            }
        }
    }
    else if (appstate == APP_FILE_SENDER)  /* If I'm sending a file.. */
    {                                        /* While room for another block.. */
        while (fhandle && buff_freelen(txb)>=BLOCKLEN+2)
        {                                        /* Get block from disk */
            lenb = (BYTE)fread(apptemp, 1, BLOCKLEN, fhandle);
            filelen += lenb;
            buff_in(txb, &lenb, 1);             /* Send length byte */
            buff_in(txb, apptemp, lenb);       /* ..and data */
            if (lenb < BLOCKLEN)               /* If end of file.. */
            {                                    /* ..send null length */
                buff_in(txb, (BYTE *)"\0", 1);
                fclose(fhandle);
                fhandle = 0;
                newappstate(APP_IDLE);
            }
        }
    }
    return(connsig);
}
```

Summary

I've looked at the elements of a protocol and how it can be slotted into the ISO standardization framework. There are a lot of decisions to be made when creating a new protocol, and I looked at the client–server model, with both modal and modeless clients. The *logical connection* is at the heart of any reliable data-transfer scheme, and connection management (opening, maintaining, and closing the connection) requires very careful organization.

In my implementation of the nonstandard SCRATCHP protocol, I looked at the issues of low-level packet storage and addressing and the strategies for buffering, byte-swapping, transmitting, and receiving packets.

The SCRATCHP utility I developed can be used to evaluate the performance of my protocol or as a test bed for the development of new protocols. It has some of the features of a "real" protocol (address resolution, reliable connection) but is implemented in a much simpler fashion.

The main weakness of my implementation is the inability to handle more than one connection at a time. In the future, I'll use the *socket* concept to group together all the information for one connection and support multiple sockets, where each may be in a different state.

Source Files

ether3c.c	3C509 Ethernet card driver
etherne.c	NE2000 Ethernet card driver
net.c	Network interface functions
netutil.c	Network utility functions
pktd.c	Packet driver (BC only)
scratchp.c	SCRATCHP protocol
serpc.c or serwin.c	Serial drivers (BC or VC)
dosdef.h	MS-DOS definitions (BC only)
ether.h	Ethernet definitions
net.h	Network driver definitions
netutil.h	Utility function and general frame definitions
scratchp.h	SCRATCHP protocol definitions
serpc.h	Serial driver definitions (BC or VC)
win32def.h	Win32 definitions (VC only)

SCRATCHP Utility

Utility	Test bed for a nonstandard protocol
Usage	`scratchp [configfile]`
	Reads `tcplean.cfg` from default directory if no file specified
Options	None
Example	`scratchp test.cfg`
Interface	Single keypress with user prompts

	[I]	Identify remote node
	[O]	Open connection to remote node
	[Q]	Quit

When connected

	[D]	Directory of remote
	[E]	Echo data test
	[G]	Get file from remote
	[P]	Put file into remote

Config	`net`	to identify network type
	`ident`	to identification string for node
Modes	Defaults to server mode unless otherwise directed	

Chapter 3

Network Addressing and Debugging

Overview

Even if you are only using TCP/IP to communicate between two computers side-by-side on the desktop, you still have to be aware of the underlying address structure, and that means understanding how the Internet works.

In this chapter, I'll briefly delve into the structure of internetworks (of which the Internet is one) and the addressing structure they use. As a practical demonstration of this, I'll build a scanning utility that searches for addresses on the network and reports back what it finds. While networks are under the microscope, I'll look at the process of detecting and tracing network activity, for help in future debugging efforts.

All Ethernet networks are not the same: there are two main standards for the format of frames transmitted on the network. I'll enhance the scanning utility to accept this other format and see what difference it makes.

Internetworks

Prior to the creation of the Internet, it was believed that all networks must use switched circuits to create a unique link between two nodes. The prime example of this was the telephone network (Figure 3.1), which could link any two telephones using electrical relays or their electronic equivalent. The relays select a unique cabling path between sender and receiver, and this is maintained for the duration of the connection.

Figure 3.1   Telephone line switching.

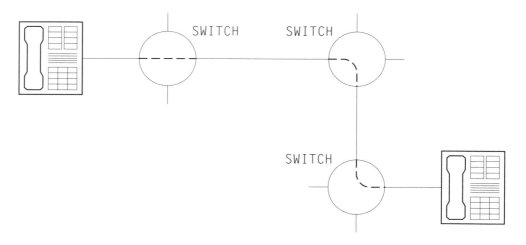

The radical step taken by the creators of the Internet was to arrange several possible paths between any two nodes and have the intervening switches (now called "routers") decide, on a *frame-by-frame* basis, which the best route was (Figure 3.2).

Figure 3.2   Network routing.

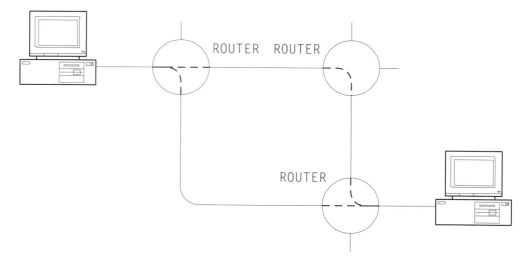

A router may have several paths for forwarding packets, and it must decide which to use on the basis of the destination address and the current network status. Because the latter changes dynamically, there is no guarantee that any two packets will use the same route, even if they have the same source and destination addresses. This means that the packets may arrive out of order, and some may not arrive at all if one of the routes is faulty. On the positive side, you gain a massive increase in flexibility and resilience. Apart from the entry and exit points, there are considerable redundancy and fault tolerance in the network.

In Figure 3.2, each router is shown as servicing only one PC. In reality, each PC would probably be on a local area network (LAN), which has various other computers attached. This LAN may be connected via a router to a wide area network (WAN) around a site, which in turn may be connected by a router to the Internet (Figure 3.3). The standard naming convention for this three-layer network hierarchy is as follows.

Subnetwork   a simple LAN

Network   a collection of subnetworks, linked by routers

Internetwork   a collection of networks, linked by routers

Figure 3.3   An internetwork.

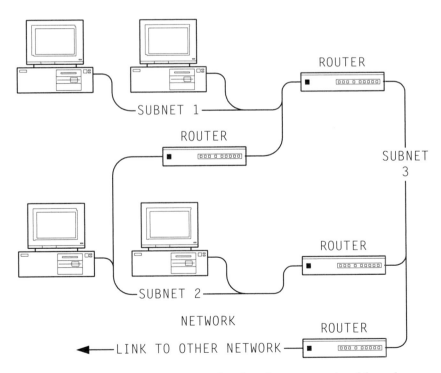

Not surprisingly, the Internet is an example of an internetwork, although not necessarily the only one, since an organization may have its own private internetwork.

In practice, the distinction between a subnetwork and a network tends to get blurred; both are called a network, and the routers may also be called *gateways*. This would be very confusing were it not that the operation of the (inter)network is almost completely transparent to your software, and you don't have to understand how gateways or routers work or what they are connected to. Your software must simply check to see if the intended destination is on your subnetwork (i.e., on your LAN). If so, the packets must be sent directly to the destination; if not, the packets must be sent to a gateway or router, which will forward them and return any replies to you or inform you that the intended destination is unreachable.

IP Addresses

To identify an individual computer on the Internet, it must have a unique address. The current version of the Internet Protocol (IPv4) uses a four-byte number, expressed in dotted decimal notation, (e.g., 123.45.67.8). This address consists of three parts.

1. A network address, which uniquely identifies an organization.
2. A subnet address, which identifies a subnet within that organization.
3. A system address, which identifies a single node on that subnet.

The size of these fields varies, depending on the size of the organization, but they must occupy a total of four bytes. The first number determines the size of the network address, using the following classes:

A. 0 to 127   one byte

B. 128 to 191   two bytes

C. 192 to 233   three bytes

So the address 123.45.67.8 has a class A network address of 123. However, if this node wants to contact another with the address 123.45.78.9, the knowledge that it is in the same *organization* is of little use; what the node really needs to know is whether the destination is on the same *subnet*; that is, whether it can be contacted directly or whether the packets must be sent to a router for forwarding to another subnet. To do this, each node is equipped with a "subnet mask"; a logical AND with this value will eliminate the system address so that the rest of the address fields can be compared. If the node 123.45.67.8 has a subnet mask of 255.255.255.0, then it would detect that 123.45.78.9 was on a different subnet (123.45.67.0 != 123.45.78.0), whereas a mask value of 255.255.0.0 would suggest it is on the same subnet (123.45.0.0 == 123.45.0.0).

The following IP address values have specific meanings:

Broadcast   A value of 255.255.255.255 is a broadcast to all nodes on the current LAN (i.e., subnetwork). A broadcast may also be directed to a remote subnet by using its address with a system address entirely of ones. It is theoretically possible to issue a broadcast to a remote network (i.e., all subnets in the network), but this is normally blocked by routers.

Loopback   Any address beginning with 127 is a loopback address, so most TCP/IP stacks bounce back the packets internally without sending them on the network. This can be useful for checking network applications on multiuser systems. Past custom dictates that only 127.0.0.1 is used for this purpose, so the rest of this large address range is unused.

Zero   If a node boots without an IP address, it assumes a value of 0.0.0.0, so this cannot be used as a node address. Zero-value system addresses are inadvisable anyway because they may not be accepted by routers.

Network not connected to the internet   One of each address class is assigned to networks (such as the one I'm developing here) that are not connected to the Internet.

- 10.0.0.0 to 10.255.255.255
- 172.16.0.0 to 172.31.255.255
- 192.168.0.0 to 192.168.255.255

These addresses are used by a large number of networks, but, because they can't intercommunicate, there is no risk of confusion.

In this book, I'll address Ethernet nodes from 10.1.1.1 to 10.1.1.20, with a subnet mask of 255.255.255.0, and serial links from 172.16.1.1 onward. These values are set in a configuration file that is read by the software at boot time. It is vital that every node has an IP address that is unique on the network (i.e., a unique configuration file), or considerable confusion will result.

Address Resolution

The software in this book supports both point-to-point serial (SLIP) links, and Ethernet networks. Although they use the same IP addressing scheme, they do require different strategies for handling those addresses.

SLIP Addressing

If there are only two nodes on a network and one emits a message, it seems obvious that the message is intended for the other node, so the IP address value would appear to be redundant. Unfortunately, SLIP links don't normally work that way, because

A SLIP link is just another network.   TCP/IP doesn't know how many nodes are actually connected, so it treats the link just like any other network, with IP addresses at both ends;

Either or both nodes might be a router.   The nodes may not be the ultimate destination of the messages, which can be forwarded to other nodes or networks; therefore, if the IP address is omitted, the router won't know where to send the message.

As a result, a SLIP node will reject a message if the IP address is incorrect, even though that node is obviously the intended recipient of the message. Just like the old storekeeper's joke, "It's no use pointing at it, the computer says it's not there."

Ethernet Addressing

The Ethernet hardware doesn't understand IP addresses; it has its own addressing scheme based on a unique six-byte address for each network adaptor manufactured; this is generally called the media access and control (MAC) address.

There are three ways for a node to send a message over Ethernet.

Broadcast Intended for all nodes on a network

Multicast Intended for a group of nodes

Unicast Intended for one specific node

Broadcasts are to be avoided if at all possible — not for security reasons (since any network adaptor can snoop on all network traffic by entering "promiscuous" mode), but because every node on the network must check the contents of the broadcast to see if a response should be generated. The more broadcasts that are sent, the greater the workload on every CPU connected to the network.

Multicast messages are intended to be received by a selected group of nodes. The applications I'll use in this book don't support multicasting, so I won't discuss this further.

Unicast addressing is for point-to-point communications across a network; if you know the unicast IP address of the node you want to communicate with, you must translate (resolve) it into the six-byte Ethernet address for that node and use that Ethernet address for all subsequent communications. The IP-to-hardware address translation protocol is called address resolution protocol (ARP) and is extremely simple in concept. A node sends a subnet broadcast containing the IP address that is to be resolved, and the node that matches that IP address sends a response with its hardware address.

Figure 3.4   ARP packet format.

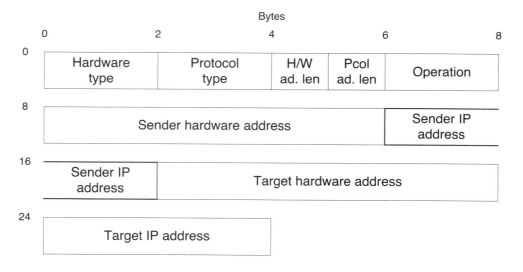

The ARP packet (Figure 3.4) was designed to accommodate other networks and protocols with different address sizes, so it includes two fields to specify the byte count of the hardware and protocol address fields.

```
/* ***** ARP (Address Resolution Protocol) packet ***** */
typedef struct
{
    WORD    hrd,                /* Hardware type */
            pro;               /* Protocol type */
    BYTE    hln,               /* Len of h/ware addr (6) */
            pln;               /* Len of IP addr (4) */
    WORD    op;                /* ARP opcode */
    BYTE    smac[MACLEN];      /* Source MAC (Ethernet) addr */
    LWORD   sip;               /* Source IP addr */
    BYTE    dmac[MACLEN];      /* Destination Enet addr */
    LWORD   dip;               /* Destination IP addr */
} ARPKT;
#define HTYPE      0x0001  /* Hardware type: ethernet */
#define ARPPRO     0x0800  /* Protocol type: IP */
#define ARPXXX     0x0000  /* ARP opcodes: unknown opcode */
#define ARPREQ     0x0001  /*              ARP request */
#define ARPRESP    0x0002  /*              ARP response */
```

ARP is generally used as a preliminary step in establishing communications: each node maintains an ARP cache, which has the current Ethernet addresses of all other nodes it is talking to. Before incorporating ARP into a larger body of code, I'll create a simple ARP-only utility that actually has a practical use — an ARP scanner.

ARP Scanner

When working with a new network or upgrading an old one, there can be some confusion over how the nodes are configured, that is, which IP addresses have been assigned and whether two nodes were given the same address by mistake. You could walk around to each machine and manually check its configuration, but it's far easier to do this over the network by sending ARP requests to each IP address in turn and checking for responses. If more than one response is received for a given IP address, then a node has been configured incorrectly. If no response is received, the IP address might be unused, but it would be wise to check for network errors (by rerunning the utility) or hardware faults (node powered down or disconnected) before reassigning it to another node.

Such an ARP scanner is useful on existing IP networks, but it is potentially useless if you don't have some TCP/IP nodes to scan. A minor addition, *server mode*, allows the same program to act like a normal TCP/IP node that simply responds to ARP requests. By setting one node as the scanner and all others as servers, you can do a simple functional check of a new network.

At the risk of stating the obvious, an ARP scanner is of no use on a SLIP network because there are no MAC addresses and no need to perform address resolution.

Implementation

A minimal ARP scanner would take the following actions:

1. Read command-line parameters (configuration filename, address range)
2. Read configuration file; open network driver
3. Main loop: for each IP address
   a. If IP address matches yours, display your Ethernet address
   b. Transmit address-resolution request
   c. Delay; if any ARP responses, display their Ethernet addresses
4. Close network driver

It is important to wait a reasonable amount of time between transmissions; ARP requests are broadcasts, and a flood of broadcast messages adversely affects the performance of all nodes on the network. A reasonable compromise between network loading and scan speed is a delay of 100ms; this is also a reasonable time for an ARP response to be received (assuming no network errors).

In server mode, step 3 is replaced with an endless loop that looks for incoming ARP requests and generates ARP responses if the IP address matches the value read from the configuration file.

Node

It is convenient to use a single structure to store the

- IP address,
- corresponding Ethernet address,
- gateway address,
- subnet mask, and
- network driver type.

A NODE structure contains these because they all apply to a single node on the network.

```
/* Structure for one node */
typedef struct
{
    WORD   dtype;                   /* Driver type */
    BYTE   mac[MACLEN];             /* MAC (Ethernet) addr */
    LWORD  ip;                      /* IP addr */
    LWORD  mask;                    /* Subnet mask */
    LWORD  gate;                    /* Gateway addr */
    WORD   port;                    /* TCP or UDP port number */
} NODE;
```

It turns out that this is also a convenient place to store another parameter, a port number, that is specific to a "logical endpoint" within the node, so perhaps the structure name isn't quite so apt. I'll discuss port numbers later in the UDP and TCP chapters.

I use the following terminology with regard to nodes:

Source   The originator of a message

Destination   The intended recipient of a message

Local   The node on which the software is running

Remote   The distant node with which you are communicating

It follows that an *incoming* message has a *remote source* and *local destination*, whereas an *outgoing* message has a *local source* and *remote destination*.

Command-Line Processing

The user specifies a starting IP address with options for an alternative configuration file, a node count, and the server mode (by omission of the starting IP address).

```c
int main(int argc, char *argv[])
{
    int args=0, err=0;
    LWORD remip=0, mstimer;
    WORD rxlen, txlen, dtype;
    GENFRAME *gfp;
    ARPKT *arp;
    char *p, temps[18];

    printf("ARPSCAN v" VERSION "\n");
    signal(SIGINT, break_handler);            /* Trap ctrl-C */
    while (argc > ++args)                      /* Process command-line args */
    {
        if (argv[args][0]=='-')
        {
            switch (toupper(argv[args][1]))
            {
            case 'C':                          /* -C: config filename */
                strncpy(cfgfile, argv[++args], MAXPATH);
                if ((p=strrchr(cfgfile, '.'))==0 || !isalpha(*(p+1)))
                    strcat(cfgfile, CFGEXT);
                break;
            case 'N':                          /* -N: num of nodes to scan */
                scancount = atoi(argv[++args]);
```

```
                        break;
                case 'V':                               /* -V: verbose packet display */
                        netdebug = 1;
                        break;
                default:
                        err = 1;
                }
            }
        else if isdigit(argv[args][0])          /* Starting IP address */
            remip = atoip(argv[args]);
        }
    if (err)                                    /* Prompt user if error */
        disp_usage();
    else if (!(dtype=read_netconfig(cfgfile, &locnode)))
        printf("Invalid configuration '%s'\n", cfgfile);
    else
    {
    ...main loop...
    ...close net driver..
    }
}
/* Display usage help */
void disp_usage(void)
{
    printf("Usage:    ARPSCAN [options] start_IP_addr\n");
    printf("Options:  -c name     Config filename (default TCPLEAN.CFG)\n");
    printf("          -n count    Number of IP addrs to scan (default %u)\n",
            SCANCOUNT);
    printf("Example:  ARPSCAN -c test.cfg 10.1.1.1\n");
}
```

Configuration File and Network Initialization

As a minimum, you must read the network driver details and local IP address from the configuration file.

```
/* Read network config file to get IP address
** Return driver type, 0 if error */
WORD read_netconfig(char *fname, NODE *np)
{
    char temps[31];
    WORD dtype=0;
```

```
    if (read_cfgstr(fname, "net", netcfg, MAXNETCFG))
    {                                              /* Get IP address */
        if (!read_cfgstr(fname, "ip", temps, 30) || (np->ip=atoip(temps))==0)
            printf("No IP address\n");
        else if (!(dtype = open_net(netcfg)))        /* Open net driver */
            printf("Can't open net driver '%s'\n", netcfg);
        else                                     /* Save ether address */
            memcpy(np->mac, ether_addr(dtype), MACLEN);
    }
    return(dtype);
}
```

The network driver string is passed to open_net() as it is read from the configuration file, which returns a driver-type identifier if successful. This is effectively a driver handle, which also contains information about the type of frames (Ethernet, SLIP) being handled. All outgoing packets must be tagged with the driver type to ensure that they are sent correctly, and all incoming packets will be automatically tagged with the driver type of the interface that received them.

Various helper functions are used for parsing the configuration file and extracting the configuration values; they do not merit a detailed description here, but are included in the source code on the CD-ROM.

Main Loop

If in scanning mode, an ARP request is sent every 100ms. The make_arp() function builds the outgoing packet, given the source and destination addresses. It is then byte swapped and transmitted. Any incoming packets are validated, using is_arp(), then checked to see whether they are a response to a previous request or a new request (server mode). If the latter, a response packet is created and sent.

… after config file and network init …

```
    remnode.dtype = genframe.g.dtype = dtype;   /* Set frame driver type */
    gfp = &genframe;                            /* Get pointer to frame */
    printf("Press ESC or ctrl-C to exit\n");
    printf("IP %s", ipstr(locnode.ip, temps));
    if (dtype & DTYPE_ETHER)
        printf(" Ethernet %s (local)\n\n", ethstr(locnode.mac, temps));
    mstimeout(&mstimer, 0);                     /* Refresh timer */
    while (!breakflag)
    {                                           /* If scanning & timeout.. */
        if (remip && mstimeout(&mstimer, DELTIME))
```

```
            {
                if (!scancount--)                    /* ..stop looping if done */
                    break;
                remnode.ip = remip++;                /* Broadcast next IP adr */
                memcpy(remnode.mac, bcast, MACLEN);
                txlen = make_arp(gfp, &locnode, &remnode, ARPREQ);
                put_frame(gfp, txlen);
            }
            poll_net(gfp->g.dtype);                  /* Keep network alive */
            if ((rxlen=get_frame(gfp)) > 0)          /* Check for incoming pkts */
            {
                if (is_arp(gfp, rxlen))
                {                                    /* ARP response? */
                    arp = getframe_datap(gfp);
                    if (arp->op==ARPRESP && arp->sip==remnode.ip)
                    {
                        printf("IP %s ", ipstr(remnode.ip, temps));
                        printf("Ethernet %s\n", ethstr(arp->smac, temps));
                    }
                    if (arp->op==ARPREQ && arp->dip==locnode.ip)
                    {                                /* ARP request? */
                        remnode.ip = arp->sip;   /* Make ARP response */
                        memcpy(remnode.mac, arp->smac, MACLEN);
                        txlen = make_arp(gfp, &locnode, &remnode, ARPRESP);
                        put_frame(gfp, txlen);
                    }
                }
            }
        }
        if (kbhit())                                 /* If user hit a key.. */
            breakflag = getch()==0x1b;               /* ..check for ESC */
    }
    close_net(dtype);                                /* Shut down net driver */
    }
    return(0);
}
```

Packet Checking and Formatting

Helper functions do the donkeywork of identifying, creating, and byte swapping the ARP packets.

```c
/* Check ARP packet, swap bytes, return -1, 0 if not ARP */
int is_arp(GENFRAME *gfp, int len)
{
    WORD pcol;
    ARPKT *arp;
    int dlen=0;

    pcol = getframe_pcol(gfp);          /* ARP only on Ether */
    if (pcol==PCOL_ARP && len>=sizeof(ARPKT))
    {                                   /* If protocol OK.. */
        arp = getframe_datap(gfp);
        swap_arp(gfp);                  /* ..check ARP data */
        if (arp->hrd==HTYPE && arp->pro==ARPPRO)
            dlen = -1;                  /* Return non-zero if OK */
        else
        {
            dlen = 0;                   /* Swap back if not OK */
            swap_arp(gfp);
        }
    }
    return(dlen);
}

/* Make an ARP packet, return its total length */
int make_arp(GENFRAME *gfp, NODE *srcep, NODE *destp, WORD code)
{
    ARPKT *arp;

    gfp->g.fragoff = 0;                 /* No fragmentation */
    arp = (ARPKT *)getframe_datap(gfp);
    memcpy(arp->smac, srcep->mac,  MACLEN);/* Srce ARP ether addr */
    memcpy(arp->dmac, destp->mac, MACLEN);  /* Dest ARP ether addr */
    arp->hrd = HTYPE;                   /* Hware & protocol types */
    arp->pro = ARPPRO;
    arp->hln = MACLEN;                  /* Hardware addr len */
    arp->pln = sizeof(LWORD);           /* IP addr len */
```

```
    arp->op  = code;                   /* ARP opcode */
    arp->dip = gate_ip(destp, srcep);  /* Dest ip addr (maybe gateway) */
    arp->sip = srcep->ip;              /* Source IP addr */
    swap_arp(gfp);
    return(make_frame(gfp, destp->mac, PCOL_ARP, sizeof(ARPKT)));
}


/* Swap byte order of ints in ARP header */
void swap_arp(GENFRAME *gfp)
{
    ARPKT *arp;

    arp = getframe_datap(gfp);
    arp->hrd = swapw(arp->hrd);
    arp->pro = swapw(arp->pro);
    arp->op = swapw(arp->op);
    arp->sip = swapl(arp->sip);
    arp->dip = swapl(arp->dip);
}
```

The `is_`, `make_`, `swap_` function hierarchy will be used for all the protocols; `is_arp()` checks an incoming frame and applies the necessary byte swaps, and `make_arp()` creates an outgoing frame, applies the necessary byte swaps, and returns the total transmit frame length.

Using ARPSCAN for Network Debugging

Assuming the system has an NE2000-compatible network card at I/O address 280h and the current directory contains the default configuration file `tcplean.cfg` with the entries

```
net     ether ne 0x280
ip      10.1.1.11
```

then the command

```
arpscan 10.1.1.1
```

scans 20 nodes from 10.1.1.1 upward. On my network, the following output is generated.

```
ARPSCAN Vx.xx
Press ESC or ctrl-C to exit
IP 10.1.1.11 Ethernet 00:c0:26:b0:0a:93 (local)

IP 10.1.1.1 Ethernet 00:c0:f0:09:bd:c3
IP 10.1.1.2 Ethernet 00:20:18:3a:76:f9
IP 10.1.1.3 Ethernet 00:20:18:3a:ed:64
```

This shows four nodes on the network: the local node is 10.1.1.11, and there are three remote nodes, with no duplication of IP addresses.

If you use the verbose (-v) option, you can see the frames as they are sent and received, though it isn't particularly informative at this low level.

```
ARPSCAN Vx.xx
Press ESC or ctrl-C to exit
IP 10.1.1.11 Ethernet 00:c0:26:b0:0a:93 (local)

Tx / len 42 ----BROADCAST---- ARP
Rx \ len 60 00:c0:f0:09:bd:c3 ARP
IP 10.1.1.1 Ethernet 00:c0:f0:09:bd:c3
Tx / len 42 ----BROADCAST---- ARP
Rx \ len 60 00:20:18:3a:76:f9 ARP
IP 10.1.1.2 Ethernet 00:20:18:3a:76:f9
Tx / len 42 ----BROADCAST---- ARP
Rx \ len 60 00:20:18:3a:ed:64 ARP
IP 10.1.1.3 Ethernet 00:20:18:3a:ed:64
Tx / len 42 ----BROADCAST---- ARP
Tx / len 42 ----BROADCAST---- ARP
Tx / len 42 ----BROADCAST---- ARP
... and so on ...
```

Note that the transmit length is reported as 42 bytes, whereas the received data, ostensibly the same size, is reported as 60 bytes. After the diagnostic display, the transmit frames are padded out to the minimum Ethernet size of 60 bytes (64 bytes including the CRC). Conversely, the low-level receiver drivers are unable to determine the actual (unpadded) length, so they must return a value of 60.

Protocol Analyzer

A protocol analyzer is a very useful yet remarkably expensive piece of software that allows you to capture and display network traffic. It checks that the software developed in this book performs as intended. Here is a log of the first ARP transaction to 10.1.1.1.

```
Packet #1
  Packet Length:64
Ethernet Header
  Destination:  FF:FF:FF:FF:FF:FF  Ethernet Broadcast
  Source:       00:C0:26:B0:0A:93
  Protocol Type:0x0806  IP ARP
ARP - Address Resolution Protocol
  Hardware:                1  Ethernet (10Mb)
  Protocol:                0x0800  IP
  Hardware Address Length: 6
  Protocol Address Length: 4
  Operation:               1  ARP Request
  Sender Hardware Address: 00:C0:26:B0:0A:93
  Sender Internet Address: 10.1.1.11
  Target Hardware Address: FF:FF:FF:FF:FF:FF  Ethernet Broadcast
(ignored)
  Target Internet Address: 10.1.1.1
```

```
Packet #2
  Packet Length:64
Ethernet Header
  Destination:  00:C0:26:B0:0A:93
  Source:       00:C0:F0:09:BD:C3
  Protocol Type:0x0806  IP ARP
ARP - Address Resolution Protocol
  Hardware:                 1  Ethernet (10Mb)
  Protocol:                 0x0800  IP
  Hardware Address Length:  6
  Protocol Address Length:  4
  Operation:                2  ARP Response
  Sender Hardware Address:  00:C0:F0:09:BD:C3
  Sender Internet Address:  10.1.1.1
  Target Hardware Address:  00:C0:26:B0:0A:93
  Target Internet Address:  10.1.1.11
```

Oscilloscope

If you aren't blessed with a protocol analyzer but are experiencing network problems, a digital storage oscilloscope can be pressed into service. Figure 3.5 shows the request and response on a 10 base 2 network, as captured on a Tektronix TDS220 DSO.

Figure 3.5   Oscilloscope trace of Ethernet transaction.

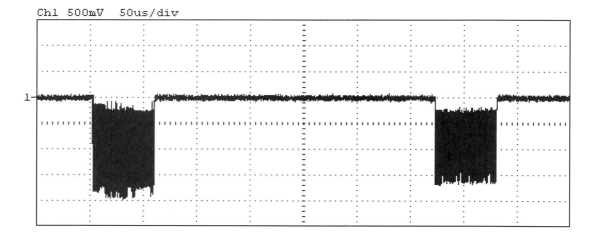

Using the serial interface and some custom software, it is even possible to decode the frame data. The upper chart of Figure 3.6 shows the start of a frame, with the preamble (62 bits), start-of-frame delimiter (2 bits), and decoded hex data. The lower chart is a magnified version of the center section, with the individual data bits shown.

Figure 3.6   Decoded oscilloscope trace of Ethernet frame.

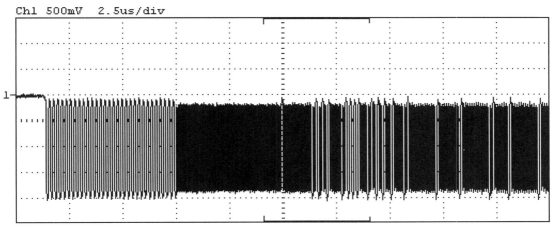

FF FF FF FF FF FF 00 C0 26 B0 0A 93 08 06 00 01 08 00 06 04 00

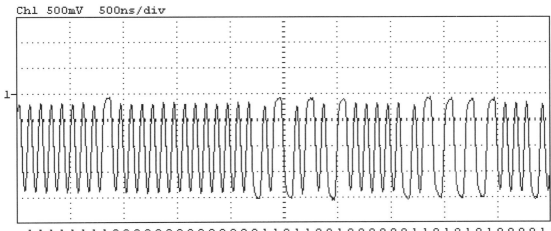

,1 1 1 1 1 1 1 1,0 0 0 0 0 0 0 0,0 0 0 0 0 0 1 1,0 1 1 0 0 1 0 0,0 0 0 0 1 1 0 1,0 1 0 1 0 0 0 0,1 :

It may seem strange using an oscilloscope to view Ethernet traffic, but it can be useful in tracking down strange network problems. If, instead of the clean traces above, you found that the waveforms were highly noisy or distorted, then you might save yourself hours of fruitless software investigation by looking at the hardware (e.g., cabling) first.

NETMON Network Monitor

As a cheap-and-cheerful alternative to an oscilloscope or protocol analyzer, I have written a simple utility to read frames off the network and decode and display them. I had intended to add features such as address filtering and logging to disk, but I ran out of time. By the time you read this, an updated version may be available from Iosoft Ltd. (see Appendix B for contact details).

Despite its simplicity, it is quite handy for checking out network transactions. The display below is of an ARP scan of four nodes.

```
NETMON vx.xx
Net 1 Ethernet
Press ESC or ctrl-C to exit
Rx0 \len 60 00:c0:26:b0:0a:93 ARP 10.1.1.11 -> 10.1.1.1
Rx0 \len 60 00:c0:f0:09:bd:c3 ARP 10.1.1.1 -> 10.1.1.11
Rx0 \len 60 00:c0:26:b0:0a:93 ARP 10.1.1.11 -> 10.1.1.2
Rx0 \len 60 00:20:18:3a:76:f9 ARP 10.1.1.2 -> 10.1.1.11
Rx0 \len 60 00:c0:26:b0:0a:93 ARP 10.1.1.11 -> 10.1.1.3
Rx0 \len 60 00:c0:26:b0:0a:93 ARP 10.1.1.11 -> 10.1.1.4
```

Although not the easiest display to read, it does show the four outgoing requests and the two responses returned. To achieve this, it puts the network card into "promiscuous mode," so it receives all frames, from whatever source. Of course, you can't use the same PC for the ARP scan and the network monitoring, unless you equip it with two network cards and load different configuration files for the two utilities running in separate DOS boxes (which I did to produce the display above).

NETMON is just a skeletal version of the TCP/IP stack I'll develop in future chapters, so I won't look at the source code here; however, it is included on the CD-ROM.

NETMON is useful for monitoring serial links, as well, and it is possible to use a PC with two spare serial ports for this purpose (one to receive each direction of serial signal), but this requires the creation of a special serial cable, and the attachment of a third device would violate the RS232 specification. Assuming you find a way around these obstacles, the configuration file for NETMON would have two entries, as in the following example.

```
net slip pc com1:38400, n,8,1
net slip pc com2:38400, n,8,1
```

The entries from the two interfaces are prefixed with RX0 and RX1. The following output is a TCP transaction recorded in this way.

```
NETMON vx.xx
Net 1 SLIP
Net 2 SLIP
Press ESC or ctrl-C to exit
Rx0 \len 64 ------SLIP------- IP 172.16.1.2 -> 172.16.1.1 TCP
    \seq 0287e09f ack 00000000 port 13<-1115 <SYN> MSS 1460 dlen 0h
Rx1 \len 40 ------SLIP------- IP 172.16.1.1 -> 172.16.1.2 TCP
    \seq 0006ffff ack 0287e0a0 port 1115<-13 <SYN><ACK> dlen 0h
Rx0 \len 40 ------SLIP------- IP 172.16.1.2 -> 172.16.1.1 TCP
    \seq 0287e0a0 ack 00070000 port 13<-1115 <ACK> dlen 0h
Rx1 \len 50 ------SLIP------- IP 172.16.1.1 -> 172.16.1.2 TCP
    \seq 00070000 ack 0287e0a0 port 1115<-13 <FIN><ACK> dlen Ah
00:12:48\r\n
Rx0 \len 40 ------SLIP------- IP 172.16.1.2 -> 172.16.1.1 TCP
    \seq 0287e0a0 ack 0007000b port 13<-1115 <ACK> dlen 0h
```

This may not be the easiest display to read, but it can be invaluable when there are protocol problems and no other diagnostic tools are available.

While I'm scrutinizing networks in detail, I should explain the two types of Ethernet frame you may encounter.

Ethernet 2

You may recall the simple frame structure I'm putting out on the Ethernet (Figure 3.7).

Figure 3.7   Ethernet 2 frame.

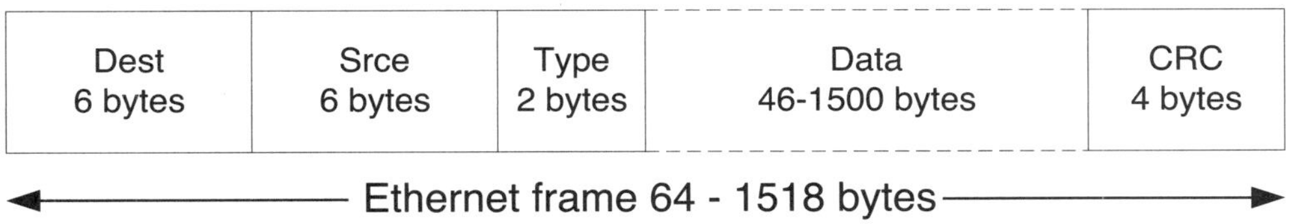

```
/* Ethernet (DIX) header */
typedef struct {
    BYTE dest[MACLEN];          /* Destination MAC address */
    BYTE srce[MACLEN];          /* Source MAC address */
    WORD ptype;                 /* Protocol type or length */
} ETHERHDR;


/* Ethernet (DIX) frame; data size is frame size minus header & CRC */
#define ETHERMTU (MAXFRAME-sizeof(ETHERHDR))
typedef struct {
    ETHERHDR h;                 /* Header */
    BYTE data[ETHERMTU];        /* Data */
    LWORD crc;                  /* CRC */
} ETHERFRAME;
```

All Ethernet frames have this general format, but unfortunately, there are differing interpretations for the type field (also known as the protocol type, Ethernet type, or E-type field) and data field.

So far, I have used the type field to signal the protocol being used (e.g., (ARP)SCRATCHP, ARP) and the data field to hold the protocol packet. For example, an ARP request from 10.1.1.11 to 10.1.1.1 has the following frame byte values.

```
Dest addr     FF FF FF FF FF FF
Srce addr     00 C0 26 B0 0A 93
Type          08 06
ARP request   00 01 08 00 06 04 00 01 00 C0 26 B0 0A 93 0A 01 01 0B
              FF FF FF FF FF FF 0A 01 01 01  ... padded to 46 bytes
```

This is the original Ethernet frame format, also known by the names

- Ethernet 2 (Ethernet 1 was an early standard, now obsolete),
- DIX Ethernet (after its originators: DEC, Intel, Xerox),
- Bluebook Ethernet, and
- RFC 894 (after the Internet standard).

IEEE 802.3 Networks

Unfortunately, the Institution of Electrical and Electronic Engineers (IEEE) decided in their 802.3 Ethernet CSMA/CD (carrier sense multiple access with collision detection) specification to change the type field into a logical link control (LLC) length field (i.e., a byte count of the rest of the frame, excluding the CRC).

Fortunately, the classic Ethernet type values are greater than the maximum possible length value (1,500 bytes), so there is no danger of one being mistaken for the other. However, how do you find out what protocol is used in an 802.3 frame?

You might think that the first two data bytes could be taken for this purpose, and there is a nonstandard format (known as raw 802.3) that does this. However, it does not conform to the IEEE 802.2 LLC standard, so it should not be used. Instead, the 802.2 standard defines the first three data bytes as an LLC header with a link layer service access point (LSAP) header containing

- a destination service access point (DSAP, one byte),
- a source service access point (SSAP, one byte), and
- a control field (one byte).

The service access points are meaningless to TCP/IP, so you must set them to AAh, which indicates that a foreign (non-IEEE) protocol is in use, which will be identified by a subnetwork access protocol (SNAP) header. The control field is equally meaningless and is normally ignored, but set it to a value of 3 (unnumbered information).

The SNAP header follows the LLC header, and has a "protocol discriminator" containing

- an organizational unit identifier (OUI, three bytes) and
- a protocol ID (two bytes).

The OUI identifies the authority for assigning the Protocol ID. If the OUI is zero, then the protocol ID takes the same value as the original Ethernet type field, for example, 806h for ARP packets.

```
/* 802.3 SNAP header */
typedef struct {
    WORD lsap;                  /* Link Service Access Point */
    BYTE ctrl;                  /* Control byte */
    BYTE oui[3];                /* Organizational Unit Identifier */
    WORD ptype;                 /* Protocol type */
} SNAPHDR;
```

```
/* 802.3 SNAP frame */
#define SNAPMTU (ETHERMTU-sizeof(SNAPHDR))
typedef struct {
    ETHERHDR e;                 /* Ethernet header (pcol is length) */
    SNAPHDR  s;                 /* 802.3 SNAP header */
    BYTE data[SNAPMTU];         /* Data */
    LWORD crc;                  /* CRC */
} SNAPFRAME;
```

The resulting frame (Figure 3.8) is known as

- 802.3 802.2 SNAP,
- 802.3 SNAP (the use of an 802.2 header is implied by the 802.3 standard), or
- RFC 1042 (after the Internet standard).

Aside from the (re)standardization issue, there is nothing to be gained by the extra eight bytes, so it is hardly surprising that most native TCP/IP LANs use the DIX standard. This is the default frame type in the software.

Figure 3.8   IEEE 802.3 SNAP frame.

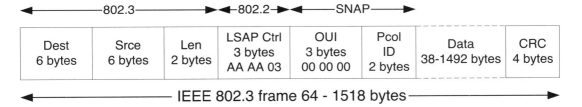

802.3 SNAP Support

The simplest way to support 802.3 SNAP on receive is to detect and delete the header so that all packets are processed internally in Ethernet 2 format. This code is included in get_net().

```
    if (gfp->g.dtype&DTYPE_ETHER && /* If Ethernet frame.. */
        gfp->buff[MACLEN*2] < 6)    /* ..and Pcol less than 600h */
    {                               /* ..might be 802.3 SNAP */
        sfp = (SNAPFRAME *)gfp->buff;
        if (len>sizeof(SNAPHDR) && sfp->s.lsap==0xaaaa)
        {                               /* If so, convert to DIX */
            len -= sizeof(SNAPHDR);
            memmove(&sfp->e.ptype, &sfp->s.ptype, len);
        }
        gfp->g.len = len;               /* Re-set length in header */
    }
```

In this way, you can accept both types of frame (even if they are intermixed) without any modifications to the rest of the receive code.

All transmissions would normally be in Ethernet 2 format, unless the user has modified the configuration file to specifically request 802.3 SNAP frames. This sets a bit in the frame-type identifier that is detected by put_net(), and the appropriate header is inserted.

```
if (dtype&DTYPE_SNAP && len+sizeof(SNAPHDR)<=MAXFRAME)
{                                         /* If 802.3 SNAP.. */
    sfp = (SNAPFRAME *)gfp->buff;         /* Make room for new header */
    memmove(&sfp->s.ptype, &sfp->e.ptype, len);
    len += sizeof(SNAPHDR);               /* Set for 802.3 802.2 & SNAP */
    sfp->e.ptype = swapw((WORD)(len-sizeof(ETHERHDR)));
    sfp->s.lsap = 0xaaaa;
    sfp->s.ctrl = 3;
    memset(sfp->s.oui, 0, 3);
}
```

The addition of the eight-byte header reduces the maximum amount of packet data that can be transferred, so the higher software levels must adjust their maximum transfer unit (MTU) value accordingly.

802.3 SNAP in Action

By using a different configuration file, you can modify ARPSCAN to use 802.3 SNAP packets.

```
ARPSCAN Vx.xx
Press ESC or ctrl-C to exit
IP 10.1.1.11 Ethernet 00:c0:26:b0:0a:93 (local)

IP 10.1.1.2 Ethernet 00:20:18:3a:76:f9
IP 10.1.1.3 Ethernet 00:20:18:3a:ed:64
```

Node 10.1.1.1 isn't responding — it's an old Linux system that is clearly unhappy with the 802.3 format. Nodes 10.1.1.2 and 10.1.1.3 (Windows 95 and 98 systems) are responding, but closer examination with the protocol analyzer shows that the response is in Ethernet 2 format.

```
Packet #2
  Packet Length:64
802.3 Header
  Destination:  FF:FF:FF:FF:FF:FF  Ethernet Broadcast
Packet #3
  Packet Length:64
Ethernet Header
  Source:        00:20:18:3A:76:F9
  Protocol Type:0x0806  IP ARP
ARP - Address Resolution Protocol
  Target Internet Address:  10.1.1.11
```

This is recommended practice for Ethernet 2 systems. If they can accept 802.3 frames, they should respond in Ethernet 2 format. If you have any doubt as to which frame format to use, I recommend you stick to Ethernet 2, unless you have specific information to the contrary.

Summary

I've looked at the following areas in this chapter.

- The fundamentals of internetworks
- IP addresses
- Special addresses for non-Internet networks
- Address resolution protocol (ARP) — conversion of IP addresses into MAC addresses
- Tracing packets on the network
- Types of network frames

To demonstrate IP addressing, I've created a scanner utility that checks which addresses are currently being used on the network, and I have also discussed the use of various packet display tools, including a simple network monitor utility.

Source Files

arpscan.c	Address scanning utility
ether3c.c	3C509 Ethernet card driver
etherne.c	NE2000 Ethernet card driver
ip.c	Low-level TCP/IP functions
net.c	Network interface functions
netmon.c	Network monitor utility
netutil.c	Network utility functions
pktd.c	Packet driver (BC only)
serpc.c or serwin.c	Serial drivers (BC or VC)
tcp.c	High-level TCP/IP functions (for NETMON only)
dosdef.h	MS-DOS definitions (BC only)
ether.h	Ethernet definitions
ip.h	TCP/IP definitions
net.h	Network driver definitions
netutil.h	Utility function and general frame definitions
serpc.h	Serial driver definitions (BC or VC)
tcp.h	TCP socket definitions (for NETMON only)
win32def.h	Win32 definitions (VC only)

ARPSCAN Utility

Utility	IP address scanner
Usage	`arpscan [`*`options`*`] [`*`start_IP_address`*`]`
	Enters server mode if no IP address given
Options	`-c` *name* Configuration filename (default `tcplean.cfg`)
	`-n` *count* Number of addresses to scan (default 20)
Example	`arpscan -n 10 10.1.1.1`
Keys	Ctrl-C or Esc to exit
Config	`net` to identify network type
	`ip` to identify IP address
Modes	Defaults to server mode (ARP responder) unless IP address given
Notes	Issues ARP requests for the given address range; displays responses
	In server mode, just responds to ARP requests

NETMON Utility

Utility	Simple network monitor; displays all network traffic
Usage	`netmon [`*`options`*`]`
Options	`-c` *name* Configuration filename (default `tcplean.cfg`)
	`-t` Display TCP segments
	`-v` Verbose packet display
Example	`netmon -c slip.cfg`
Keys	Ctrl-C or Esc to exit
Config	`net` to identify network type (multiple entries allowed)
Notes	Can be overloaded by moderate network traffic

Chapter 4

The Network Interface: IP and ICMP

Overview

A journey through TCP/IP starts with the lower layers and works upward so that you can experiment en route to check your understanding. I'll start with Internet Protocol (IP) and the associated Internet control message protocol (ICMP), which are low-level interfaces between TCP/IP and the network drivers.

First, I'll place IP and ICMP into the context of the overall TCP stack and sketch the roles of other protocols (such as TCP) that I will describe in future chapters. Then I'll look at the details of IP and ICMP, the way in which the IP messages (datagrams) are stored within software, and implement the standard diagnostic utility, Ping.

To consolidate your understanding of IP and the concept of routing, I'll develop a simple router that can support multiple Ethernet and serial interfaces.

TCP/IP Stack

I have already discussed the concept of a protocol stack containing protocol layers, and I have referred to TCP/IP as a protocol family. It is now time to be more specific and start working the way up a TCP/IP stack (Figure 4.1).

Figure 4.1   TCP/IP protocol stack.

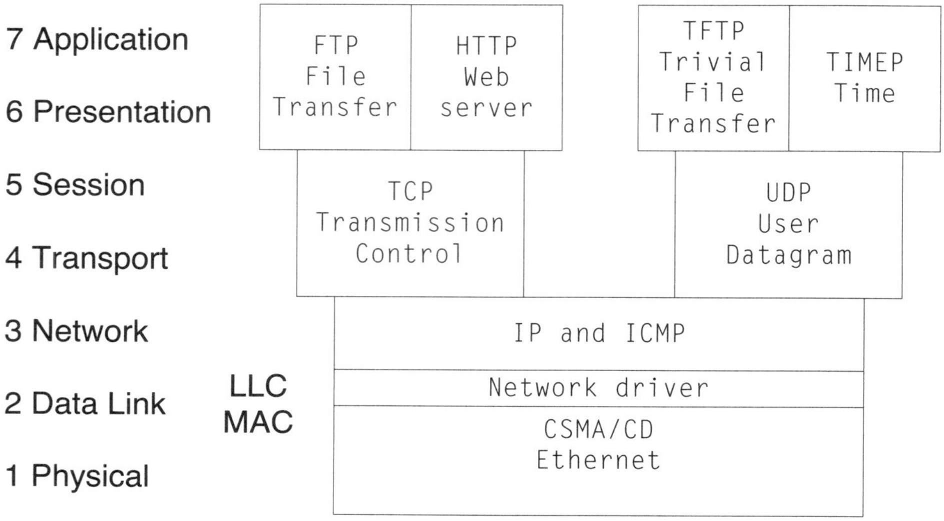

7 Application	FTP File Transfer	HTTP Web server	TFTP Trivial File Transfer	TIMEP Time
6 Presentation				

```
7 Application     ┌─────────┬─────────┐   ┌──────────┬─────────┐
                  │  FTP    │  HTTP   │   │  TFTP    │         │
                  │  File   │  Web    │   │ Trivial  │ TIMEP   │
6 Presentation    │Transfer │ server  │   │  File    │ Time    │
                  │         │         │   │Transfer  │         │
                  └─────────┴─────────┘   └──────────┴─────────┘
5 Session            ┌──────────────┐      ┌──────────────────┐
                     │     TCP      │      │      UDP         │
                     │ Transmission │      │     User         │
4 Transport          │   Control    │      │   Datagram       │
                     └──────────────┘      └──────────────────┘
3 Network            ┌──────────────────────────────────────┐
                     │            IP and ICMP               │
                     ├──────────────────────────────────────┤
2 Data Link   LLC    │           Network driver             │
              MAC    ├──────────────────────────────────────┤
                     │             CSMA/CD                  │
1 Physical           │             Ethernet                 │
                     └──────────────────────────────────────┘
```

The first software layer above the network drivers is IP and its partner ICMP. Above these, there is a split: connection-oriented applications use transmission control protocol (TCP), whereas connectionless applications use user datagram protocol (UDP). I have illustrated this using two examples of each type, though the actual number of applications is far greater.

If you employ TCP, you get a logical connection between your application and the remote application, with automatic retries in the event of an error. Connectionless UDP is often known as "send and pray" because there is no guarantee that your data will get through.

What is the point of IP and ICMP? In a word, *routing*. You will recall that the Internet is an arbitrary collection of networks connected by routers. It would be very unfortunate if these routers had to perform a detailed examination of the contents of each packet in order to decide what to do with it. It is far better to concentrate all the information the router needs into a single header at the front of each packet. This is the IP header — it contains all the *static* information the router needs to perform its job, such as the source and destination IP addresses.

However, static information is not enough; for example, if a route is unavailable, its associated router needs a way of dynamically informing the sender of this fact so that measures can be taken to redirect the traffic. This function is performed by ICMP, which carries routing control and diagnostic information. An IP packet is known as a "datagram."

IP Datagram

An IP datagram occupies the data area of an Ethernet frame or the whole of a SLIP frame (Figure 4.2). As implied by its name, SLIP stands for serial line IP.

Figure 4.2   Ethernet and SLIP datagrams.

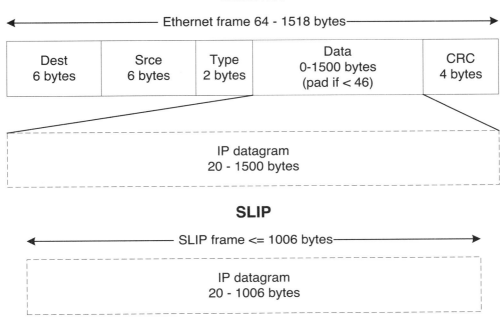

Before discussing the datagram structures, an important problem must be addressed, namely the differing frame header sizes for Ethernet (14 bytes) and SLIP (zero bytes, assuming the framing characters are deleted by the device driver). The generic frame has already been defined.

```
/* General-purpose frame header, and frame including header */
typedef struct {
    WORD len;                    /* Length of data in genframe buffer */
    WORD dtype;                  /* Driver type */
    WORD fragoff;                /* Offset of fragment within buffer */
} GENHDR;

typedef struct {
    GENHDR g;                    /* General-pupose frame header */
    BYTE buff[MAXGEN];           /* Frame itself (2 frames if fragmented) */
} GENFRAME;
```

However, depending on the frame type, a datagram may start at a buffer offset of zero or 14 bytes. There are various solutions.

- Chop off the Ethernet header so that the frame is the same as with SLIP. This has the disadvantage of losing the protocol type and sender's address, making it impossible to return a message to the sender.

- Chop off the Ethernet header and store it elsewhere. You'd have to retain some linkage between the chopped frames and their headers, which might get complicated.

- Pad the SLIP frame with a dummy header so that it is the same length as the Ethernet frame. This is temptingly simple, if slightly wasteful, and potentially very confusing if you introduced another physical transport with a different header length.

- Use function calls to determine the header length based on the driver-type flags.

I use the last of these methods because it offers the most versatile solution. I've created a function that returns a pointer to the datagram start, and this is used every time a datagram pointer is required. For example, assume IPKT is an IP datagram pointer (described in the next section), and gfp points to an incoming frame.

```
IPKT *ip;
GENFRAME *gfp;
...
ip = getframe_datap(gfp);
...
```

You now have a pointer to the datagram in the incoming frame, regardless of the frame type. The actual code to do this is simple.

```
/* Get pointer to the data area of the given frame */
void *getframe_datap(GENFRAME *gfp)
{
    return(&gfp->buff[dtype_hdrlen(gfp->g.dtype)]);
}
/* Return frame header length, given driver type */
WORD dtype_hdrlen(WORD dtype)
{
    return(dtype&DTYPE_ETHER ? sizeof(ETHERHDR) : 0);
}
```

It is worthwhile to create functions that access the address and protocol type fields because they don't exist in SLIP frames. These functions can return harmlessly when attempting to access a nonexistent field.

```
/* Get pointer to the source address of the given frame, 0 if none */
BYTE *getframe_srcep(GENFRAME *gfp)
{
    ETHERHDR *ehp;
    BYTE *srce=0;
```

```
    if (gfp->g.dtype & DTYPE_ETHER)          /* Only Ethernet has address */
    {
        ehp = (ETHERHDR *)gfp->buff;
        srce = ehp->srce;
    }
    return(srce);
}
/* Copy the source MAC addr of the given frame; use broadcast if no addr */
BYTE *getframe_srce(GENFRAME *gfp, BYTE *buff)
{
    BYTE *p;

    p = getframe_srcep(gfp);
    if (p)
        memcpy(buff, p, MACLEN);
    else
        memcpy(buff, bcast, MACLEN);
    return(p);
}
```

... and ditto for the destination address ...

```
/* Get the protocol for the given frame; if unknown , return 0 */
WORD getframe_pcol(GENFRAME *gfp)
{
    ETHERHDR *ehp;
    WORD pcol=0;

    if (gfp->g.dtype & DTYPE_ETHER)          /* Only Ethernet has protocol */
    {
        ehp = (ETHERHDR *)gfp->buff;
        pcol = ehp->ptype;
    }
    return(pcol);
}
```

Having sorted this out, you can now define the IP structures.

IP Header

An IP datagram consists of a variable-length header and an optional data field (Figure 4.3).

Figure 4.3   IP datagram.

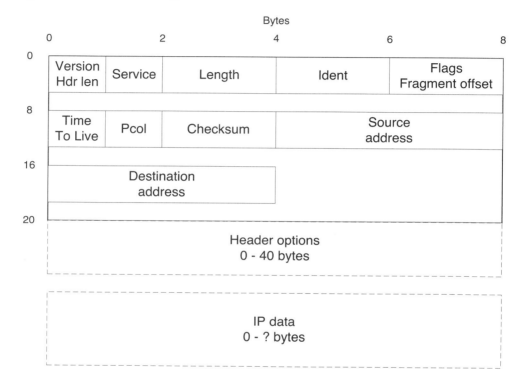

```
/* ***** IP (Internet Protocol) header ***** */
typedef struct
{
    BYTE  vhl,              /* Version and header len */
          service;         /* Quality of IP service */
    WORD  len,             /* Total len of IP datagram */
          ident,           /* Identification value */
          frags;           /* Flags & fragment offset */
    BYTE  ttl,             /* Time to live */
          pcol;            /* Protocol used in data area */
    WORD  check;           /* Header checksum */
    LWORD sip,             /* IP source addr */
          dip;             /* IP dest addr */
```

```
} IPHDR;
#define PICMP    1          /* Protocol type: ICMP */
#define PTCP     6          /*                 TCP */
#define PUDP     17         /*                 UDP */
```

Version and header length. Two four-bit fields. I'll use IP version 4, and the usual header size (measured in 32-bit words) is five, so this field is normally 45h.

Service. The sender of the datagram can use this field in an attempt to prioritize one datagram over others. However, this information is largely ignored by routers, so I set it to zero, which is "normal" precedence.

Length. The total datagram length, including the IP header, in bytes.

Ident and fragmentation. IP is designed to support a wide range of physical media, each of which will have its own limit on the number of bytes that a datagram can hold (the maximum transmission unit, or MTU). If a datagram is larger than the MTU, it will be *fragmented* (i.e., broken into smaller parts), and it must be reassembled at the destination. This is discussed in more detail later.

Time to live. Each datagram circulating on the Internet is equipped with a time value (in seconds), which is gradually reduced as it travels from router to router. If the value is reduced to zero, the datagram is assumed to be undeliverable and is discarded. A typical starting time is 100 seconds.

Protocol. This identifies the protocol that is used in the data area of the datagram. Typical values are 1 for ICMP, 6 for TCP, and 17 for UDP. I present more details on these protocols later.

Checksum. This is a simple checksum of the IP header only, to ensure that it hasn't been corrupted. The checksum algorithm is given later.

Source and destination addresses. These are four-byte IP address values, as described in the previous chapter.

Options. The header field may be extended by certain options to allow tighter control over the routing process and to add security. I will not be using any of these options, but you must accept that the IP header may be greater than the default 20 bytes. All extra information may safely be ignored.

IP Implementation

In line with the standard is_, make_, swap_ ARP implementation, you need

is_ip()   to check an incoming packet,

make_ip()   to build an outgoing packet, and

swap_ip()   to byte swap an incoming or outgoing packet.

```
/* Check frame is IP/SLIP, checksum & byte-swap, return data len */
int is_ip(GENFRAME *gfp, int len)
{
    int ver, dlen=0, hlen;
    WORD pcol, sum;
    IPKT *ip;

    pcol = getframe_pcol(gfp);
    if ((pcol==PCOL_IP || pcol==0) && len>=sizeof(IPHDR))
    {
        ip = getframe_datap(gfp);          /* Get pointer to IP frame */
        ver = ip->i.vhl >> 4;              /* Get IP version & hdr len */
        hlen = (ip->i.vhl & 0xf) << 2;
        sum = ~csum(&ip->i, (WORD)hlen);   /* Do checksum */
        if (ver==4 && len>=hlen && sum==0) /* If OK.. */
        {
            swap_ip(gfp);                  /* Do byte-swaps */
            dlen = mini(ip->i.len, len) - hlen;
            if (hlen > sizeof(IPHDR))      /* If IP options present.. */
            {                              /* ...delete them, move data down */
                memmove(ip->data, &ip->data[hlen-sizeof(IPHDR)], len);
                dlen -= hlen-sizeof(IPHDR);
            }
            if ((ip->i.frags & 0x3fff)!=0)  /* If a fragment.. */
                dlen = defrag_ip(ip, dlen); /* ...call defragmenter */
        }
    }
    return(dlen);
}
```

It is possible, though not probable, that the IP header is oversized because of the presence of IP options, which are concerned with routing and security and can safely be ignored by the software. However, an oversized IP header means that all subsequent network data will be offset, causing a mismatch with structures.

The previous code employs an easy way around this by copying the data into the correct place, which is similar to the way I deleted the IEEE 802.3 header.

Making the datagram would be a relatively simple fill-in-the-blanks exercise were it not for the checksum and fragmentation, which I'll explore in the next two sections.

```c
/* Make an IP packet, if greater than the MTU, also make fragment (subframe) in
** this frame. Return total length of frame and subframes (if any) */
int make_ip(GENFRAME *gfp, NODE *srcep, NODE *destp, BYTE pcol, WORD dlen)
{
    IPKT *ip, *ip2;
    int len, sublen=0, fhlen;
    static WORD ident=1, oset=0;
    GENFRAME *sfp;

    ip = getframe_datap(gfp);           /* Get pointer to IP datagram */
    ip->i.ident = ident;                /* Set datagram ident */
    ip->i.frags = oset >> 3;            /* Frag offset in units of 8 bytes */
    gfp->g.fragoff = 0;                 /* ...assuming no more frags */
    len = mini(dlen, getframe_mtu(gfp)-sizeof(IPHDR));
                                        /* Size of this frame */
```

... fragmentation check ...

```c
    ip->i.vhl = 0x40+(sizeof(IPHDR)>>2);  /* Version 4, header len 5 LWORDs */
    ip->i.service = 0;                    /* Routine message */
    ip->i.ttl = IP_TTL;                   /* Time To Live */
    ip->i.pcol = pcol;                    /* Set IP protocol */
    ip->i.sip = srcep->ip;                /* Srce, dest IP addrs */
    ip->i.dip = destp->ip;
    ip->i.len = len + sizeof(IPHDR);      /* Data length */
    swap_ip(gfp);                         /* Do byte-swaps (for checksum) */
    ip->i.check = 0;                      /* Clear checksum */
    ip->i.check = ~csum(ip, sizeof(IPHDR));  /* ..then set to calc value */
    ident++;                              /* Increment datagram ident */
    oset = 0;                             /* Clear fragment offset */
    len += sizeof(IPHDR) + sublen;        /* Bump up length */
    return(make_frame(gfp, destp->mac, PCOL_IP, (WORD)len));
}
```

```
/* Swap byte order of ints in IP header */
void swap_ip(GENFRAME *gfp)
{
    IPHDR *iph;

    iph = getframe_datap(gfp);
    iph->len = swapw(iph->len);
    iph->ident = swapw(iph->ident);
    iph->frags = swapw(iph->frags);
    iph->sip = swapl(iph->sip);
    iph->dip = swapl(iph->dip);
}
```

Checksum

As a precaution against corruption by routers, there is a checksum for the IP header. The steps for transmission are as follows.

1. Prepare and byte swap the header.
2. If it is an odd length, pad with a zero byte.
3. Clear the checksum field.
4. Sum the 16-bit words in the header.
5. Put the one's complement of the result in the checksum field.

The addition is somewhat unusual, in that any carry bits are added back on.

```
long sum;
WORD *data;

sum += *data++;
if (sum & 0x10000L)
    sum = (sum & 0xffffL) + 1;
```

This addition method is particularly suitable for implementation in assembly language because it is just a repeated add-with-carry, irrespective of whether a big-endian or little-endian processor is being used. There is a useful optimization to the C version, which involves saving up the carry bits and then adding them on at the end.

```
/* Do checksum. Improved algorithm is from RFC 1071 */
WORD csum(void *dp, WORD count)
{
    register LWORD total=0L;
    register WORD n, *p, carries;
```

```
    n = count / 2;
    p = (WORD *)dp;
    while (n--)
        total += *p++;
    if (count & 1)
        total += *(BYTE *)p;
    while ((carries=(WORD)(total>>16))!=0)
        total = (total & 0xffffL) + carries;
    return((WORD)total);
}
```

This also caters to an odd length by doing an extra single-byte addition (little-endian processors only). On reception, the datagram header is verified by applying the checksum across it; the result should be FFFFh.

Fragmentation

IP has the facility for sending packets that are larger than the maximum size permitted by the network hardware (the MTU) by splitting them into fragments on transmit and then reassembling them on receipt. It is very tempting to ignore fragmentation altogether, on the assumption that it can be avoided by sending the maximum data size the network will bear. Indeed, fragmentation is generally to be avoided because it carries a significant performance penalty. There are two reasons why it is worth considering the inclusion of fragmentation:

- Your network may include an old link with a small MTU value, and throughput on the newer links will suffer if they are constrained to this size.
- Some connectionless, higher-level protocols make unfortunate assumptions about the minimum UDP datagram size, so these protocols might not work unless there is support for IP fragmentation.

I suggest you bite the bullet and make provision for at least one level of fragmentation (i.e., a total of two fragments). Chapter 1GENFRAME is large enough to store two Ethernet frames and their associated headers, so on transmit you will have to split the single oversized frame into two smaller ones, not forgetting to insert the extra headers (Figure 4.4).

Figure 4.4 **Transmit fragmentation.**

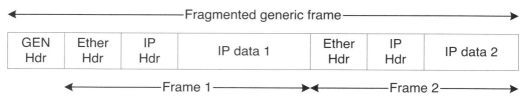

The fragmentation code in make_ip() uses a block copy and recursion technique to make one frame into two.

```
if (dlen > len)                     /* If fragmentation required.. */
{                                   /* Create new frag within this frame */
    fhlen = dtype_hdrlen(gfp->g.dtype);         /* Frame hdr len */
    gfp->g.fragoff = len + sizeof(IPHDR) + fhlen; /* Subframe offset */
    sfp = (GENFRAME *)&gfp->buff[gfp->g.fragoff]; /* Subframe ptr */
    ip->i.frags = (oset>>3)+0x2000;             /* Show there is frag */
    oset += len;                                /* New data offset */
    ip2 = (IPKT*)((BYTE*)sfp+sizeof(GENHDR)+fhlen);
                                                /* Ptr to 2nd IP frag */
    memmove(ip2->data, &ip->data[oset], dlen-len);
                                                /* Copy data 1st->2nd */
    sfp->g.dtype = gfp->g.dtype;    /* Copy driver type into subframe */
    sublen = make_ip(sfp, srcep, destp, pcol, (WORD)(dlen-len));
}                                               /* Recursive call to make frag */
```

The low-level function put_frame() checks for the presence of two fragments and sends them both. There is a compile-time option to allow the fragments to be sent in reverse order, which is useful for testing defragmentation code.

```
/* Put frame out onto the network; if sub-frame (fragment), send it as well */
int put_frame(GENFRAME *gfp, int len)
{
    int ret=0, len1, len2;
    GENFRAME *sfp;

    len1 = gfp->g.fragoff ? gfp->g.fragoff : len; /* Get len of 2 fragments */
```

```
    len2 = len - len1;
    sfp = (GENFRAME *)&gfp->buff[gfp->g.fragoff]; /* ..and ptr to 2nd frag */
#if SUBFIRST
    if (len2 > 0)                               /* Send sub-frame first.. */
        ret = put_net(sfp, (WORD)len2);
    if (len1 > 0)                               /* ..then main frame */
        ret += put_net(gfp, (WORD)len1);
#else                                           /* Or send main frame first */
    if (len1 > 0)
        ret = put_net(gfp, (WORD)len1);
    if (len2 > 0)
        ret += put_net(sfp, (WORD)len2);
#endif
    return(ret);
}
```

For receive, it's just a question of saving one fragment until its partner comes along and then stitching them together into a single, large datagram. Unfortunately, the network is unreliable, so you can't guarantee in which order the fragments will arrive or that both fragments will arrive at all. Instead of a regular sequence of A1, A2, B1, B2, ..., you might receive A1, B2, B1, B2, C1, ... and have to reassemble the resulting jigsaw puzzle without mismatching any of the pieces (Figure 4.5).

Figure 4.5   Receive defragmentation.

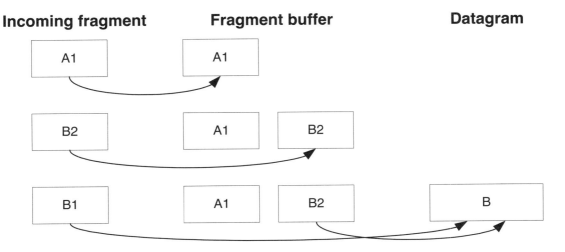

The fragment buffer holds the fragments awaiting reassembly.

```
typedef struct {                  /* Fragment buffer structure */
    int tries;                    /* Number of times to attempt a match */
    WORD ident;                   /* IP ident field */
    LWORD sip;                    /* Source IP address */
    WORD oset;                    /* Offset in IP data area */
    WORD len;                     /* Length of fragment */
    BYTE data[MAXIP];             /* Fragment data */
} FRAG;
FRAG frags[NFRAGS];               /* Fragment buffer */
```

I've included a match count, which will be decremented every time a match is attempted; when zero, the fragment is deleted. This is to ensure that the buffer doesn't gradually get clogged up with unmatched fragments from long-dead connections. The values in the IP header concerned with fragmentation are listed in the code fragment below.

```
    WORD  ident,                  /* Identification value */
          frags;                  /* Flags & fragment offset */
```

The identification value is the same for all fragments that belong together. It is incremented by the sender every time a new datagram (or set of datagram fragments) is sent.

The top three bits of the next word are used as flags.

Bit 15:   0
Bit 14:   0 = may fragment
          1 = won't fragment
Bit 13:   0 = last fragment
          1 = more fragments

Bit 14 can disable fragmentation for this datagram, in which case it may be rejected by a router if it's too big for a single frame. Bit 13 signals that there are more fragments to come.

The remaining 13 bits form a fragment offset, measured in eight-byte blocks. The maximum offset is 8,191 by 8, or 65,528 bytes. Conveniently, a `frags` value of zero means no offset and no more fragments (i.e., no fragmentation of the datagram).

This defragmenter checks for a matching half in the buffer. If found, the complete datagram is returned; if not, the incoming fragment is added to the buffer to await matching.

```
/* Defragment an incoming IP datagram by matching with existing fragments
** This function handles a maximum of 2 fragments per datagram
** Return total IP data length, 0 if datagram is incomplete */
int defrag_ip(IPKT *ip, int dlen)
{
    int n=0, match=0;
    WORD oset;
    FRAG *fp, *fp2=0;
```

```
    oset = (ip->i.frags & 0x1fff) << 3; /* Get offset for imcoming frag */
    while (n<NFRAGS && !match)               /* Search for matching half */
    {
        fp = &frags[n++];
        if (fp->tries)
            {                                   /* ..by checking ident */
            if (!(match = (ip->i.ident==fp->ident && ip->i.sip==fp->sip)))
                fp->tries--;            /* If no match, reduce attempts left */
            }
        else
            fp2 = fp;
    }
    if (match)
    {                                   /* Matched: check isn't a duplicate */
        if ((oset+dlen == fp->oset || fp->oset+fp->len == oset) &&
            dlen+fp->len <= MAXGEN)        /* ..and length is OK */
        {
            if (oset)                      /* Move old data as necessary */
                memmove(&ip->data[oset], ip->data, dlen);
            ip->i.len = dlen += fp->len;/* ..and add in new data */
            memcpy(&ip->data[fp->oset], fp->data, fp->len);
            fp->tries = 0;
        }
        else
        {
            if (netdebug)
                printf("Mismatched frag oset %u buff len %u\n", oset, fp->len);
            match = 0;
        }
    }
    else if (fp2)                           /* No match, but there is spare space */
    {
        fp2->tries = FRAGTRIES;             /* Save frag for matching later */
        fp2->ident = ip->i.ident;
        fp2->sip = ip->i.sip;
        fp2->oset = oset;
        fp2->len = dlen;
        memcpy(fp2->data, ip->data, dlen);
    }
    return(match ? dlen : 0);
}
```

Internet Control Message Protocol

ICMP is an adjunct to IP that gives all nodes on the network the ability to perform simple diagnostics and return error messages. For example, if you ask a router to forward a datagram to an address it can't reach, it will return an ICMP "destination unreachable" message. ICMP messages are contained within the data field of an IP datagram using IP protocol number 1 (Figure 4.6).

Figure 4.6   ICMP header.

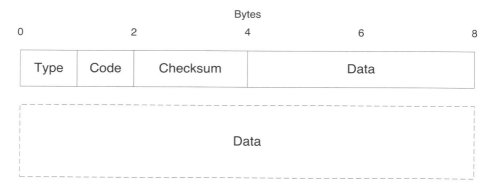

The message length depends on the message type, but it is always a minimum of eight bytes, even though four of these may be unused. The checksum method is the same as for the IP header and covers the header plus all of the data.

The type and code fields identify the ICMP operation required. The most commonly used values are as follows:

```
code 0    Network unreachable
     type 0    Echo reply
          3    Destination unreachable
          8    Echo request
     1    Host unreachable
     2    Protocol unreachable
     4    Port unreachable
     5    Fragmentation needed but not allowed
     6    Destination network unknown
     7    Destination host unknown
```

For now, I'll concentrate on the simple diagnostic message "Echo request" and the corresponding "Echo reply." These are commonly known by the name of the utility that sends the requests: Ping.

The ICMP structure definition reflects the parameters needed for an Echo request.

```
/* ***** ICMP (Internet Control Message Protocol) header ***** */
typedef struct
{
    BYTE    type,           /* Message type */
            code;           /* Message code */
    WORD    check,          /* Checksum */
            ident,          /* Identifier (possibly unused) */
            seq;            /* Sequence number (possibly unused) */
} ICMPHDR;
#define ICREQ           8   /* Message type: echo request */
#define ICREP           0   /*                  echo reply */
#define ICUNREACH       3   /*                  destination unreachable */
#define ICQUENCH        4   /*                  source quench */
#define UNREACH_NET     0   /* Destination Unreachable codes: network */
#define UNREACH_HOST    1   /*                                    host */
#define UNREACH_PORT    3   /*                                    port */
#define UNREACH_FRAG    4   /*   fragmentation needed, but disable flag set */
```

This header is followed by a block of undefined data; the recipient should echo back the Ident and sequence numbers and all the data. The Ping utility checks the response and prints an appropriate message on the console.

All TCP/IP computers have the ability to respond to a ping, although it might be disabled in external gateways for security reasons, so it is a good candidate for your first IP utility.

ICMP Message Functions

As with previous protocols, you need functions for checking incoming ICMP messages, making outgoing messages, and swapping bytes.

```
/* Return ICMP data length (-1 if no data), 0 if not ICMP */
int is_icmp(IPKT *ip, int len)
{
    ICMPKT *icmp;
    WORD sum;
    int dlen=0;

    if (ip->i.pcol==PICMP && len>=sizeof(ICMPHDR))
    {
        icmp = (ICMPKT *)ip;
        if ((sum=csum(&icmp->c, (WORD)len)) == 0xffff)
        {
```

```
            swap_icmp(icmp);
            dlen = len>sizeof(ICMPHDR) ? len-sizeof(ICMPHDR) : -1;
        }
        else
            printf("\nICMP checksum error: %04X\n", sum);
    }
    return(dlen);
}
```

The `make_icmp()` function only creates the message framework; it is assumed that the appropriate data has been copied into place before the function is called.

```
/* Make an ICMP packet */
int make_icmp(GENFRAME *gfp, NODE *srcep, NODE *destp, BYTE type, BYTE code,
    WORD dlen)
{
    ICMPKT *icmp;
    WORD len;

    icmp = getframe_datap(gfp);
    icmp->c.type = type;
    icmp->c.code = code;
    icmp->c.check = 0;
    swap_icmp(icmp);
    len = (WORD)(dlen + sizeof(ICMPHDR));
    icmp->c.check = ~csum(&icmp->c, len);
    return(make_ip(gfp, srcep, destp, PICMP, len));
}

/* Swap byte order of ints in ICMP header */
void swap_icmp(ICMPKT *icmp)
{
    icmp->c.ident = swapw(icmp->c.ident);
    icmp->c.seq = swapw(icmp->c.seq);
}
```

Ping Implementation

A Ping utility needs to take the following actions:

1. If not using a SLIP link, resolve the IP address.
   a. Send ARP request.
   b. Wait for ARP response.

2. Send ICMP Echo request with some data.

3. Wait for ICMP response.

While waiting for the responses, the utility might as well respond to any ICMP or ARP requests it receives, which would allow two nodes, each pinging the other. It is also worthwhile creating a "server mode," in which the utility doesn't initiate any messages but simply responds to incoming ones.

The usual delay between ICMP requests is one second, though it can be useful to shorten this time when investigating problems. The ultimate shortening is a "flood" ping, where requests are sent as fast as the recipient can accept them. The flood technique is very useful for stress-testing network drivers, but it should not be used on a real network because of the amount of CPU resources and network bandwidth it may consume.

Main Program

Command-Line Options

This is the usual process of checking option switches.

```
int main(int argc, char *argv[])
{
    int i, args=0, len;
    LWORD mstimer;
    WORD dtype;
    GENFRAME *gfp;
    char *p, temps[18];

    printf("PING v" VERSION "\n");              /* Sign on */
    signal(SIGINT, break_handler);              /* Trap ctrl-C */
    while (argc > ++args)                        /* Process command-line args */
    {
        if (argv[args][0]=='-')
        {
            switch (toupper(argv[args][1]))
            {
            case 'C':                           /* -C: config filename */
                strncpy(cfgfile, argv[++args], MAXPATH);
                if ((p=strrchr(cfgfile, '.'))==0 || !isalpha(*(p+1)))
                    strcat(cfgfile, CFGEXT);
                break;
            case 'L':                           /* -L: length of data */
                datalen = maxi(atoi(argv[++args]), 1);
                break;
```

```
            case 'W':                              /* -W: waiting time in msec */
                waitime = maxi(atoi(argv[++args]), MINWAIT);
                break;
            case 'V':                              /* -V: verbose (debug) mode */
                netdebug = 1;
                break;
            case 'S':                              /* -S: server mode */
                servermode = 1;
                break;
            case 'F':                              /* -F: flood mode */
                floodmode = 1;
                break;
            }
        }
        else if isdigit(argv[args][0])             /* Destination IP address */
            remip = atoip(argv[args]);
    }
```

Initialization

An addition to the usual network initialization is the creation of a buffer with the source data for the Ping. Two options are available: straightforward ASCII characters (to emulate a DOS ping) or a pseudorandom byte sequence, which is more likely to show up overrun and underrun errors.

```
    for (i=0; i<datalen*2; i++)                 /* Test block is 2x data size */
#if ASCDATA
        testdata[i] = (BYTE)(i%23 + 'a');       /* ..same data as DOS ping */
#else
        testdata[i] = (BYTE)rand();             /* ..or random test data.. */
#endif
    if (remip==0L && !servermode)               /* Prompt user if no IP addr */
        disp_usage();                           /* Read net config */
    else if (!(dtype=read_netconfig(cfgfile, &locnode)))
        printf("Invalid configuration '%s'\n", cfgfile);
    else
    {
        remnode.ip = remip;                     /* Set remote addr */
        memcpy(remnode.mac, bcast, MACLEN);     /* ..as broadcast */
        genframe.g.dtype = dtype;               /* Set frame driver type */
        gfp = &genframe;                        /* Get pointer to frame */
        printf("IP %s", ipstr(locnode.ip, temps));
```

```
        printf(" mask %s", ipstr(locnode.mask, temps));
    if (locnode.gate)
        printf(" gate %s", ipstr(locnode.gate, temps));
    if (dtype & DTYPE_ETHER)
        printf(" Ethernet %s", ethstr(locnode.mac, temps));
    if (gfp->g.dtype & DTYPE_SLIP)          /* No ARP if SLIP */
        arped = 1;
    if (datalen > (len=icmp_maxdata(gfp)*2))/* Don't exceed 2 frames */
    {
        printf("\nWarning: data length reduced to %u bytes", len);
        datalen = len;
    }
    if (servermode)
        printf("\nServer mode");
    else
    {
        printf("\n%s ", arped ? "Pinging" : "Resolving");
        printf("%s", ipstr(gate_ip(&remnode, &locnode), temps));
    }
```

Main Loop

By delegating the frame transmission and reception into separate functions, little is left in the main program.

```
        printf(" - ESC or ctrl-C to exit\n");
    mstimeout(&mstimer, 0);                      /* Refresh timer */
    while (!breakflag)
    {
        if (!servermode)
        {
            if (!arped)                          /* If not ARPed.. */
            {                                    /* ..and timeout.. */
                if (mstimeout(&mstimer, ARPTIME))
                    do_transmit(gfp);            /* ..send ARP */
            }
            else if (floodmode)                  /* If flood ping.. */
            {                                    /* ..and response or timeout */
                if (txseq==rxseq || mstimeout(&mstimer, waitime))
                {
                    mstimeout(&mstimer, 0); /* ..refresh timer */
```

```
                    do_transmit(gfp);          /* ..transmit next packet */
                }
            }
            else                               /* If normal pinging.. */
            {                                  /* ..and timeout */
                if (mstimeout(&mstimer, waitime))
                    do_transmit(gfp);          /* ..transmit next packet */
            }
        }
        do_receive(gfp);                       /* Check responses */
        do_poll();                             /* Poll net drivers */
        if (kbhit())                           /* If user hit a key.. */
            breakflag = getch()==0x1b;         /* ..check for ESC */
    }
    close_net(dtype);                          /* Shut down net driver */
}
free(testdata);                                /* Free test data memory */
printf("ICMP echo: %lu sent, %lu received, %lu errors\n",
    txcount, rxcount, errcount);               /* Print stats */
return(0);
}
```

Transmit Function

Transmitting the echo request is made simple by the previously described make_icmp(). The source data is either alphabetic or pseudorandom from the buffer prepared above.

To make the pseudorandom test more rigorous, it is worthwhile introducing a data offset within the buffer so that the same data isn't sent twice. To permit the returned message to be checked, this offset is tied in to the message sequence number.

The steps in transmitting a message are common to all protocols.

1. Copy the data into the correct place in the frame buffer.
2. Make the frame; save the resulting total frame length.
3. Put the frame into the outgoing frame buffer.

Note that this process won't necessarily transmit the frame on the network: this is normally done by the polling cycle do_poll() above.

```
/* Do next transmission cycle */
void do_transmit(GENFRAME *gfp)
{
    ICMPKT *icmp;
    BYTE *data;
```

```
    int txlen;

    if (!arped)                                /* If not arped, send ARP */
    {
        printf("ARP ");                        /* Make packet */
        txlen = make_arp(gfp, &locnode, &remnode, ARPREQ);
    }
    else
    {
        icmp = getframe_datap(gfp);            /* Send echo req */
        icmp->c.seq = ++txseq;
#if ASCDATA
        data = testdata;                       /* ..using plain data */
#else
        data = &testdata[txseq%datalen];       /* ..or random */
#endif
        memcpy(icmp->data, data, datalen);
        icmp->c.ident = 1;                     /* Make packet */
        txlen = make_icmp(gfp, &locnode, &remnode, ICREQ, 0, datalen);
        txcount++;
    }
    put_frame(gfp, txlen);                      /* Transmit packet */
}
```

Receive Function

The basic structure of the receive function is to

- get the incoming frame,
- check the protocol type using nested is_ functions, and
- take the appropriate action for the protocol.

The first half of the function is concerned with ARP: either a response to your request or a request from another node. In the former case, extract the sender's Ethernet address; in the latter case, send your Ethernet address as a response.

```
/* Check for incoming packets, send response if required */
void do_receive(GENFRAME *gfp)
{
    NODE node;
    ICMPKT *icmp;
    IPKT *ip;
    ARPKT *arp;
```

```
    BYTE *data;
    int rxlen, txlen, len;
    char temps[18];

    if ((rxlen=get_frame(gfp)) > 0)                    /* Any incoming frames? */
    {
        ip = getframe_datap(gfp);
        if (is_arp(gfp, rxlen))
        {                                              /* ARP response? */
            arp = getframe_datap(gfp);
            if (arp->op==ARPRESP && arp->sip==remip)
            {
                memcpy(remnode.mac, arp->smac, MACLEN);
                printf("OK\n");
                arped = 1;
            }
            else if (arp->op==ARPREQ && arp->dip==locnode.ip)
            {                                          /* ARP request? */
                node.ip = arp->sip;                    /* Make ARP response */
                memcpy(node.mac, arp->smac, MACLEN);
                txlen = make_arp(gfp, &locnode, &node, ARPRESP);
                put_frame(gfp, txlen);                 /* Send packet */
            }
        }
    }
```

The other half is concerned with incoming ICMP messages. It is essential to preserve the protocol hierarchy by calling is_ip() and then is_icmp(); otherwise, frame checking won't work.

```
        else if ((rxlen=is_ip(gfp, rxlen))!=0 &&      /* IP datagram? */
                ip->i.dip==locnode.ip || ip->i.dip==BCASTIP)
        {
            if ((rxlen=is_icmp(ip, rxlen))!=0)         /* ICMP? */
            {
                icmp = (ICMPKT *)ip;
                if (icmp->c.type == ICREP)             /* Echo response? */
                {
                    printf("Reply from %s seq=%u len=%u ",
                        ipstr(icmp->i.sip, temps), icmp->c.seq, rxlen);
                    rxseq = icmp->c.seq;               /* Check response */
#if ASCDATA
```

```
                    data = testdata;
#else

                    data = &testdata[rxseq%datalen];
#endif

                    if (rxlen==datalen && !memcmp(icmp->data, data, rxlen))
                    {
                        printf("OK\n");
                        rxcount++;
                    }
                    else
                    {
                        printf("ERROR\n");
                        errcount++;
                    }
                }
                else if (icmp->c.type==ICREQ)        /* Echo request? */
                {

                    getip_srce(gfp, &node);
                    len = (WORD)maxi(rxlen, 0);      /* Make response */
                    txlen = make_icmp(gfp, &locnode, &node, ICREP,
                                        icmp->c.code, (WORD)len);
                    put_frame(gfp, txlen);           /* Send packet */
                }
            }
        }
    }
}
```

Sender's Address

When replying to an unsolicited request, you obviously need to obtain the sender's address so that you can reply to it.

```
/* Get the frame driver type, source IP and Ethernet addresses
** Returned data does not include port number, netmask or gateway addr */
void getip_srce(GENFRAME *gfp, NODE *np)
{
    IPHDR *iph;
```

```
    np->dtype = gfp->g.dtype;
    getframe_srce(gfp, np->mac);
    iph = getframe_datap(gfp);
    np->ip = iph->sip;
}
```

This shortcut assumes that whichever node sent the request will want to receive the response. This is certainly true for nodes on your subnetwork, but isn't necessarily true if the message was sent via a gateway — there may be a better return path via another gateway.

Strictly speaking, you should look up the sender's IP address, and if it's on a different subnetwork, you should send the response to the router specified in the configuration file. I'll be examining routing in greater detail later in this chapter.

Ping in Action

To Ping node 10.1.1.1, type the following line:

```
ping 10.1.1.1
```

Assuming 10.1.1.1 is alive, the following display is produced (key pressed after three cycles):

```
PING Vx.xx
IP 10.1.1.11 mask 255.0.0.0 gate 10.1.1.111 Ethernet 00:c0:26:b0:0a:93
Resolving 10.1.1.1 - press any key to exit
ARP OK
Reply from 10.1.1.1 seq=1 len=32 OK
Reply from 10.1.1.1 seq=2 len=32 OK
Reply from 10.1.1.1 seq=3 len=32 OK
ICMP echo: 3 sent, 3 received, 0 errors
```

Adding the verbose option produces the following display:

```
PING Vx.xx
IP 10.1.1.11 mask 255.0.0.0 gate 10.1.1.111 Ethernet 00:c0:26:b0:0a:93
Resolving 10.1.1.1 - press any key to exit
ARP Tx to   ff:ff:ff:ff:ff:ff 0806 (ARP) len 42
Rx from 00:c0:f0:09:bd:c3 0806 (ARP) len 60
OK
Tx to   00:c0:f0:09:bd:c3 0800 (IP) len 74
Rx from 00:c0:f0:09:bd:c3 0800 (IP) len 74
Reply from 10.1.1.1 seq=1 len=32 OK
Tx to   00:c0:f0:09:bd:c3 0800 (IP) len 74
Rx from 00:c0:f0:09:bd:c3 0800 (IP) len 74
Reply from 10.1.1.1 seq=2 len=32 OK
Tx to   00:c0:f0:09:bd:c3 0800 (IP) len 74
Rx from 00:c0:f0:09:bd:c3 0800 (IP) len 74
Reply from 10.1.1.1 seq=3 len=32 OK
ICMP echo: 3 sent, 3 received, 0 errors
```

Frame Sizes

A useful cross-check is to verify the frame sizes.

```
ARP request 42 bytes:
14 bytes Ethernet header (excl. CRC)
28 bytes ARP header

ARP response 60 bytes:
Minimum size for Ethernet 2 (excl. CRC)

Ping request or response 74 bytes:
14 bytes Ethernet header (excl. CRC)
20 bytes IP header
8 bytes ICMP header
32 bytes data
```

The ARP response is constrained by the minimum Ethernet frame size and the inability of the low-level drivers to deduce the amount of padding added to the real message.

Fragmentation

To check fragmentation, you can increase the ping data size beyond the maximum for one frame (1,472 bytes for Ethernet 2). The following verbose display comes from sending a ping with 2,000 bytes of data to a Linux system.

```
PING Vx.xx
IP 10.1.1.11 mask 255.0.0.0 gate 10.1.1.111 Ethernet 00:c0:26:b0:0a:93
Resolving 10.1.1.1 - press any key to exit
ARP Tx to   ff:ff:ff:ff:ff:ff 0806 (ARP) len 42
Rx from 00:c0:f0:09:bd:c3 0806 (ARP) len 60
OK
Tx to   00:c0:f0:09:bd:c3 0800 (IP) len 1514
Tx to   00:c0:f0:09:bd:c3 0800 (IP) len 562
Rx from 00:c0:f0:09:bd:c3 0800 (IP) len 562
Rx from 00:c0:f0:09:bd:c3 0800 (IP) len 1514
Reply from 10.1.1.1 seq=1 len=2000 OK
```

Interestingly, the Linux system returns the fragments in the reverse order. It's a good thing the software can accommodate out-of-order arrival. The following output is from the same exercise with SLIP, using a 1,000-byte ping from 172.16.1.2 to 172.16.1.3.

```
IP 172.16.1.2 mask 255.255.0.0 SLIP
Pinging 172.16.1.3 - ESC or ctrl-C to exit
Tx / len 1004 ------SLIP------- IP 172.16.1.2 -> 172.16.1.3 ICMP
Tx / len 44 ------SLIP------- IP 172.16.1.2 -> 172.16.1.3 ICMP
Rx \ len 1004 ------SLIP------- IP 172.16.1.3 -> 172.16.1.2 ICMP
Rx \ len 44 ------SLIP------- IP 172.16.1.3 -> 172.16.1.2 ICMP
Reply from 172.16.1.3 seq=1 len=1000 OK
```

At first sight, this seems wrong; the MTU for SLIP is 1,006 bytes, yet the first frame length is 1,004 bytes.

The reason for the discrepancy lies with the fact that the datagram fragment size is measured in units of eight bytes, so the first fragment data length must be divisible by eight.

Unfortunately, the usual SLIP MTU (1,006 bytes) minus the default IP header (20 bytes) isn't divisible by eight, so it is rounded down to 1,004 bytes.

Router Implementation

When experimenting with TCP/IP, it is handy to have a router that provides flexibility when interconnecting random items of network equipment with different network interfaces and network configurations. It is also useful for demonstrating the decision-making process when handling a frame destined for another network. The router I've developed has the ability to deliberately introduce network faults by dropping packets. This will help in checking the error-handling algorithms in higher-level protocols.

Routers are very complicated. It will come as no surprise that our router is a bare-bones design with only the minimum of intelligence, but it will provide a useful framework to support further experimentation in this area.

Interfaces

The router must be installed in a computer with at least two network interfaces, which need not necessarily be of the same type. The standard configuration file has to be extended to cover these interfaces, as shown below.

```
net     ether ne 0x280
ip      10.1.1.111

net     slip pc com2:38400,n,8,1
ip      172.16.1.111

net     ether 3c 0x300
ip      192.168.1.111
```

Here I have defined three network interfaces, each on a completely different subnetwork. Two use Ethernet cards, and one is a SLIP link on a PC serial port.

The Router utility scans the file and associates the first net entry with the first IP address, the second with the second, and so on. This means that the following configuration file is functionally equivalent:

```
net     ether ne 0x280
net     slip pc com2:38400,n,8,1
net     ether 3c 0x300

ip      172.16.1.111
ip      10.1.1.111
ip      192.168.1.111
```

For each of the interfaces, create a local node structure using a variant of read_netconfig().

```
NODE locnodes[MAXNETS];              /* My local node addrs for each net */
int nnets;                           /* Number of nets in use */


...main program...
    while (nnets<MAXNETS &&                        /* Load net drivers */
          (dt=read_netconfig_n(cfgfile, nnets, &locnodes[nnets]))!=0)
    {
        np = &locnodes[nnets++];                   /* Print net addresses etc. */
        np->dtype = dt;
        printf ("Net %u %-15s ", nnets, ipstr(np->ip, temps));
        printf ("mask %-15s ", ipstr(np->mask, temps));
        if (dt & DTYPE_ETHER)
            printf("Ethernet %s\n", ethstr(np->mac, temps));
        else
            printf("SLIP\n");
    }
```

This explains why all the packet-creation functions are given both the source and destination nodes as parameters.

```
int make_arp(GENFRAME *gfp, NODE *srcep, NODE *destp, WORD code);
```

It would have been simpler to dispense with the source-node parameter and use some globally defined address values, but then you would have lost the ability to support the multiple local nodes that are needed for routing.

Routing Algorithm

Generally, the router will only speak when spoken to, so start with an incoming datagram, taking the following steps (Figure 4.7):

1. When an ARP request is received, send the usual ARP response.
2. When a nonbroadcast datagram is received, take one of the following steps.
    a. Check if the destination is on the local network (more specifically, the local network of the interface it arrived on). If so, ignore the datagram, since it doesn't require routing.
    b. Check if the destination is on the local network of another interface. If not, ignore the datagram, since you don't know how to reach the destination.
    c. Check if the destination IP address is in your ARP cache (i.e., whether you know the destination Ethernet address or not). If not, then
        i. buffer the datagram and send an ARP request to the destination and
        ii. receive the ARP response. If there is none, discard the datagram to be routed.
3. Send the datagram to the destination.

Figure 4.7 **Datagram routing.**

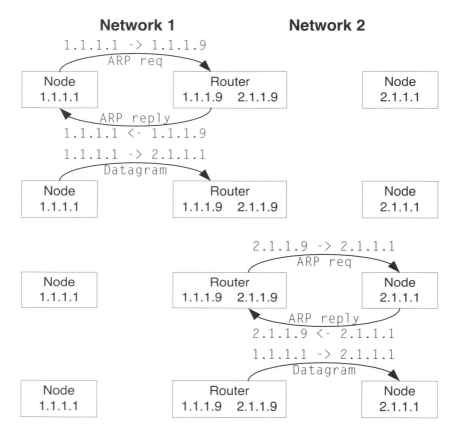

It is important to realize that the routing process is symmetrical, so node 2.1.1.1 will have to go through a similar process (possibly without the ARP cycles) when sending a response datagram. Suppose node 2.1.1.1 has no gateway entries in its configuration file; it won't then be able to find a return route to 1.1.1.1, so the communication will fail. One may wish that TCP/IP had a simple "return to sender" mode, but the reality is that one must know the router IP address through which the sender is reachable, in order to establish two-way communications.

You will note that the router must receive datagrams where the destination IP address doesn't match its own and send datagrams using a source IP address that is equally alien. Hence, the router must filter incoming datagrams based on their Ethernet addresses, rather than their IP addresses. Any datagram sent to the router's Ethernet address that doesn't match its own IP address is considered a potential candidate for routing.

ARP Cache

Of course, routing would be extremely inefficient if an ARP cycle were performed every time a datagram is received. The router should cache all ARP responses so it can forward the datagram immediately.

```c
/* Remote node store is an ARP cache, retaining recently accessed nodes */
NODE remnodes[MAXREMS];                  /* Cache of remote node addresses */
int remidx;                              /* Index number of latest entry */


/* Add a node to the ARP cache */
void add_cache(NODE *np)
{
    char temps[21];

    if (!lookup_cache(np))
    {
        remnodes[remidx] = *np;
        remidx = (remidx+1) % MAXREMS;
        if (netdebug)
            printf ("Adding %s to ARP cache\n", ipstr(np->ip, temps));
    }
}
/* Look up a node in the ARP cache. If found, fill in MAC addr, return non-0 */
int lookup_cache(NODE *np)
{
    int i=remidx, n=MAXREMS, ok=0;

    while (!(ok=remnodes[i=(i-1)%MAXREMS].ip == np->ip) && n>0)
        n--;
    if (ok)
        *np = remnodes[i];
    return(ok);
}
```

This cache has a fixed capacity, and old entries are overwritten when it is full. A real ARP cache would also have an aging mechanism, to refresh or delete old entries.

Datagram Storage

Unfortunately, you'll need to temporarily store datagrams that are awaiting ARP resolution of the destination address. In the diagram, I've shown only one such datagram, but, at any moment in time, you could have a few stacked up awaiting forwarding and a few more that (unbeknownst to you) can never, ever be forwarded, since the destination is, say, powered down.

As with the fragmentation buffer, you need to use a temporary buffer, without permitting the steady accumulation of unrouted datagrams. It just so happens that, in the previous utilities, I've used only one frame buffer for transmission and reception.

```
GENFRAME genframe;                          /* Frame for network Tx/Rx */
```

A simple way of providing a datagram buffer is to provide an array of generic frames with an index pointer to mark the current frame. Every time a datagram must be stored, increment the array index (wrapping the value as necessary), and use the next generic frame for all the subsequent transactions. When an ARP reply arrives, check all the destination IP addresses in the frame array, and send out any datagrams that match the newly contacted node.

```
/* Handle an incoming ARP packet, given local node address */
void do_arp(GENFRAME *gfp, NODE *locp)
{
    ARPKT *arp;
    NODE rem, srce;
    IPKT *ip;
    int txlen, i, n;

    arp = getframe_datap(gfp);
    if (arp->dip == locp->ip)               /* ARP intended for me? */
    {
        rem.ip = arp->sip;                  /* Get remote (source) addrs */
        memcpy(rem.mac, arp->smac, MACLEN);
        rem.dtype = gfp->g.dtype;           /* ..and driver type */
        if (arp->op == ARPREQ)              /* If ARP request.. */
        {                                   /* ..make ARP response */
            txlen = make_arp(gfp, locp, &rem, ARPRESP);
            put_frame(gfp, txlen);          /* ..send packet */
        }
        else if (arp->op == ARPRESP)        /* If ARP response.. */
        {
            add_cache(&rem);                /* ..add to cache */
            for (n=0, i=frameidx; n<NFRAMES-1; n++)
            {                               /* Search framestore */
                gfp = &framestore[i=(i+1)%NFRAMES];
                ip = getframe_datap(gfp);
                if (rem.ip == ip->i.dip)    /* If matching IP addr.. */
                {
                    gfp->g.dtype = rem.dtype;       /* ..make datagram.. */
                    srce = locnodes[rem.dtype & NETNUM_MASK];
```

```
                        srce.ip = ip->i.sip;
                        txlen = make_ip(gfp, &srce, &rem, ip->i.pcol, gfp->g.len);
                        put_frame(gfp, txlen);              /* ..and send it */
                        ip->i.dip = 0;
                    }
                }
            }
        }
    }
}
```

Datagram Reception

The Datagram reception function is significantly more complicated than in previous utilities in order to provide the decision-making process described above. An additional complication is that the interfaces may be of different types (SLIP and Ethernet), with different network header sizes, so you have to translate between the two frame formats.

```
/* Check for incoming packets, send response if required */
void do_receive(void)
{
    int txlen;
    WORD rxlen;
    NODE *locp, srce, dest;
    IPKT *ip, *ip2;
    GENFRAME *gfp;

    gfp = &framestore[frameidx];                      /* Ptr to current frame */
    gfp->g.dtype = 0;                                 /* Has any net driver.. */
    if ((rxlen=get_frame(gfp)) > 0)                   /* ..got incoming frame? */
    {
        locp = &locnodes[gfp->g.dtype&NETNUM_MASK];  /* Get my local address */
        if (is_arp(gfp, rxlen))                        /* If ARP.. */
        {
            do_arp(gfp, locp);                         /* ..handle it */
        }
        else if ((rxlen=is_ip(gfp, rxlen))!=0 &&      /* If IP datagram.. */
                !is_bcast(gfp))                        /* ..but not broadcast.. */
        {
            ip = getframe_datap(gfp);                  /* If ICMP  pkt to me.. */
            if (ip->i.dip==locp->ip && (rxlen=is_icmp(ip, rxlen))!=0)
```

```
        {
            do_icmp(gfp, rxlen, locp);           /* Send ICMP response */
        }
        else                                     /* Datagram needs routing */
        {
            getip_srce(gfp, &srce);              /* Get source addr */
            add_cache(&srce);                    /* ..and add to cache */
            getip_dest(gfp, &dest);              /* Get dest addr */
            gfp->g.len = gfp->g.fragoff = 0;
            if ((locp=find_subnet(&dest))!=0)    /* Find correct subnet.. */
            {
                gfp->g.dtype = locp->dtype;      /* ..and its driver type */
                ip2 = getframe_datap(gfp);       /* Get datagram posn */
                if (ip != ip2)                   /* If new posn in frame.. */
                    memmove(ip2, ip, rxlen+sizeof(IPHDR)); /* ..move it */
                ip = ip2;
                if (locp->dtype & DTYPE_SLIP ||  /* If SLIP.. */
                    lookup_cache(&dest))         /* ..or dest in cache */
                {                                /* Make IP datagram */
                    txlen = make_ip(gfp, &srce, &dest, ip->i.pcol, rxlen);
                    put_frame(gfp, txlen);       /* ..and send it */
                }
                else
                {
                    gfp->g.len = rxlen;          /* Save current frame */
                    frameidx = (frameidx+1) % NFRAMES;
                    gfp = &framestore[frameidx];
                    gfp->g.dtype = locp->dtype; /*Broadcast ARP request*/
                    memcpy(dest.mac, bcast, MACLEN);
                    txlen = make_arp(gfp, locp, &dest, ARPREQ);
                    put_frame(gfp, txlen);       /* Send packet */
                }
            }
        }
    }
}
```

Note the use of a frame index to determine which generic frame is currently in use and how it is incremented to save the current frame (which is the current datagram).

```
GENFRAME *gfp;


gfp = &framestore[frameidx];                  /* Ptr to current frame */
...
gfp->g.len = rxlen;                           /* Save current frame */
frameidx = (frameidx+1) % NFRAMES;
...
```

Main Loop

There is very little left to do in the main loop — just handle incoming packets, and poll the network interface to keep it alive.

```
        printf("Press ESC or ctrl-C to exit\n");
        while (!breakflag)                    /* Main loop.. */
        {
            do_receive();                     /* ..handle Rx packets */
            do_poll();                        /* ..keep net drivers alive */
            if (kbhit())                      /* If key hit, check for ESC */
                breakflag = getch()==0x1b;
        }
        for (n=0; n<nnets; n++)
            close_net(locnodes[n].dtype);     /* Shut down net drivers */
```

Router in Action

When run from the command line, the router reads a router.cfg by default. I'll use the following configuration:

```
net     ether ne 0x280
ip      10.1.1.111

net     slip pc com2:38400,n,8,1
ip      172.16.1.111

net     ether 3c 0x300
ip      192.168.1.111
```

The router digests this and produces the following display.

```
ROUTER v0.05
Net 1 10.1.1.111      mask 255.0.0.0       Ethernet 00:c0:26:b0:0a:93
Net 2 172.16.1.111    mask 255.255.0.0     SLIP
Net 3 192.168.1.111   mask 255.255.255.0   Ethernet 00:50:04:f7:7c:ca
Press ESC or ctrl-C to exit
```

The mask values have been derived from the network class. They could have been overridden by explicit mask entries in the configuration file.

Routing Test

Imagine you've set up a machine on network 3 as 192.168.1.3 with a gateway address of 192.168.1.111, and, in attempting to ping 10.1.1.1, you get no response, even though the target machine is fully functional. Selecting the verbose router option, you get the following display:

```
Rx \ len 74 00:20:18:3a:ed:64 IP 192.168.1.3 -> 10.1.1.1 ICMP        (1)
Adding 192.168.1.3 to ARP cache
Tx / len 42 ----BROADCAST---- ARP   10.1.1.111 -> 10.1.1.1           (2)
Rx \ len 60 00:c0:f0:09:bd:c3 ARP   10.1.1.1 -> 10.1.1.111           (3)
Adding 10.1.1.1 to ARP cache
Tx / len 74 00:c0:f0:09:bd:c3 IP 192.168.1.3 -> 10.1.1.1 ICMP        (4)
```

You can see

1. the incoming datagram,
2. the router ARP request to the destination,
3. the destination ARP response, and
4. the datagram forwarded to the destination.

Unfortunately, there is no ICMP response from 10.1.1.1. Since it responded to the ARP request, it isn't off-line, so the most likely cause is that its gateway address hasn't been set, so it doesn't know where to send the response. If two systems are communicating via a router, it is essential that they both know of the router's existence; node 192.168.1.3 must be configured to use the router interface at 192.168.1.111, while 10.1.1.1 must be configured to use the interface 10.1.1.111.

If the system is running Linux, you can make the following addition to its routing table relatively easily.

```
route add -net 192.168.1.0 gw 10.1.1.111
```

If it is a Windows system, you'd have to add a gateway to the network configuration and reboot.

Now all datagrams from 10.1.1.1 for 192.168.1.x will be routed through 10.1.1.111. Rerunning the ping test produces the following display.

```
Rx \ len 60 00:20:18:3a:ed:64 ARP 192.168.1.3 -> 192.168.1.111       (1)
Tx / len 42 00:20:18:3a:ed:64 ARP 192.168.1.111 -> 192.168.1.3       (2)
Rx \ len 74 00:20:18:3a:ed:64 IP 192.168.1.3 -> 10.1.1.1 ICMP        (3)
Adding 192.168.1.3 to ARP cache
Tx / len 42 ----BROADCAST---- ARP 10.1.1.111 -> 10.1.1.1             (4)
Rx \ len 60 00:c0:f0:09:bd:c3 ARP 10.1.1.1 -> 10.1.1.111             (5)
Adding 10.1.1.1 to ARP cache
Tx / len 74 00:c0:f0:09:bd:c3 IP 192.168.1.3 -> 10.1.1.1 ICMP        (6)
Rx \ len 60 00:c0:f0:09:bd:c3 ARP 10.1.1.1 -> 10.1.1.111             (7)
Tx / len 42 00:c0:f0:09:bd:c3 ARP 10.1.1.111 -> 10.1.1.1             (8)
Rx \ len 74 00:c0:f0:09:bd:c3 IP 10.1.1.1 -> 192.168.1.3 ICMP        (9)
Tx / len 74 00:20:18:3a:ed:64 IP 10.1.1.1 -> 192.168.1.3 ICMP        (10)
```

The ping has succeeded, but with a remarkably high level of network traffic.

1, 2	ARP cycle from 192.168.1.3 (checking router address)
6	Ping request datagram to be routed
4, 5	ARP cycle from router to destination
6	Datagram forwarded to 10.1.1.1
7,8	ARP cycle from 10.1.1.1 (checking router address)
7	Ping response datagram to be routed
8	Datagram forwarded to 192.168.1.3

The router can also handle pings between the Ethernet and SLIP interfaces. What is less obvious is that it will also fragment or defragment as appropriate for the particular interface. A 1,000-byte ping fits within one Ethernet frame, yet it must be fragmented for SLIP, as in the following example:

```
Rx \ len 1042 00:20:18:3a:ed:64 IP 192.168.1.3 -> 172.16.1.3 ICMP
Tx / len 1004 ------SLIP------- IP 192.168.1.3 -> 172.16.1.3 ICMP
Tx / len 44 ------SLIP------- IP 192.168.1.3 -> 172.16.1.3 ICMP
Rx \ len 1004 ------SLIP------- IP 172.16.1.3 -> 192.168.1.3 ICMP
Rx \ len 44 ------SLIP------- IP 172.16.1.3 -> 192.168.1.3 ICMP
Tx / len 1042 00:20:18:3a:ed:64 IP 172.16.1.3 -> 192.168.1.3 ICMP
```

Strictly speaking, this is incorrect behavior for a router. Once fragmented, a datagram should not be reassembled until the destination is reached. However, this behavior happens to be useful for network testing, so it has been allowed to remain.

Summary

I've looked at the role of IP and ICMP in handling the low-level aspects of the TCP/IP protocol. The IP datagram format is straightforward, but fragmentation is a tricky issue, and its implementation requires significant forethought.

ICMP is mainly known for the ping diagnostic it provides, although other ICMP messages are very useful for diagnosing network problems. The implementation of Ping provides an essential diagnostic tool and the ability to check frame sizes and fragmentation issues in the networks you'll be using.

A do-it-yourself router is a rather unusual development, but is a very useful item in the network developer's toolbox, since it can be used as a temporary bridge between various networks under test. It also allows you to explore some simple routing techniques and could form the basis of a much more comprehensive implementation.

Source Files

ether3c.c	3C509 Ethernet card driver
etherne.c	NE2000 Ethernet card driver
ip.c	Low-level TCP/IP functions
net.c	Network interface functions
netutil.c	Network utility functions
ping.c	Ping utility
pktd.c	Packet driver (BC only)
router.c	Router utility
serpc.c or serwin.c	Serial drivers (BC or VC)
dosdef.h	MS-DOS definitions (BC only)
ether.h	Ethernet definitions
ip.h	TCP/IP definitions
net.h	Network driver definitions
netutil.h	Utility function and general frame definitions
serpc.h	Serial driver definitions (BC or VC)
win32def.h	Win32 definitions (VC only)

Ping Utility

Utility:	Emulation of standard Ping utility with server capabilities	
Usage:	ping [*options*] [*IP_address*]	
	Enters server mode if no IP address given	
Options:	-c *name*	Configuration filename (default tcplean.cfg)
	-f	Flood mode
	-l *xxx*	Length of ICMP data in bytes (default 32)
	-v	Verbose display mode
	-w *xxx*	Waiting time in milliseconds (default 1000)
Example:	ping -c slip -v 172.16.1.1	
Keys:	Ctrl-C or Esc to stop pings	
Config:	net	to identify network type
	ip	to identify IP address
	mask	optional subnet mask
	gate	optional gateway IP address
Modes:	Defaults to server mode (Ping responder) unless IP address given	

Router Utility

Utility	Simple router for two or more networks	
Usage	`router [`*`options`*`]`	
Options	`-c` *`name`*	Configuration filename (default `router.cfg`)
	`-v`	Verbose display mode
Example	`router -v`	
Keys	Ctrl-C or Esc to exit	
Config	At least two networks must be defined with	
	`net`	to identify network type
	`ip`	to identify IP address
	`mask`	optional subnet mask
	`gate`	optional gateway IP address
Notes	The default configuration file is `router.cfg`, not `tcplean.cfg`	
	Networks can be a mixture of Ethernet and SLIP	

Chapter 5

User Datagram Protocol: UDP

Overview

While traveling up the TCP stack, you might have wondered when you'd get to the point of doing something useful. Pinging and routing are all very fine, but you can't use them to exchange meaningful data with another system. What is really needed is a simple protocol that sits on top of IP and allows you to launch data and receive replies on your network. That protocol already exists: it is user datagram protocol (UDP). In this chapter, I'll look at how it works, and how you can send simple requests to a system.

Surprisingly, there isn't a standard general-purpose UDP utility to emulate, so I'll have to invent one. I'll be implementing a utility called Datagram, which can be used as a general-purpose UDP workhorse and a platform for further exploration of UDP-based protocols.

Ports and Sockets

What is this "meaningful" data that can be exchanged, and who defines the meaning? The answer is twofold. First are standard applications that run on most network servers and give a standard response to a UDP request, and second are user-defined applications, which may initiate or respond to communications in an entirely nonstandard way. So long as all of these applications encapsulate their data in UDP datagrams, there is no risk of confusion. Each UDP datagram is equipped with source and destination port numbers, which identify the source and destination applications for the data so that one application can't accidentally receive another's data.

Some port numbers are "well-known" (i.e., predefined). If you send a datagram to one of these, you will receive a standard response. For example, port 13 is reserved for the "Daytime" protocol: whatever data you send, a date and time string is received in return, as below.

```
Thu Apr  6 17:24:03 2000
```

If I want to use this service, I must set the destination port number on my outgoing datagram to 13, and the source port number to some temporary (ephemeral) value that is currently unused on my system — a value between 1,024 and 5,000 is generally appropriate. The response is then returned using port 13 as the source and my port as the destination (Figure 5.1).

Figure 5.1   Simultaneous UDP requests and responses.

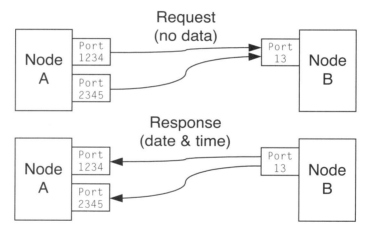

At first sight, it might seem more sensible to return the reply to port 13 on my system, but this would cause two problems.

1. If two separate applications on my system happened to make simultaneous requests, it would be impossible to determine which application should receive which response.

2. The returned response would be indistinguishable from a request to my system, so my system might produce a response, which would be interpreted by the remote as another request, and so on ad infinitum.

For these reasons, the response is usually returned to the originating port and, by implication, to the originating application, since there is usually a one-to-one correspondence between ports and applications (or in the case of ephemeral ports, instances of the application).

Well-Known Ports

The following are common, predefined (well-known) UDP port numbers.

7	Echo	echoes incoming data
9	Discard	discards incoming data
13	Daytime	returns date and time string
19	Chargen	generates character strings
37	Time	four-byte time value
69	TFTP	simple file transfer
161	SNMP	simple network management

For historic reasons, these are all odd numbers. It was originally intended that all requests use even numbers and responses use odd numbers, but (for the reasons described above) this was found to be unworkable, so the concept of ephemeral port numbers was introduced instead.

All the applications are relatively simple, as befits the simple, no-frills nature of UDP. There is no guarantee that a datagram will arrive at all, and if it arrives, there is no guarantee that the response will be received. As a result, you may think UDP is useless and wonder why anyone would bother with it. However, a short exposure to the intricacies of TCP may convince you that UDP is quite adequate for your (simple) application. If UDP is enhanced by a lightweight retry scheme, it can be reliable yet flexible, consume fewer resources, and be much easier to debug than its TCP alternative.

Sockets

A combination of a specific network address and a port is often referred to as a "logical endpoint" or a "socket." There is a good analogy between this (virtual) network socket and a physical plug-and-socket connection. If networks didn't exist, you'd have to find a cable and plug it into a socket on the back of the remote computer. The socket might even be labeled with the service it supported There would be different sockets for terminal login, printing, and peripheral communication. Having concluded your use of the socket, you'd unplug (disconnect) to allow others to use it.

The analogy falls down in that the virtual sockets can support multiple simultaneous connections: so long as the sender possesses a unique socket, a response can be returned.

If two systems happen to choose the same port number when communicating with a node, it doesn't matter because their sockets are defined by a port number *and* the node address, so the (potentially dissimilar) requests can be distinguished from each other (Figure 5.2).

Figure 5.2   UDP requests from the same port number.

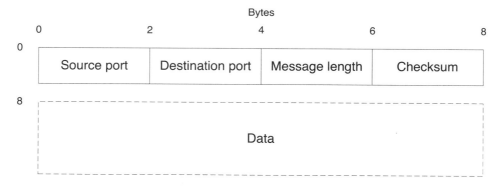

Datagram Format

A UDP message is known by the same name as an IP message, namely a *datagram*.

Figure 5.3   Datagram format.

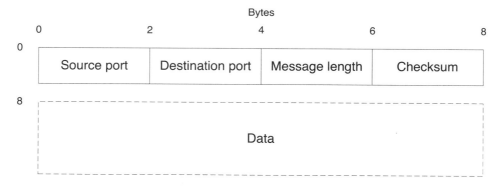

Aside from the source and destination port numbers, there is a length value (length in bytes of this header plus the data) and a checksum field. All values are in the standard network (big endian) byte order.

```
/* ***** UDP (User Datagram Protocol) header ***** */
typedef struct udph
{
    WORD   sport,             /* Source port */
           dport,             /* Destination port */
           len,               /* Length of datagram + this header */
           check;             /* Checksum of data, header + pseudoheader */
} UDPHDR;


#define MAXUDP (MAXIP-sizeof(UDPHDR))
/* ***** UDP packet ('datagram') ***** */
typedef struct udp
{
    IPHDR   i;                /* IP header */
    UDPHDR  u;                /* UDP header */
    BYTE    data[MAXUDP];     /* Data area */
} UDPKT;
```

Thanks to IP fragmentation, the maximum UDP data size is not restricted by the Ethernet or SLIP frame size. If the datagram is too large for one frame, it will be broken up into several fragments (Figure 5.4), each with its own IP header, but only the first possessing a UDP header.

Figure 5.4   Datagram fragmentation.

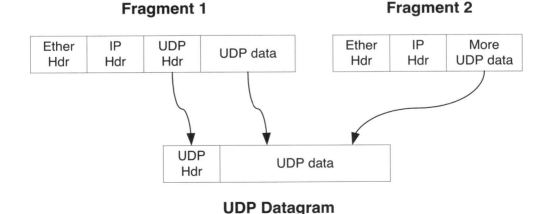

Although it may be tempting to encapsulate all your data in a single, huge datagram, a word of warning: all systems have limits on the number of fragments that can be accommodated, since they occupy precious space during reassembly. The software in this book can accept a total of only two fragments, and many implementations don't accept any fragmentation at all.

UDP Checksum

This uses the same addition method as IP, but with two new twists. First, the UDP checksum re-checks part of the IP header (by incorporating an IP "pseudoheader" in the calculation), and second, the checksum process can be disabled by inserting a zero value.

IP Pseudoheader

The creators of UDP wanted to extend the checksum protection to the IP header. Rather than adding on a checksum of the complete IP header, they opted to extract the important parts of the IP header into a smaller pseudoheader (Figure 5.5) and then add this checksum onto the value obtained from the UDP header and data.

Figure 5.5   IP pseudoheader.

It must be stressed that this header is an internal software fiction, used purely for computing the UDP checksum. It is never seen on the network.

```
/* ***** Pseudo-header for UDP or TCP checksum calculation ***** */
/* The integers must be in hi-lo byte order for checksum */
typedef struct             /* Pseudo-header... */
{
    LWORD srce,            /* Source IP address */
          dest;            /* Destination IP address */
    BYTE  z,               /* Zero */
          pcol;            /* Protocol byte */
    WORD  len;             /* UDP length field */
} PHDR;
```

Computing the checksum is the usual fill-in-the-blanks exercise, taking care to ensure that correct length values are used and that data is in the unswapped (big endian) order.

```
/* Return TCP checksum, given UDP (header + data) length.
** The values must be in network byte-order */
WORD check_udp(UDPKT *udp, LWORD sip, LWORD dip, int ulen)
{
    PHDR tph;
    LWORD sum;

    sum = csum(&udp->u, (WORD)ulen);          /* Checksum TCP segment */
    tph.len = swapw((WORD)ulen);              /* Make pseudo-header */
    tph.srce = sip;
    tph.dest = dip;
    tph.z = 0;
    tph.pcol = udp->i.pcol;
    sum += csum(&tph, sizeof(tph));           /* Checksum pseudo-header */
    return((WORD)(sum + (sum >> 16)));        /* Return total plus carry */
}
```

The value inserted in the checksum field is the one's complement of this value.

```
udp->u.check = 0;
udp->u.check = ~check_udp(udp, udp->i.sip, udp->i.dip, ulen);
```

On receive, the datagram checksum should be FFFFh.

```
sum = check_udp(udp, swapl(ip->i.sip), swapl(ip->i.dip), len);
if (sum==0xffff)
```

                                     ... datagram is OK ...

Disabling the UDP Checksum

If the sender of the datagram doesn't want to compute a checksum, a zero value can be sent instead, which indicates that the value is to be ignored on reception. If the computed value happens to be zero, but you do want to send a checksum, send a value of FFFFh.

```
udp->u.check = ~check_udp(udp, udp->i.sip, udp->i.dip, ulen);
if (udp->u.check == 0)                 /* Change sum of 0 to FFFF */
    udp->u.check = 0xffff;
```

On receive, you now have a quandary. If the checksum field contains FFFFh, how do you know whether this is actually FFFFh or really zero adjusted to FFFFh? Surprisingly, this is a problem that goes away if you ignore it. Just compute the checksum as usual, and everything is OK.

```
sum = check_udp(udp, swapl(ip->i.sip), swapl(ip->i.dip), len);
if (!udp->u.check || sum==0xffff)
```

... datagram is OK ...

This seems totally counterintuitive, so an example is in order. Assume that the data bytes to be checked are 67h, 89h, 98h, 76h. The transmit checksum would be computed as follows:

1st byte: ~(0x67 + 0x98) = 0
2nd byte: ~(0x89 + 0x76) = 0

Hence, the transmitted data might be 67h, 89h, 98h, 76h, 00h, 00h, which checks out OK.

1st byte: 0x67 + 0x98 + 0x00 = 0xff
2nd byte: 0x89 + 0x76 + 0x00 = 0xff

However, you should change the checksum value from zero to FFh as described above, so the actual data sent should be 67h, 89h, 98h, 76h, FFh, FFh. Remarkably, this still checks out OK.

1st byte: 0x67 + 0x98 + 0xff + carry = 0xff
2nd byte: 0x89 + 0x76 + 0xff + carry = 0xff

Because there is a carry from byte 1 to byte 2 *and* from byte 2 to byte 1, adding FFFFh to any non-zero value will leave it unchanged — a great feature or a significant weakness; decide for yourself. Either way, you don't have to make special provisions for the translation from zero to FFFFh when checking incoming datagrams.

UDP Utility

In order to experiment with UDP, I created Datagram, a general-purpose datagram send/receive utility. To interrogate port 13 (daytime) on node 10.1.1.1, simply type

```
datagram 10.1.1.1 13
```

and a reply like

```
Fri Apr  7 10:45:16 2000
```

will be returned. You can also try the echo port (7) by sending a character string, such as

```
datagram 10.1.1.1 7 "Hello there"
```

receiving the following response:

```
Hello there
```

If no IP address is specified on the command line, the utility goes into server mode, where it only responds to daytime or echo requests.

Implementation

Transmission and Reception

Aside from the checksum peculiarities, there is little to be added to the IP transmit and receive functions.

```c
/* Return UDP data length (-1 if no data), 0 if not UDP */
int is_udp(IPKT *ip, int len)
{
    UDPKT *udp;
    WORD sum;
    int dlen=0;
                                        /* Check protocol & minimum length */
    if (ip->i.pcol==PUDP && len>=sizeof(UDPHDR))
    {
        udp = (UDPKT *)ip;              /* Do checksum */
        sum = check_udp(udp, swapl(ip->i.sip), swapl(ip->i.dip), len);
        if (!udp->u.check || sum==0xffff)
        {                               /* If zero or correct.. */
            swap_udp(udp);              /* Do byte-swaps */
            len -= sizeof(UDPHDR);      /* Subtract header len */
            if (udpdebug)               /* Display segment if in debug mode */
                disp_udp(udp, len, 0);
            dlen = len>0 ? len : -1;    /* Return -1 if data len=0 */
        }
        else if (udpdebug)             /* Display error */
            printf("  ERROR: UDP checksum %04X\n", sum);
    }
    return(dlen);
}


/* Make a UDP datagram given the source & destination, data len */
int make_udp(GENFRAME *gfp, NODE *srcep, NODE *destp, WORD dlen)
{
    UDPKT *udp;
    int ulen, ilen;

    udp = getframe_datap(gfp);
    udp->u.sport = srcep->port;         /* Set ports */
    udp->u.dport = destp->port;
```

```
    udp->u.len = ulen = dlen + sizeof(UDPHDR);
    udp->u.check = 0;
    if (udpdebug)                        /* Display datagram if in debug mode */
        disp_udp(udp, dlen, 1);
    swap_udp(udp);                       /* Byte-swap */
    ilen = make_ip(gfp, srcep, destp, PUDP, (WORD)(ulen));
    udp->u.check = ~check_udp(udp, udp->i.sip, udp->i.dip, ulen);
    if (udp->u.check == 0)               /* Change sum of 0 to FFFF */
        udp->u.check = 0xffff;
    return(ilen);                        /* Return IP length */
}


/* Swap byte order of ints in TCP header */
void swap_udp(UDPKT *udp)
{
    udp->u.sport = swapw(udp->u.sport);
    udp->u.dport = swapw(udp->u.dport);
    udp->u.len = swapw(udp->u.len);
}


/* Send a UDP datagram, given destination node, data and length */
void udp_transmit(GENFRAME *gfp, NODE *sp, NODE *dp, void *dat, int len)
{
    UDPKT *udp;

    udp = getframe_datap(gfp);
    memmove(udp->data, dat, len);
    put_frame(gfp, make_udp(gfp, sp, dp, (WORD)maxi(len, 0)));
}
```

The source- and destination-node structures contain the necessary addresses and port numbers, so they might more properly be called socket structures. The source-node structure might seem superfluous, since the UDP utility possesses only one IP address, which could be hard-coded; however, there are various circumstances in which it is useful for a node to possess multiple identities (known as multihoming). One instance was the router described in the previous chapter. Also, a multihomed host can be useful for testing a large network using a small amount of hardware.

Local Node and Multihoming

For this scheme to work, an incoming packet must be matched to the appropriate local node (local socket) before any response can be sent. This is achieved by calling getudp_locdest(), which in turn does an upcall via a `get_locnode_n()` function pointer to interrogate the configuration file data that was loaded at startup.

```c
/* Get complete UDP local node data corresponding to frame dest IP address
** Return 0 if no matching node */
int getudp_locdest(GENFRAME *gfp, NODE *np)
{
    UDPKT *udp;
    int ok;

    ok = getip_locdest(gfp, np);        /* Get addresses, dtype & netmask */
    udp = getframe_datap(gfp);          /* Get dest port */
    np->port = udp->u.dport;
    return(ok);
}
/* Get local node data corresponding to a frame destination IP address
** Data does not include port number. Return 0 if no matching local node */
int getip_locdest(GENFRAME *gfp, NODE *np)
{
    IPHDR *iph;
    NODE *locp;
    int ok=0;

    iph = getframe_datap(gfp);
    ok = (locp = findloc_ip(iph->dip)) != 0;
    if (ok)
        *np = *locp;
    return(ok);
}
/* Find local node corresponding to given IP addr, return 0 if not found */
NODE *findloc_ip(LWORD locip)
{
    NODE *np=0;
    int n=0;

    while (get_locnode_n && (np=get_locnode_n(n))!=0 && np->ip!=locip)
        n++;
    return(np);
}
```

Only the IP address of the incoming datagram is checked, not the port number. This is acceptable for a lightweight UDP implementation, since `udp_receive()` can simply check the port number and take the appropriate action, but it's worth pointing out that a proper socket scheme (such as the one I'll use for TCP in the next chapter) would match the port number to a socket and reject the datagram if no match was found.

To make this work, all you have to do on startup is to set the function pointer to a handler that will return your local-node information.

```
NODE  locnode;                        /* My Ethernet and IP addresses */
...
extern NODE *(*get_locnode_n)(int n);     /* Get local node */
...
get_locnode_n = locnode_n;            /* Set upcall ptr to func */
...
/* Return ptr to local node 'n' (n=0 for first), return 0 if doesn't exist
** Used by IP functions to get my netmask & gateway addresses */
NODE *locnode_n(int n)
{
    return(n==0 ? &locnode : 0);
}
```

This is somewhat over the top for a humble UDP utility, but it allows you to potentially emulate a whole battery of UDP-controlled data collection systems, for example, by creating multiple local nodes and tweaking locnode_n() accordingly.

Main Processing Loop

Processing of the initialization and configuration files is very similar to previous utilities, so I will not describe them here. The main processing loop has a little state machine to do the following:

- Send the ARP request to the intended host.
- Wait for the ARP response; if none, retry a few times; then exit.
- Send the UDP data.
- Wait for the UDP response; if none, retry a few times; then exit.
- Display the UDP response.

```
printf("\nPress ESC or ctrl-C to exit\n");
if (client)
{
    printf("Contacting %s...\n\n", ipstr(remnode.ip, temps));
    memcpy(remnode.mac, bcast, MACLEN);
}
mstimeout(&mstime, 0);
while (!breakflag)                        /* Main loop.. */
{
    if (client && (mstimeout(&mstime, TRYTIME) || cstate!=lastcstate))
```

```
    {
        if (tries++ > MAXTRIES)          /* Giving up? */
            breakflag = 1;
        else if (cstate == ARP_TX)       /* (Re)transmit ARP? */
            put_frame(gfp, make_arp(gfp, &locnode, &remnode, ARPREQ));
        else if (cstate == ARP_RX)       /* ARP response? */
        {
            if (in)
                udp_transmit_file(gfp, &locnode, &remnode, in);
            else
                udp_transmit(gfp, &locnode, &remnode, cmd, strlen(cmd));
        }
        else if (cstate == CLIENT_DONE) /* UDP response? */
            breakflag = 1;
        lastcstate = cstate;             /* Record state-change */
    }
    st = do_receive(gfp);                /* Receive frames */
    cstate = st ? st : cstate;           /* ..maybe change state */
    do_poll(gfp);                        /* Poll net drivers */
    if (kbhit() && getch()==0x1b)        /* Check keyboard */
        breakflag = 1;
}
close_net(dtype);                        /* Shut down net driver */
```

As usual, do_receive() does the demultiplexing for the incoming packet (which may or may not be a datagram) and takes the appropriate action.

```
/* Check for incoming packets, send response if required
** Return state-change value if ARP response or datagram received */
int do_receive(GENFRAME *gfp)
{
    NODE node;
    ARPKT *arp;
    IPKT *ip;
    ICMPKT *icmp;
    int rxlen, txlen, len, ret=0;

    if ((rxlen=get_frame(gfp)) > 0)                      /* Any incoming frames? */
    {
        ip = getframe_datap(gfp);
        if (is_arp(gfp, rxlen))
```

```
                                                       /* ARP? */
        arp = getframe_datap(gfp);
        if (arp->op==ARPREQ && arp->dip==locnode.ip)
        {                                              /* ARP request? */
            node.ip = arp->sip;                        /* Make ARP response */
            memcpy(node.mac, arp->smac, MACLEN);
            txlen = make_arp(gfp, &locnode, &node, ARPRESP);
            put_frame(gfp, txlen);                     /* Send packet */
        }
        if (arp->op==ARPRESP && arp->dip==locnode.ip)
        {                                              /* ARP response? */
            memcpy(remnode.mac, arp->smac, MACLEN);
            ret = ARP_RX;
        }
    }
    else if ((rxlen=is_ip(gfp, rxlen))!=0 &&     /* IP datagram? */
            ip->i.dip==locnode.ip || ip->i.dip==BCASTIP)
    {
        getip_srce(gfp, &node);
        if ((len=is_icmp(ip, rxlen))!=0)         /* ICMP? */
        {
            icmp = (ICMPKT *)ip;
            if (icmp->c.type == ICREQ)           /* Echo request? */
            {
                len = (WORD)maxi(len, 0);        /* Make response */
                txlen = make_icmp(gfp, &locnode, &node, ICREP,
                                 icmp->c.code, (WORD)len);
                put_frame(gfp, txlen);           /* Send packet */
            }
            else if (icmp->c.type == ICUNREACH)
                printf("ICMP: destination unreachable\n");
        }
        else if ((len=is_udp(ip, rxlen))!=0)     /* UDP? */
        {
            ret = udp_receive(gfp, maxi(len, 0));
        }
    }
}
return(ret);
}
```

UDP Services

As previously mentioned, `udp_receive()` checks the port number and applies the appropriate service.

```
/* Receive a UDP datagram: return non-0 if client state-change */
int udp_receive(GENFRAME *gfp, int len)
{
    UDPKT *udp;
    int ret=0;
    NODE loc, rem;
    char temps[30];
    time_t t;

    udp = getframe_datap(gfp);
    getudp_srce(gfp, &rem);                         /* Get srce & dest nodes */
    getudp_locdest(gfp, &loc);
    if (loc.port == locnode.port)                   /* Client response */
    {
        disp_data(udp->data, len);                  /* Display data.. */
        ret = CLIENT_DONE;                          /* ..and exit */
    }
    else if (loc.port == ECHOPORT)                  /* Echo req: resend data */
        udp_transmit(gfp, &loc, &rem, udp->data, len);
    else if (loc.port == DAYPORT)                   /* Daytime req: get date */
    {
        time(&t);                                   /* Get standard string */
        strcpy(temps, ctime(&t));
        strcpy(&temps[24], "\r\n");                 /* Patch newline chars */
        udp_transmit(gfp, &loc, &rem, (BYTE *)temps, 26);
    }
    else                                            /* Unreachable send ICMP */
    {
        swap_udp(udp);
        put_frame(gfp, icmp_unreach(gfp, &loc, &rem, UNREACH_PORT));
    }
    return(ret);
}
```

If the destination port number matches your local client port, then the datagram must be a response to your request, so the data is displayed. If not, and the port doesn't match an Echo or Daytime request, an ICMP "destination unreachable" response is sent.

Experimenting with the Datagram Utility

Echo

The Datagram utility can be used as both a client and a server. Run the server using

```
datagram
```

with no command-line arguments. The utility should sign on, giving its IP and Ethernet addresses.

```
DATAGRAM Vx.xx
IP 10.1.1.12 mask 255.0.0.0 gate 10.1.1.111 Ethernet 00:80:66:03:00:09
Press ESC or ctrl-C to exit
```

Now run the client on another machine, making sure you aren't using the same values in the default configuration file and, hence, the same IP address.

```
datagram 10.1.1.12 echo "Hello"
```

If all is well, the string should be echoed back.

```
DATAGRAM Vx.xx
IP 10.1.1.11 mask 255.0.0.0 gate 10.1.1.111 Ethernet 00:c0:26:b0:0a:93
Press ESC or ctrl-C to exit
Contacting 10.1.1.12...

Hello
```

To help in diagnosing problems, you can use the -v (verbose), -u (UDP display), or both flags.

```
datagram -u -v 10.1.1.12 echo "Hello"

DATAGRAM Vx.xx
IP 10.1.1.11 mask 255.0.0.0 gate 10.1.1.111 Ethernet 00:c0:26:b0:0a:93
Press ESC or ctrl-C to exit
Contacting 10.1.1.12...

Tx0 /len 42 ----BROADCAST---- ARP 10.1.1.11 -> 10.1.1.12
Rx0 \len 60 00:80:66:03:00:09 ARP 10.1.1.12 -> 10.1.1.11


    /port 1024->7  dlen 5
Tx0 /len 47 00:80:66:03:00:09 IP 10.1.1.11 -> 10.1.1.12 UDP
Rx0 \len 60 00:80:66:03:00:09 IP 10.1.1.12 -> 10.1.1.11 UDP
    \port 1024<-7  dlen 5

Hello
```

This shows the initial ARP request and response, followed by the UDP transaction. If the host can't be contacted, there would be no ARP responses.

```
Tx0 /len 42 ----BROADCAST---- ARP 10.1.1.11 -> 10.1.1.13
Tx0 /len 42 ----BROADCAST---- ARP 10.1.1.11 -> 10.1.1.13
... and so on ...
```

If the host doesn't support that service, it returns an ICMP "destination unreachable" message.

```
Tx0 /len 42 ----BROADCAST---- ARP 10.1.1.11 -> 10.1.1.12
Rx0 \len 60 00:80:66:03:00:09 ARP 10.1.1.12 -> 10.1.1.11
     /port 1024->123  dlen 5
Tx0 /len 47 00:80:66:03:00:09 IP 10.1.1.11 -> 10.1.1.12 UDP
Rx0 \len 70 00:80:66:03:00:09 IP 10.1.1.12 -> 10.1.1.11 ICMP
ICMP: destination unreachable
     /port 1024->123  dlen 5
Tx0 /len 47 00:80:66:03:00:09 IP 10.1.1.11 -> 10.1.1.12 UDP
Rx0 \len 70 00:80:66:03:00:09 IP 10.1.1.12 -> 10.1.1.11 ICMP
ICMP: destination unreachable
```

... and so on ...

Use the -u and -v options to produce similar diagnostics on the server.

```
datagram -u -v
```

Test the echo service with longer strings using the -f option to send data from a file.

```
datagram -u -v -f myfile.txt 10.1.1.12 echo
```

Perform the test with nontext files using the -b (binary) option, which displays the response in hexadecimal.

```
datagram -b -i \temp\test.bin 10.1.1.12 echo
DATAGRAM Vx.xx
IP 10.1.1.11 mask 255.0.0.0 gate 10.1.1.111 Ethernet 00:c0:26:b0:0a:93
Press ESC or ctrl-C to exit
Contacting 10.1.1.12...

0000: 01 12 23 34 45 56 67 78 89 9A AB BC CD DE EF
          #  4  E  V  g  x
```

Time

There are two UDP time protocols: Daytime and Time. The former returns a string, and the latter returns a four-byte binary value.

```
datagram 10.1.1.1 daytime
DATAGRAM Vx.xx
IP 10.1.1.11 mask 255.0.0.0 gate 10.1.1.111 Ethernet 00:c0:26:b0:0a:93
Press ESC or ctrl-C to exit
Contacting 10.1.1.1...
Fri Apr  7 17:21:50 2000
```

To display a binary value, use the -b option.

```
datagram -b 10.1.1.1 time
DATAGRAM Vx.xx
IP 10.1.1.11 mask 255.0.0.0 gate 10.1.1.111 Ethernet 00:c0:26:b0:0a:93
Press ESC or ctrl-C to exit
Contacting 10.1.1.1...

0000: BC 98 89 A8
```

This value represents the number of seconds since the birth of Unix on 1 January 1970.

Summary

User datagram protocol (UDP) is a simple way of conveying data from one node to another. Although it inherits all the inherent unreliability of the parent network, it is excellent for low-level network housekeeping tasks or as a basis for a more complex user-defined protocol. I developed a general-purpose UDP utility, Datagram, that can be used to explore existing UDP protocols or as a basis for new ones, and I looked at one area for expansion, multihoming, that can emulate several nodes using a single utility.

Many network developers favor TCP (as discussed in Chapter 6) over UDP, because it offers a reliable connection, but you'll see that reliability comes at a heavy price in terms of code complexity. If it's simplicity you're after, UDP is hard to beat. Even after error-handling code has been added, it is still much smaller and easier to debug than the corresponding TCP implementation. Read the next chapter, and judge for yourself.

Source Files

`datagram.c`	UDP utility
`ether3c.c`	3C509 Ethernet card driver
`etherne.c`	NE2000 Ethernet card driver
`ip.c`	Low-level TCP/IP functions
`net.c`	Network interface functions
`netutil.c`	Network utility functions
`pktd.c`	Packet driver (BC only)
`router.c`	Router utility
`serpc.c` or `serwin.c`	Serial drivers (BC or VC)
`udp.c`	UDP functions
`dosdef.h`	MS-DOS definitions (BC only)
`ether.h`	Ethernet definitions
`ip.h`	TCP/IP definitions
`net.h`	Network driver definitions
`netutil.h`	Utility function and general frame definitions
`serpc.h`	Serial driver definitions (BC or VC)
`udp.c`	UDP definitions
`win32def.h`	Win32 definitions (VC only)

Datagram Utility

Utility	General-purpose UDP interface
Usage	datagram [*options*] [*IP_address* [*port* [*data*]]]
	Enters server mode if no IP address given
Options	-b Binary mode (hex data display)
	-c *name* Configuration filename (default tcplean.cfg)
	-i *name* Data input filename
	-u UDP datagram display mode
	-v Verbose display mode
	-x Hex packet display
Example	datagram -v 10.1.1.1 echo "hello there"
Keys	Ctrl-C or Esc to exit
Config	net to identify network type
	ip to identify IP address
	mask optional subnet mask
	gate optional gateway IP address
Modes	Defaults to server mode (Echo and Daytime) if no IP address
Notes	The port can be a number or one of the following:
	echo
	daytime
	time
	snmp

Chapter 6

Transmission Control Protocol: TCP

Overview

The next step up the TCP/IP stack is TCP (transmission control protocol). This isn't an easy protocol to implement; it has to perform a multitude of tasks in pursuit of its ultimate objective: creating a reliable connection between two points on the network.

This chapter explores the elements of TCP, how they can be translated into software, and how that software can be tested. I'll be creating my own implementation of the general-purpose TCP communications program, TELNET.

Why is TCP so difficult? Fundamentally, it is trying to do several jobs at once.

1. Initiate a connection between two nodes.
2. Send data bidirectionally between the nodes.
3. Handle network datagram loss.
4. Handle network datagram duplication.
5. Handle out-of-order arrival of datagrams.
6. Handle network failure.
7. Handle all data rates, from occasional single characters to bulk transfer of large files.

8. Provide flow control to avoid data overload.
9. Provide the ability to hasten urgent data.
10. Close the connection between two nodes.
11. Support "half-closure," in which one participant wants to close, but the other doesn't.
12. Handle datagram arrival after the connection is closed.

To add to your problems, the situation between any two nodes changes dynamically and very rapidly, so any failure situations can be difficult to reproduce, making bugs hard to find and fix.

The key to creating a solid TCP implementation is to find a clear expression of the underlying concepts and perform exhaustive testing. To help in this, I'll include significant diagnostic capability in the code, with the ability to deliberately generate errors, in order to analyze how the software handles them.

TCP Concepts

The key word in understanding TCP is *connection*. TCP establishes the logical equivalent of a physical connection between two points. Data then passes bidirectionally along this connection. Both sides must keep track of the data sent and received so that they can detect any omissions or duplications in the data stream (Figure 6.1).

Data is sent in blocks (IP datagrams), and, as already discussed in Chapter 2, the nodes can track in various ways the quantity of data sent and received.

A *lockstep* method (the sender waits until each block has been acknowledged before sending the next one) is a relatively inefficient way of transferring bulk data. *Block sequencing*, where each block is individually numbered, is better for bulk data transfers because the recipient need not acknowledge each block as it arrives but can delay until several blocks have arrived and acknowledge them all with one response.

The *byte-sequencing* scheme adopted by TCP goes one stage further, in that both sides track the byte count of the transmitted and received data. After an appropriate amount of data has been received, the recipient acknowledges that byte count, irrespective of the number of blocks it represents. This gives greater flexibility in avoiding congestion, as you will see later.

Sequence and Acknowledgment Numbers

I've already used these terms informally, but now is the time to tie down their definitions.

The *sequence number* indicates the position of the current data block in the data stream. If the first data block has a sequence number of zero and is 10 bytes long, then you'd expect the next data block to have a sequence number of 10. The sequence number is often abbreviated to seq.

The *acknowledgment number* indicates the total amount of data received. If the initial sequence number is zero and 10 bytes of data are to be acknowledged, the response would have an acknowledgment value of 10. The acknowledgment number is often abbreviated to ack.

Because TCP data transfer is bidirectional and symmetrical, each side has a sequence and acknowledgment number for its own transmissions, and each side tracks the sequence and acknowledge numbers it receives from the other node.

How big is a sequence or acknowledge number, and does this limit the maximum size of data that can be transferred? The numbers are 32 bits, so they wrap around when 4.3Gb of data is transferred. This isn't a problem, because the sequence and acknowledgment numbers are used to track the *relative*, not the absolute, amount of data transferred. There is no compulsion to start with a sequence number of zero; indeed, there are good reasons not to do so. If you start with an arbitrary sequence number, you only need to check the *difference* between the last and the current sequence numbers received in order to detect missing or duplicated data. So long as there aren't several gigabytes of data in transit between sender and receiver, there is no possibility of the wraparound causing any problems.

Figure 6.1   Sequencing methods.

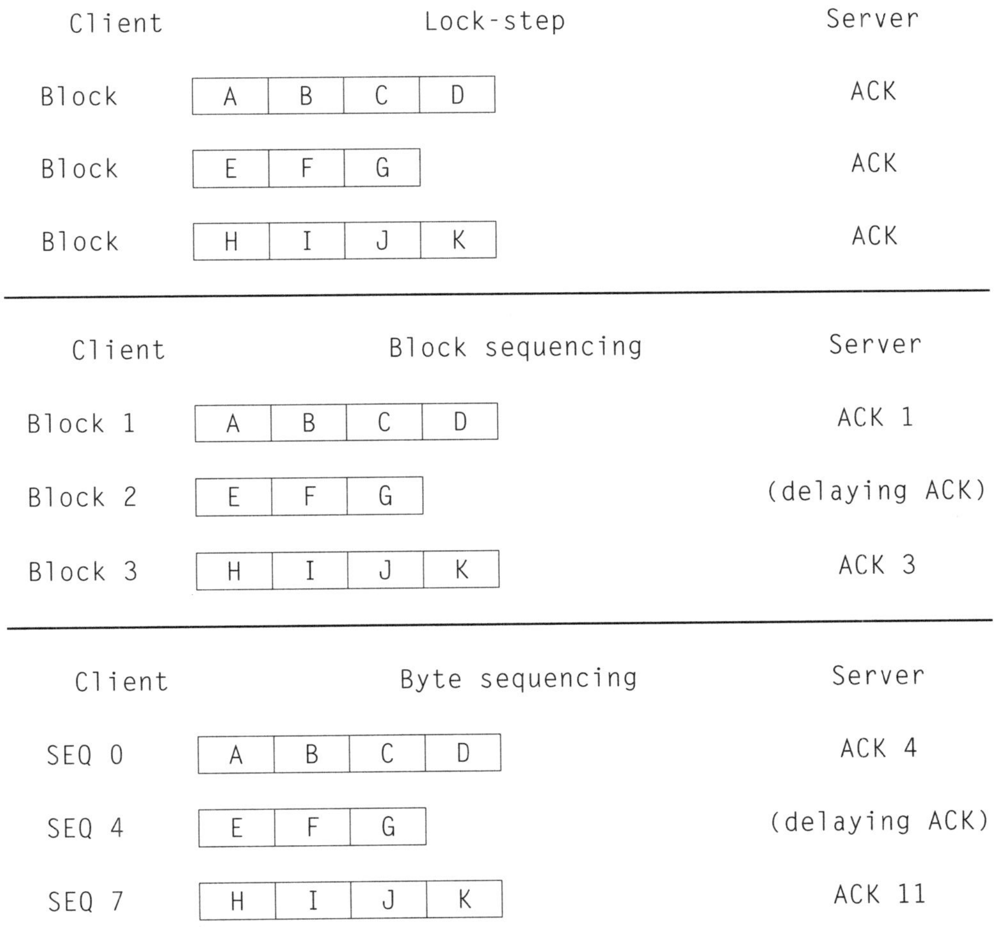

Client	Lock-step	Server
Block	A B C D	ACK
Block	E F G	ACK
Block	H I J K	ACK

Client	Block sequencing	Server
Block 1	A B C D	ACK 1
Block 2	E F G	(delaying ACK)
Block 3	H I J K	ACK 3

Client	Byte sequencing	Server
SEQ 0	A B C D	ACK 4
SEQ 4	E F G	(delaying ACK)
SEQ 7	H I J K	ACK 11

TCP Sequence Space

When coming to grips with sequencing, it can help to think in terms of a *sequence space*, which is the 4.3Gb area covered by a sequence number and a small data transfer within this space with a start marker, data block, and end marker (Figure 6.2).

Figure 6.2   Sequence space.

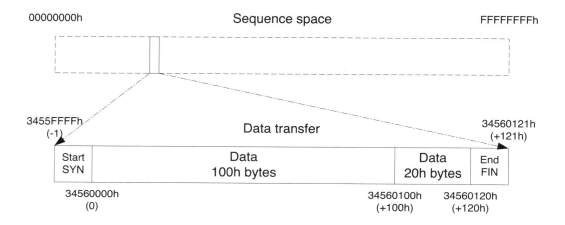

The TCP start marker is known as SYN because it is used to synchronize the two nodes; the end marker is known as FIN because it finishes the connection. Between them lies a data transfer, which I'll assume to be 288 (120h) bytes, in two blocks of 100h and 20h bytes. TCP counts the SYN and FIN markers as one byte each (even though they don't contain data), so this transfer occupies a total of 122h bytes of sequence space.

The simplest way to grasp the underlying concept of TCP is to realize that it has only two primary objectives:

• to send out a SYN, some optional data, and a FIN or
• to receive a SYN, some optional data, and a FIN.

If a marker or data block has gone astray, the sender must repeat the transmission using the same sequence number as before. Any attempt to use a different sequence number would cause the recipient to misplace it within the sequence of markers and data.

The actual starting sequence number will be assigned when the connection is opened, and it is impossible to predict in advance. All subsequent sequence numbers in the data transfer will be measured with reference to the starting value. To help in creating and debugging the code, I find it convenient to think of the initial SYN as having a relative sequence number of minus one (one byte before the start of the data) and of the closing FIN as having the relative sequence number of the data length plus one (one byte after the end of the data).

All incoming data must be validated by ensuring that it fits into the sequence space, but this matching can be quite complicated. Imagine that you've received the SYN marker and the first 100h data block, and then another block arrives. Here are some possibilities, depending on the relative sequence number.

-10h	Unknown data; before the start of the current transfer
-1	The same relative sequence number as the SYN — a duplicate SYN?
0	Duplicate of the first data block?
10h	Duplicate of part of the first data block?
100h	Second data block
110h	Future data block received out of order?
1000h	Future data block or unknown data?
10000h	Unknown data

The complexity arises because the recipient doesn't know how much data is being transferred (and, in some applications, neither does the sender). All it can do is attempt to fit each incoming block into the sequence space and judge whether it looks sensible.

For this scheme to work, the sender and receiver must agree on a *window size*, which limits the amount of (unacknowledged) data in transit at any one time. The sender won't transmit data outside the window, and the receiver won't accept it. The above example assumes a window size of 16Kb (4000h), so the receiver can reject the relative sequence number of 10000h, knowing it is outside the window. The values of 1000h and 110h are within the window, so they might be future data that has been received out of order. Unfortunately, the receiver won't know if this is true until the intervening data blocks arrive and doesn't know how many blocks have yet to arrive (or how many are lost in transit), because there is no compulsion on the sender to keep the block sizes uniform.

The relative sequence number of zero carries a hidden trap. It is tempting to assume it is a repeat of the first data block, and discard it, but imagine the following scenario. The sender transmits the first 100h bytes and waits for a response. Receiving none, it retransmits the data in its buffer, which happens to be more than 100h bytes (because the user is frantically hammering on the keyboard hoping something will happen). Therefore, the repeat contains the old 100h of data plus some new data. The receiver must check the data length of the new block to ensure that it really has received a duplicate and not a mixture of old and new data.

The relative sequence number 10h is yet more complex, so feel free to ignore it. For completeness, here is a possible scenario that assumes that the first 100h block wasn't really the first block sent; instead, the data blocks were sent in the following order:

- First data block of 10h data bytes, relative seq zero, is lost in transit.
- Second data block of F0h data bytes, relative seq 10h, is delayed.
- Third data block of 100h data bytes, relative seq zero, is a resend of the first and second data blocks.

If the third block arrives first, then the second arrives, you would get the unexpected relative sequence number of 10h. This may be highly unlikely, but it isn't impossible, so it must be catered for. The ultimate fallback is to ignore the block, which happens to be the correct action in this case.

TCP data transfers are completely symmetrical, so both sides must perform these sequence-matching tests. Are you still wondering why TCP is so difficult to implement?

Acknowledgments

Acknowledgments are sent by a recipient to indicate it has received one or more messages, which may contain data, SYN, or FIN markers. The important principle is that an ACK number acknowledges everything before it in the sequence space. If I have received 120h bytes of data, then I must send a relative ACK value of 120h.

At any stage of the transfer, whether any data has been received or not, an acknowledgment *may* be sent. An acknowledgment *must* be sent when the quantity of unacknowledged data approaches the window size or if it appears that no more data will be received for a while. The difficulty with acknowledgments is knowing when to send them. Too many, and network bandwidth will be wasted; too few, and the sender will assume the data has been lost and start resending it.

You can add the acknowledgments to the sequence space diagram, assuming a simple transaction initiated by a client with 80h bytes of data to send to a server with 120h bytes of data to send, and close the connection (Figure 6.3).

Client:	SYN
Server:	SYN + ACK
Client:	ACK
Client:	80h bytes of data + ACK
Server:	100h bytes of data + ACK
Server:	20h bytes of data + ACK
Server:	FIN + ACK
Client:	FIN + ACK
Server:	ACK

This is how a typical Web client–server connection is structured, although the data sizes, block counts, and ACK counts will differ for each transaction.

TCP State Machine

I have already implied the existence of startup, data transfer, and close-down phases of a TCP connection, and now these need to be explained in terms of the standard state diagram (Figure 6.4).

Figure 6.3   Transaction in sequence space.

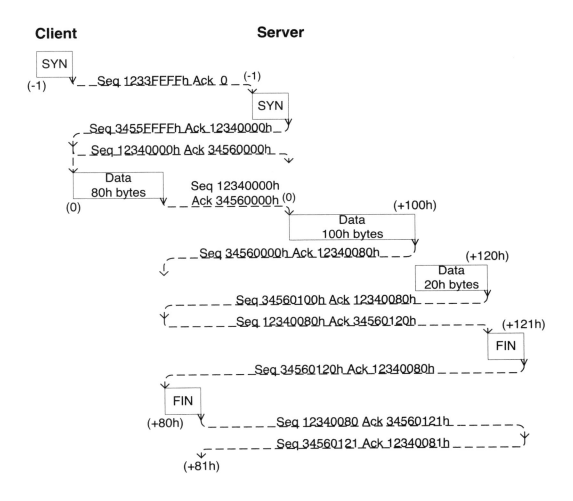

Figure 6.4   TCP state diagram.

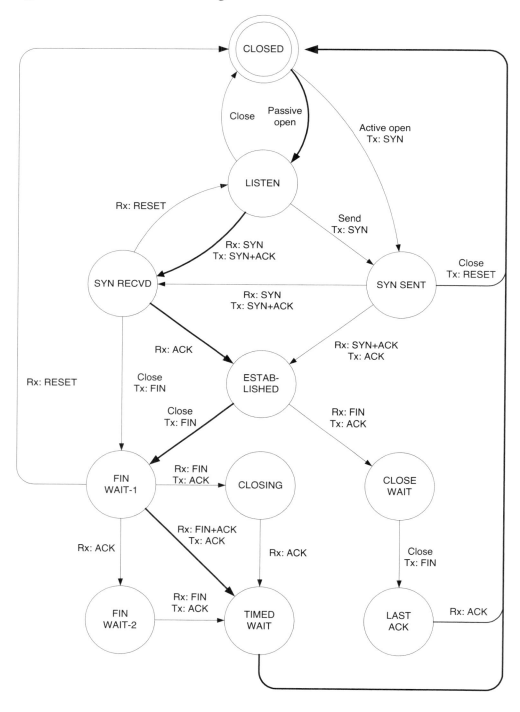

Before wending your way through the various states, some general words of explanation are in order. Figure 6.4 shows all the possible TCP states and what network signals are received (Rx) and transmitted (Tx) for each state transition. The signals are labeled as follows:

SYN   initial synchronizing message
ACK   acknowledgment
FIN   final closure message
Reset   forced closure signal

You need to be slightly careful in interpreting these signals; the ACK label is particularly misleading, since all packets except the first will have valid SEQ and ACK numbers and will be checked at all stages of the connection. See my earlier comments about sequence space in "TCP Sequence Space" on page 158.

The diagram incorporates all the possible state changes, but only a subset is applicable to any one transaction. The typical state changes for a Web server are indicated by the bold connecting lines. To establish a path through the state machine, you must answer the main following questions:

- At the start, which side, client or server, opens the connection?
- When all data has been sent, which side closes the connection?

Either side can choose an active or passive role. Just because the client opens the connection, there is no compulsion for it to start the closure. The Web servers I'll describe have limited resources, so start closure as soon as all the data has been sent (Figure 6.5).

The states and transitions most frequently encountered in a Web client follow:

Closed   Initial state when booting. Socket is unable to send or receive network packets.

SYN Sent   Client software requests connection.

Established   Connected; able to send and receive data.

Close Wait   Server has requested closure.

Last ACK   Client acknowledges closure; awaiting final acknowledgment.

The states and transitions for a Web server follow:

Closed   Initial state when booting. Socket is unable to send or receive network packets.

Listen   Internal state transition when server software is run. Allows connection requests to be accepted. No packets sent or received yet.

SYN Received   Received first message (SYN) from browser; requesting a connection.

Established   Connected; able to send and receive data.

FIN Wait 1   All data sent; server requests closure.

Timed Wait   Closure acknowledged by client.

Figure 6.5   Typical Web client and server state changes.

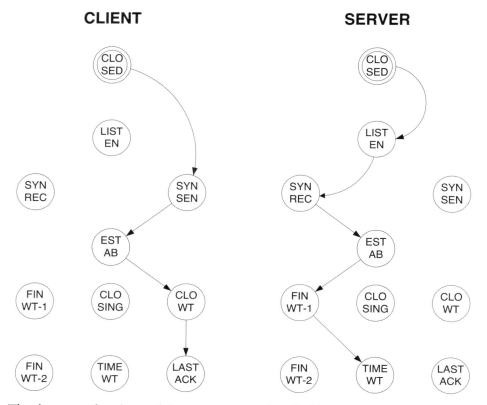

The data transfer phase of the connection is handled by a single state, Established, which is synonymous with "connected." There are no state transitions while the bulk of data is being transferred, but what happens if one side starts closure while the other still has data to send? The answer is "half closure," which accounts for the two states Close Wait and Last ACK. On receipt of a FIN, the TCP stack goes to Close Wait and can then respond with just an ACK and keep on sending data. When all data is sent, it sends a FIN and waits for a final ACK. In practice, a Web client hasn't any extra data to send, so these two steps are generally compressed into one. On receipt of a FIN ACK, the client responds FIN ACK immediately. This explains why I've shown the server going from FIN Wait 1 to Timed Wait; in the half-Closed scenario, the server would hang around in FIN Wait 2 until it received a FIN from the client. You can now add the state transitions to your sample client–server transaction (Table 6.1).

Table 6.1   State transitions.

Client State				Server State
Closed				Listen
SYN Sent				
	SEQ 1233FFFFh	SYN >	ACK 00000000h	
				SYN Received
	ACK 12340000h	< SYN + ACK	SEQ 3455FFFFh	
Established				
	SEQ 12340000h	ACK >	ACK 34560000h	
				Established
Established (data Tx)				
	SEQ 12340000h	DATA + ACK >	ACK 34560000h	
				Established (data Rx)
				Established (data Tx)
	ACK 12340080h	< DATA + ACK	SEQ 34560000h	
Established (data Rx)				
				Established (data Tx)
	ACK 12340080h	< DATA + ACK	SEQ 34560100h	
Established (data Rx)				
	SEQ 12340080h	ACK >	ACK 34560120h	
				FIN Wait 1
	ACK 12340080h	< FIN	SEQ 34560120h	
Close Wait				
Last ACK				
	SEQ 12340080h	FIN + ACK >	ACK 34560121h	
				Timed Wait
	ACK 12340081h	< ACK	SEQ 34560121h	
Closed				

Segment Format

A TCP header plus data block is called a *segment*, which fits into the data field of an IP datagram (Figure 6.6).

Figure 6.6   TCP segment format.

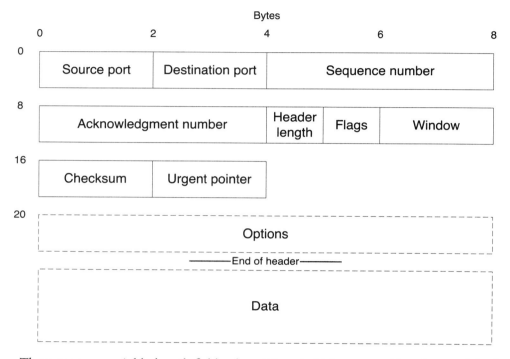

There are two variable-length fields; the *options* (which are considered part of the header) and the data. Because a C structure can't accommodate variable-length fields, I declare a single data[] field and separate out the options at run time.

```
/* ***** TCP (Transmission Control Protocol) header ***** */
typedef struct tcph
{
    WORD   sport,          /* Source port */
           dport;          /* Destination port */
    LWORD  seq,            /* Sequence number */
           ack;            /* Ack number */
    BYTE   hlen,           /* TCP header len (num of bytes << 2) */
           flags;          /* Option flags */
    WORD   window,         /* Flow control credit (num of bytes) */
           check,          /* Checksum */
           urgent;         /* Urgent data pointer */
```

```
} TCPHDR;
#define TFIN       0x01    /* Option flags: no more data */
#define TSYN       0x02    /*            sync sequence nums */
#define TRST       0x04    /*            reset connection */
#define TPUSH      0x08    /*            push buffered data */
#define TACK       0x10    /*            acknowledgment */
#define TURGE      0x20    /*            urgent */


/* ***** TCP packet ('segment') ***** */
typedef struct tcp
{
    IPHDR   i;             /* IP header */
    TCPHDR  t;             /* TCP header */
    BYTE    data[MAXDATA]; /* Data area */
} TCPKT;
```

The source and destination ports are very similar to their UDP counterparts. There are well-known port numbers for standard services (such as port 80 for an HTTP server) and ephemeral port numbers for the client to use when opening connections to the server. Typical well-known port numbers are listed in Table 6.2.

Table 6.2   Well-known port numbers.

Port	Service	
7	Echo	Echoes back all incoming characters
9	Discard	Discards all incoming characters
13	Daytime	Returns time and date string
21	FTP	File transfer protocol
23	Telnet	Remote login
25	SMTP	Simple mail transfer protocol
37	Time	Binary value of date and time
80	HTTP	Hypertext transfer protocol

The sequence and acknowledgment numbers are 32-bit values for tracking the transmit and receive data, as already described.

The header length gives the total byte count of the standard header plus options, and the flags field contains one-bit flags for FIN, SYN, Reset, Push, ACK, and Urgent. I define these as two eight-bit values. This differs from the standard, which defines a four-bit header length (giving length in units of four bytes), then six reserved bits, then six code bits. A small amount of shifting and masking converts my byte-wide fields into the standard values.

The push flag encourages the recipient to handle the incoming data now, rather than waiting for more data to arrive. The classic usage is in the Telnet terminal emulation, where it may be important that each keystroke is handled as soon as it is received.

The urgent flag and urgent data pointer give TCP the capability of sending data out of band, as if there were a completely separate channel for urgent data. Relatively few utilities require this capability, so my software does not support it.

The window size gives an indication of how much buffer space the sender has for incoming data. If its buffers are getting full, it will reduce the window size; in extremis, a window size of zero can be sent, in which case, no more data can be sent to that socket. If more buffer space becomes available, the window size is increased.

The checksum is similar to the UDP checksum, in that a pseudoheader is created from the IP header and added in to the checksum of the TCP header and data (Figure 6.7).

Figure 6.7   IP pseudoheader.

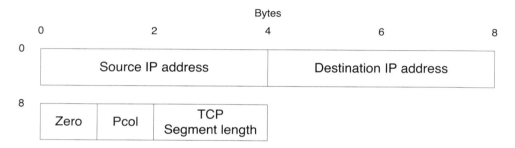

Unlike UDP, there is no provision for disabling the checksum. A value of zero indicates the checksum is zero and does not mean the checksum should be ignored.

TCP Options

The variable-length options field in the header may be present in any TCP segment, though it generally only appears in the first SYN of each connection. Each option within the field starts with a one-byte identifier, as follows.

0   End of option list
1   No operation
2   Maximum segment size
3   Window scale factor
8   Time stamp

Options 2 and above are followed by a one-byte length field, giving the total length of the option, including its identifier. This allows the recipient to skip through the options, ignoring any it does not recognize.

The only option I want to address is maximum segment size (MSS), which usually accompanies the first SYN sent. It restricts the total amount of data that can be sent in any one segment and has the following format:

- one-byte identifier = 2
- one-byte length = 4
- two bytes MSS (usual network hi–lo byte order)

You would expect the MSS to be the maximum total length of the TCP header plus data, but, unfortunately, the name is highly misleading. It really indicates the maximum amount of TCP data that can be carried *assuming the TCP and IP headers are at their minimum sizes*. Because these headers are a minimum of 20 bytes each, the actual limit on the TCP data size is

MSS + 40 – (actual size of TCP and IP headers).

The sender may transmit less data than this to avoid fragmentation if an intermediate link in the network has a lower maximum transfer unit (MTU).

TCP Implementation

TCP States

Each TCP state is defined as an enumeration and as a string for diagnostic printouts.

```
/* TCP states */
#define TSTATE_STRINGS \
    "closed",       "SYN rcvd",    "established", \
    "close wait",   "last ACK",    "reset rcvd",    \
    "active open",  "ARP sent",    "ARP rcvd",     "SYN sent", \
    "active close", "FIN wait 1",  "FIN wait 2",   "closing",\
    "timed wait"
typedef enum {
    /* Passive open & close */
    TCP_CLOSED,             /* Closed */
    TCP_SYNR,               /* SYN recvd: send SYN ACK */
    TCP_EST,                /* Established: connection OK */
    TCP_CLWAIT,             /* Close wait: FIN recvd, send data then FIN ACK */
    TCP_LASTACK,            /* Last ACK: waiting for ACK of my FIN */
    TCP_RSTR,               /* Reset received */
    /* Active open and close */
    TCP_AOPEN,              /* Active open: send ARP */
    TCP_ARPS,               /* ARP sent, awaiting reply */
    TCP_ARPR,               /* ARP reply received */
    TCP_SYNS,               /* SYN sent: awaiting SYN ACK */
                            /* Established: same as for passive open */
    TCP_ACLOSE,             /* Active close: send FIN */
    TCP_FINWT1,             /* FIN wait 1: FIN sent, waiting for ACK */
    TCP_FINWT2,             /* FIN wait 2: receiving data, waiting for FIN */
    TCP_CLING,              /* Closing: awaiting final ACK */
    TCP_TWAIT               /* Timed wait */
} TCP_STATE;
```

Segment Reception and Transmission

The functions for checking an incoming TCP segment are complicated by the variable-length options header, so I have created a function specifically to decode this. The only option of interest is the MSS, so the function optionally returns this value, too, if it is present in the header.

```
/* Return TCP options field length; if Max Seg Size option, get value */
WORD gettcp_opt(TCPKT *tcp, WORD *mssp)
{
    int olen;

    olen = ((tcp->t.hlen & 0xf0) >> 2) - sizeof(TCPHDR);
    if (mssp && olen>=4 && tcp->data[0]==2 && tcp->data[1]==4)
        *mssp = swapw(*(WORD *)&tcp->data[2]);
    return(olen);
}
```

In common with the IP and UDP equivalents, is_tcp() returns a non-zero value if the packet is TCP. The return value is the data length, or -1 if there is no data.

```
/* Return TCP data length (excl. options, -1 if no data), 0 if not TCP */
int is_tcp(IPKT *ip, int len)
{
    TCPKT *tcp;
    WORD sum;
    int dlen=0;
                                    /* Check protocol & minimum length */
    if (ip->i.pcol==PTCP && len>=sizeof(TCPHDR))
    {
        tcp = (TCPKT *)ip;              /* Do checksum */
        sum = check_tcp(tcp, swapl(ip->i.sip), swapl(ip->i.dip), len);
        if (tcp->t.hlen < 0x50)
        {
            if (tcpdebug)
                printf("  ERROR: TCP header len %u\n", (tcp->t.hlen&0xf0)>>2);
        }
        if (sum == 0xffff)                  /* If correct.. */
        {
            swap_tcp(tcp);                  /* Do byte-swaps */
            len -= sizeof(TCPHDR);          /* Subtract header len */
            if (tcpdebug)                   /* Display segment if in debug mode */
                disp_tcp(tcp, len, 0);
```

```
                len -= gettcp_opt(tcp, 0);   /* Subtract options len */
                dlen = len>0 ? len : -1;     /* Return -1 if data len=0 */
            }
        else if (tcpdebug)                   /* Display error */
            printf("  ERROR: TCP checksum %04X\n", sum);
    }
    return(dlen);
}
```

The checksum calculation uses an IP pseudoheader in the same way as UDP.

```
/* ***** Pseudo-header for UDP or TCP checksum calculation ***** */
/* The integers must be in hi-lo byte order for checksum */
typedef struct                 /* Pseudo-header... */
{
    LWORD srce,                /* Source IP address */
          dest;                /* Destination IP address */
    BYTE  z,                   /* Zero */
          pcol;                /* Protocol byte */
    WORD  len;                 /* UDP length field */
} PHDR;


/* Return TCP checksum, given segment (TCP header + data) length.
** The TCP segment and both IP addrs must be in network byte-order */
WORD check_tcp(TCPKT *tcp, LWORD sip, LWORD dip, int tlen)
{
    PHDR tph;
    LWORD sum;

    sum = csum(&tcp->t, (WORD)tlen);         /* Checksum TCP segment */
    tph.len = swapw((WORD)tlen);             /* Make pseudo-header */
    tph.srce = sip;
    tph.dest = dip;
    tph.z = 0;
    tph.pcol = tcp->i.pcol;
    sum += csum(&tph, sizeof(tph));          /* Checksum pseudo-header */
    return(WORD)(sum + (sum>>16));           /* Return total plus carry */
}


/* Swap byte order of ints in TCP header */
void swap_tcp(TCPKT *tcp)
```

```
{
    tcp->t.sport = swapw(tcp->t.sport);
    tcp->t.dport = swapw(tcp->t.dport);
    tcp->t.window = swapw(tcp->t.window);
    tcp->t.urgent = swapw(tcp->t.urgent);
    tcp->t.seq = swapl(tcp->t.seq);
    tcp->t.ack = swapl(tcp->t.ack);
}
```

The make_tcp() function has many more arguments than its UDP equivalent. Most of the values will be obtained from the socket storage described in the next section, but there are occasions when you'll need to use other values (e.g., when responding to a rogue segment that doesn't belong to the current transaction).

```
/* Make a TCP segment given the socket, flags, data len */
int make_tcp(GENFRAME *gfp, NODE *srcep, NODE *destp, BYTE flags,
    LWORD seq, LWORD ack, WORD window, WORD dlen)
{
    TCPKT *tcp;
    int hlen, tlen, ilen, olen=0;

    tcp = getframe_datap(gfp);
    tcp->t.seq = seq;                       /* Set seq and ack values */
    tcp->t.ack = ack;
    tcp->t.window = window;                 /* Window size srce & dest port nums */
    tcp->t.sport = srcep->port;
    tcp->t.dport = destp->port;
    hlen = sizeof(TCPHDR);                  /* TCP header len */
    if (flags&TSYN && dlen==0)              /* Add 4 bytes for options if SYN.. */
    {
        hlen += (olen = 4);
        tcp->data[0] = 2;                   /* ..and send max seg size */
        tcp->data[1] = 4;
        *(WORD*)&tcp->data[2] = swapw((WORD)tcp_maxdata(gfp));
    }
    tcp->t.hlen = (BYTE)(hlen<<2);          /* Set TCP header len, and flags */
    tcp->t.flags = flags;
    tcp->t.urgent = tcp->t.check = 0;
    if (tcpdebug)                           /* Display segment if in debug mode */
        disp_tcp(tcp, dlen+olen, 1);
    swap_tcp(tcp);                          /* Do byte-swaps, encapsulate in IP */
```

```
    tlen = hlen + dlen;
    ilen = make_ip(gfp, srcep, destp, PTCP, (WORD)(tlen));
    tcp->t.check = ~check_tcp(tcp, tcp->i.sip, tcp->i.dip, tlen);
    return(ilen);                        /* Checksum final packet */
}


/* Make a TCP RESET response to incoming segment */
int make_reset_resp(GENFRAME *gfp, int rdlen)
{
    TCPKT *tcp;
    NODE loc, rem;
    LWORD ack;

    gettcp_srce(gfp, &rem);              /* Get source & dest nodes */
    gettcp_locdest(gfp, &loc);           /* (including port numbers) */
    tcp = getframe_datap(gfp);
    ack = tcp->t.seq + maxi(rdlen, 0);
    if (tcp->t.flags & (TSYN+TFIN))
        ack++;
    return(make_tcp(gfp, &loc, &rem, TRST+TACK, tcp->t.ack, ack, 0, 0));
}
```

Socket Data

As with UDP, you'll be able to support multiple connections through multiple sockets. Each socket has its own storage, with a pointer to an application-specific data area (defined later).

```
/* Well-known TCP port numbers */
#define ECHOPORT    7        /* Echo */
#define DAYPORT     13       /* Daytime */
#define CHARPORT    19       /* Character generator */
#define TELPORT     23       /* Telnet remote login */
#define HTTPORT     80       /* HTTP */
#define MINEPORT    1024     /* Start of ephemeral (temporary) port numbers */
#define MAXEPORT    5000     /* Max ephemeral port number */

/* Dummy values for the segment data length */
#define DLEN_NODATA -1       /* Segment received, but no data */

/* Storage structure for a single TCP socket
** The positions of the first 3 items are fixed to simplify initialization */
```

```
typedef struct {
    int         index;          /* Index number - must be first */
    CBUFF       rxb,            /* Receive & transmit circular buffers */
                txb;            /* (must be 2nd and 3rd) */
    NODE        loc,            /* Local and remote nodes */
                rem;
    TCP_STATE   state;          /* Current connection state */
    int         server;         /* Flag to identify server socket */
    LWORD       time;           /* Time at last state change (msec) */
    int         timeout,        /* Timeout value (0=no timeout) */
                retries;        /* Number of retries left */
    WORD        txmss,          /* Max seg data size for my transmit */
                rxwin,          /* Current Rx & Tx window sizes */
                txwin;
    LWORD       rxseq,          /* Seq & ack values in latest Rx segment */
                rxack;
    BYTE        txflags,        /* Latest Tx flags */
                connflags;      /* Extra Tx connection flags (push/urgent) */
    int         txdlen;         /* Latest transmit data length */
    void        *app;           /* Pointer to application-specific data */
} TSOCK;
```

Sequence and Acknowledgment Values

Standard circular buffers are defined using 32-bit pointers, so you can map the sequence and acknowledgment numbers directly onto them. The sequence number is mapped onto the circular buffer `trial` pointer, since it reflects the start of the data that has tentatively been sent out, whereas the acknowledgment number is the circular buffer input pointer, since it reflects the amount of data that has been successfully received. These circular buffer values are only correct when connected (Established); when setting up or closing down, they must be adjusted to allow for the SYN and FIN markers.

This mapping has the slightly strange side effect that the Transmit buffer pointers must be initialized to the outgoing sequence value and the Receive buffer pointers initialized to the incoming sequence value. It may seem counterintuitive to initialize a circular buffer to a non-zero value, but this is perfectly sensible because the buffers are circular.

```
/* Make a TCP segment given the socket state, flags, data len */
int make_sock_tcp(GENFRAME *gfp, TSOCK *ts, BYTE flags, WORD dlen)
{
    WORD len;
    BYTE st;
    TCPKT *tcp;
    LWORD tseq, tack;
```

```
    tcp = getframe_datap(gfp);
    tseq = ts->txb.trial;                /* Seq and ack values if connected */
    tack = ts->rxb.in;
    ts->txflags = flags;
    ts->txdlen = dlen;
    if ((st=ts->state)==TCP_SYNR || st==TCP_SYNS)
        tseq--;                          /* Decrement SEQ if sending SYN */
    else if (st==TCP_CLING || st==TCP_FINWT2 || st==TCP_TWAIT)
        tseq++;                          /* ..or increment if sending FIN */
    if (st==TCP_LASTACK || st==TCP_CLWAIT || st==TCP_CLING || st==TCP_TWAIT)
        tack++;                          /* Increment ACK if FIN received */
    if (dlen > 0)                        /* Get the Tx data */
        dlen = buff_try(&ts->txb, tcp->data, dlen);
    len = make_tcp(gfp, &ts->loc, &ts->rem, flags, tseq, tack, ts->rxwin,dlen);
    return(len);
}
```

TCP State Machine

The TCP state machine is in tsock_rx() and is fairly complicated because it includes all possible states. Each state change is accompanied by a function call, to help in creating a diagnostic printout and to refresh the activity timer.

```
/* Change TCP socket state; if closed/opening/closing/established, do upcall
** If upcall returns error, don't change state, and return 0 */
void new_state(TSOCK *ts, int news)
{
    if (news != ts->state)
    {
        if (statedebug)
            printf("    (%u) new state '%s'\n", ts->index, tstates[news]);
        ts->state = news;
    }
    mstimeout(&ts->time, 0);
    ts->retries = TCP_RETRIES;
    ts->timeout = TCP_TIMEOUT;
}
```

The tsock_rx() function is a slight misnomer because it is not only called every time a TCP segment is received, but it is also polled on a routine basis to keep the TCP stack alive (since not all TCP state changes are triggered by the arrival of a TCP segment).

This first code fragment handles an incoming TCP segment by

- isolating the flags;
- saving the SEQ, ACK, and window values;
- aborting the current connection if Reset is received; and
- moving the Transmit buffer pointer to reflect acknowledged data.

Then comes the start of a large switch statement for the current state.

```
/* Receive an incoming TCP seg into a socket, given data length (-1 if no data)
** Also called with length=0 to send out Tx data when connected.
** Returns transmit length, 0 if nothing to transmit */
int tsock_rx(TSOCK *ts, GENFRAME *gfp, int dlen)
{
    BYTE rflags=0;
    TCPKT *tcp;
    int txlen=0, tx=0;
    static WORD eport=MINEPORT;
    int (*upcall)(TSOCK *ts, CONN_STATE conn);

    tcp = getframe_datap(gfp);
    upcall = ts->server ? server_upcall : client_upcall;
    if (dlen>0 || dlen==DLEN_NODATA)
    {
        rflags = tcp->t.flags & (TFIN+TSYN+TRST+TACK);
        ts->rxseq = tcp->t.seq;
        ts->rxack = tcp->t.ack;
        ts->txwin = tcp->t.window;
        if (rflags & TRST)
            new_state(ts, TCP_RSTR);
        else if (rflags&TACK &&
                    in_limits(ts->rxack, ts->txb.out, ts->txb.trial))
            ts->txb.out = ts->rxack;
    }
    switch (ts->state)
    {
```

Closed and Listen States

For present purposes, it is convenient to ignore the distinction between the Closed and Listen states. If a socket is unused, it is available to be opened, without the necessity of undergoing the administrative step of a Passive Open. On arrival of a SYN segment, the Transmit buffer is initialized to reflect the sequence number and the Receive buffer is initialized using a new acknowledgment number. The latter is derived from the system millisecond timer to ensure that two successive connections will not have the same values. The SYN ACK response is then prepared [it is sent out by the caller of `tsock_rx()`].

```
/* Passive (remote) open a connection */
case TCP_CLOSED:
    if (rflags == TSYN)                        /* If closed & SYN recvd.. */
    {
        buff_setall(&ts->rxb, ++ts->rxseq); /* Load my ACK value */
        buff_setall(&ts->txb, mstime()*0x100L); /* ..and my SEQ */
        ts->rxwin = TCP_RXWIN;                /* Default Rx window */
        ts->connflags = 0;                    /* No special flags */
        if (server_upcall && !server_upcall(ts, TCP_OPEN)) /*Upcall err?*/
            txlen = make_reset_resp(gfp, dlen); /* ..don't accept SYN */
        else
        {
            new_state(ts, TCP_SYNR);          /* If OK, send SYN+ACK */
            txlen = make_sock_tcp(gfp, ts, TSYN+TACK, 0);
            ts->server = 1;                   /* Identify as server socket */
        }
    }
    else                                      /* If not SYN.. */
        txlen = make_reset_resp(gfp, dlen);/* ..send reset */
    break;
```

In this section, as in many others, you will note there is an `upcall` (callback) to the higher-level code. This is to give the application code an opportunity to respond to changes in TCP state. The application can also influence the TCP state machine by returning a zero value to the `upcall`. In the above code, this causes the new connection to be refused by sending a Reset signal.

A more detailed description of the `upcall` mechanism is given later, when discussing the application interface in "TCP Callbacks" on page 190.

SYN Received

Having received a SYN and sent SYN ACK, you must wait for an ACK to be connected (Established).

```
case TCP_SYNR:                            /* SYN+ACK sent, if ACK Rx.. */
    tsock_estab_rx(ts, gfp, dlen);        /* Fetch Rx data */
    if (rflags==TACK && ts->rxseq==ts->rxb.in && ts->rxack==ts->txb.out)
    {
        if (upcall && !upcall(ts, TCP_CONN))     /* If upcall not OK.. */
            new_state(ts, TCP_ACLOSE);    /* ..close after sending dat */
        else                              /* If OK.. */
```

```
                new_state(ts, TCP_EST);          /* ..go established */
        }
        else if (dlen && ts->rxseq==ts->rxb.in-1)    /* If repeat SYN.. */
            txlen = make_sock_tcp(gfp, ts, TSYN+TACK, 0);
        break;                                    /* ..repeat SYN+ACK */
```

Of course, you can't accept an ACK at face value, but you must check the sequence and acknowledgment values to make sure it is really intended for this connection and is not a stray packet from a past connection. Another precautionary measure is to check for incoming data. It may seem strange to receive data before you're actually connected, but it is possible the transmitting TCP stack is somewhat overeager to get its data out, and most TCP stacks tolerate this behavior. However, you still have to check the position of any such data in the sequence space.

```
/*Put Rx data into an established socket, return non-zero if Tx ACK required*/
int tsock_estab_rx(TSOCK *ts, GENFRAME *gfp, int dlen)
{
    TCPKT *tcp;
    int oset=0, oldlen, tx=0;
    WORD rdlen=0;
    long rxdiff;

    if (dlen > 0)                                 /* If any data received.. */
    {
        tcp = getframe_datap(gfp);
        oset = gettcp_opt(tcp, 0);
        rxdiff = ts->rxseq - ts->rxb.in;          /* Find posn w.r.t. last dat */
        if (rxdiff == 0)                          /* If next block, accept it */
            rdlen = (WORD)dlen;
        else if (rxdiff < 0)                      /* If part or all is repeat */
        {
            oldlen = -(int)rxdiff;                /* ..read in new part */
            if (oldlen<=ts->rxwin && dlen>oldlen)
            {
                rdlen = dlen - oldlen;
                oset += (int)oldlen;
            }
        }
        if (rdlen)                                /* Read the data in */
        {
            if (statedebug)
                printf("    (%u) Rx data %u bytes\n", ts->index, rdlen);
```

```
                  buff_in(&ts->rxb, &tcp->data[oset], rdlen);
          }                                     /* Tx if rpt or half full */
          tx = tcp->t.flags&TPUSH || rxdiff<0 || buff_dlen(&ts->rxb)>ts->rxwin/2;
      }
      return(tx);
  }
```

This is an implementation of the sequence space logic described earlier. It looks at the difference between the incoming sequence number and the amount of data received so far. If zero, then the new data is appended to the old. If negative, then some (or all) of the incoming data is a repeat of old data. A notable omission is the case where the difference is greater than zero, that is, a future data block has been received out of order. The software relies on the out-of-order data being retransmitted, which is less than ideal for Internet operation and should be improved in a future software release.

Established

In the Established state, data is exchanged with the other node. You need to check for a FIN flag from the other side; if your application isn't happy about the closure you must half-Close (so you can continue sending data). If the application status is OK, you can acknowledge the closure with FIN ACK. As a precaution, still check for incoming data, even if there is a FIN. It is terribly embarrassing if your TCP stack fails to respond to a slightly unusual, but perfectly comprehensible, packet sequence.

```
      /* Connection established */
      case TCP_EST:
          if (rflags)                           /* Refresh timer if Rx */
              new_state(ts, ts->state);
          if (rflags&TFIN && ts->rxseq==ts->rxb.in)  /* If remote close.. */
          {
              tx = tsock_estab_rx(ts, gfp, dlen);    /* Fetch Rx data */
              if (!upcall || upcall(ts, TCP_CLOSE))
              {
                  new_state(ts, TCP_LASTACK);         /* ..FIN+ACK if OK */
                  txlen = make_sock_tcp(gfp, ts, TFIN+TACK, 0);
              }
              else
              {
                  new_state(ts, TCP_CLWAIT);          /* ..or send data if not */
                  txlen = tsock_estab_tx(ts, gfp, 1);
              }
          }
          else
          {
```

```
        tx = tsock_estab_rx(ts, gfp, dlen);     /* Fetch Rx data */
        if (upcall && !upcall(ts, dlen>0 ? TCP_DATA : TCP_NODATA))
        {
            new_state(ts, TCP_ACLOSE);      /* If upcall 0, start close */
            tx = 1;
        }
        else if (dlen > 0)                  /* If Rx data, send ack */
            tx = 1;
        txlen = tsock_estab_tx(ts, gfp, tx);/* Send pkt, maybe Tx data */
    }
    break;
```

The decision whether or not to send data, acknowledgments, or both is complex and should really involve detailed measurements of network transit time to maximize throughput. My code makes a slightly over-simplistic decision, which nevertheless serves quite well.

```
/* Prepare Tx frame containing outgoing data for connection, return frame len
** If 'force' is non-zero, make frame even if there is no data */
int tsock_estab_tx(TSOCK *ts, GENFRAME *gfp, int force)
{
    int tdlen, txlen=0;

    tdlen = mini(buff_untriedlen(&ts->txb), tcp_maxdata(gfp));
    tdlen = mini(tdlen, ts->txwin-buff_trylen(&ts->txb));
    if (tdlen>0 && !force)
        force = buff_trylen(&ts->txb)==0 || tdlen<tcp_maxdata(gfp)/2;
    if (force)
    {
        if (tdlen>0 && statedebug)
            printf("    (%u) Tx data %u bytes\n", ts->index, tdlen);
        txlen = make_sock_tcp(gfp, ts, (BYTE)(TACK+ts->connflags),(WORD)tdlen);
    }
    return(txlen);
}
```

Close Wait

This half-Closed state is caused when the other node wants to close, but your application still wants to send data. You don't need to check for incoming data (because there can't be any after a FIN), so just keep shovelling the data out, until the application allows you to close the connection by sending a FIN.

```
    /* Passive (remote) close a connection */
    case TCP_CLWAIT:                            /* Do upcall to application */
        if (!upcall || upcall(ts, TCP_CLOSE))   /* If OK, send FIN */
        {
            new_state(ts, TCP_LASTACK);
            txlen = make_sock_tcp(gfp, ts, TFIN+TACK, 0);
        }
        else                                    /* If not, keep open */
            txlen = tsock_estab_tx(ts, gfp, 0);
        break;
```

Last ACK

Wait until the FIN is acknowledged; then close.

```
    case TCP_LASTACK:                           /* If ACK of my FIN.. */
        if (rflags==TACK && ts->rxseq==ts->rxb.in+1)
            new_state(ts, TCP_CLOSED);          /* ..connection closed */
        break;
```

Active Open

Although not on the standard TCP state diagram, it is useful to have an extra state for local (Active) open. The local application only needs to initialize the socket data with the identity and port number of the remote node and set the socket to this state, and the state machine will take over and establish a connection with that node. For simplicity, the TCP stack assumes there is no address-resolution cache, so it sends out an ARP request.

```
    /* Active (local) open a connection */
    case TCP_AOPEN:
        ts->server = 0;                         /* Identify as client socket */
        if (gfp->g.dtype & DTYPE_SLIP)
            new_state(ts, TCP_ARPR);            /* If SLIP, don't ARP */
        else
        {                                       /* If Ether, do ARP */
            memcpy(ts->rem.mac, bcast, MACLEN); /* Set broadcast addr */
            new_state(ts, TCP_ARPS);            /* Send ARP */
            txlen = make_arp(gfp, &ts->loc, &ts->rem, ARPREQ);
        }
        break;

    case TCP_ARPS:                              /* Idle until ARP response */
        break;                                  /* (see ARP_receive func) */
```

The `arp_receive()` function receives the ARP response and moves on to the next state.

```
/* Find socket(s) for incoming ARP response */
void arp_receive(TSOCK tss[], int nsocks, GENFRAME *gfp)
{
    ARPKT *arp;
    TSOCK *ts;
    int n, txlen;

    arp = getframe_datap(gfp);
    for (n=0; n<nsocks; n++)                    /* Try matching to socket */
    {
        ts = &tss[n];
        if (ts->state==TCP_ARPS && arp->sip==ts->rem.ip)
        {                                       /* If matched, change state */
            memcpy(ts->rem.mac, arp->smac, MACLEN); /* ..copy Ethernet addr */
            new_state(ts, TCP_ARPR);     /* Send SYN */
            if ((txlen = tsock_rx(ts, gfp, 0))>0)
                put_frame(gfp, txlen);
        }
    }
}
```

ARP Received

ARP Received is another nonstandard state that takes action after an Active Open ARP response has been received. You need to choose a new sequence value and ephemeral port number and then send a SYN.

```
case TCP_ARPR:
    buff_setall(&ts->txb, mstime()*100L);   /* ARPed: set my SEQ value */
    ts->rxwin = TCP_RXWIN;                   /* Default window size */
    ts->connflags = 0;                       /* No special flags */
    ts->loc.port = ++eport>=MAXEPORT ? MINEPORT : eport;/* New port num */
    if (upcall)                              /* Do upcall */
        upcall(ts, TCP_OPEN);
    new_state(ts, TCP_SYNS);
    txlen = make_sock_tcp(gfp, ts, TSYN, 0);
    break;
```

Unlike most of the `upcalls`, this one does not check the return status. There are two reasons for this.

- My application must have requested the opening of a connection, so it shouldn't want to close it this early.

- It is inadvisable to send a SYN without establishing a connection. Some systems will assume that your application is attempting a Denial Of Service hacking attack and will ignore any further connection attempts.

SYN Sent

Having sent a SYN, wait until the SYN ACK response arrives; then respond with an ACK and go Established.

```
    case TCP_SYNS:                                /* Sent SYN, if SYN+ACK Rx */
        if (rflags==TSYN+TACK && ts->rxack==ts->txb.out)
        {
            buff_setall(&ts->rxb, ts->rxseq+1); /* Set my ACK value */
            if (upcall)                           /* Do upcall */
                upcall(ts, TCP_CONN);
            new_state(ts, TCP_EST);               /* ..send ACK, go estab */
            txlen = make_sock_tcp(gfp, ts, TACK, 0);
        }
        else if (rflags)                          /* If anything else.. */
            txlen = make_reset_resp(gfp, dlen); /* ..send reset */
        break;
```

The Established state for an Active Open is the same as that for a Passive Open, described in "Closed and Listen States" on page 176.

Active Close

As with the Active Open, the Active Close state has been added to simplify the interface between an application and the TCP stack. When it has finished sending data, the application changes the state to Active Close, and the TCP state machine performs an orderly disconnection, starting with a FIN segment.

```
    /* Active (local) close a connection */
    case TCP_ACLOSE:
        tsock_estab_rx(ts, gfp, dlen);            /* Fetch Rx data */
        if (buff_dlen(&ts->txb))                  /* If any Tx data left.. */
            txlen = tsock_estab_tx(ts, gfp, 0);  /* If unsent, send it */
        else                                      /* All data sent: close conn */
        {
            if (upcall)                           /* Do upcall */
                upcall(ts, TCP_CLOSE);
            new_state(ts, TCP_FINWT1);
            txlen = make_sock_tcp(gfp, ts, TFIN+TACK, 0);
        }
        break;
```

FIN Wait 1

Having sent a FIN, there are three possible responses from the remote system.

FIN ACK   Immediate acknowledgment of the closure.

ACK   Half-closure; remote system continues to send data.

FIN   Simultaneous closure. The remote system has requested closure at the same time as the local system, and the requests have crossed in transit.

Of course, the response may contain more data, so this must be read in before checking the flags.

```
case TCP_FINWT1:
    tsock_estab_rx(ts, gfp, dlen);            /* Fetch Rx data */
    if (rflags&TFIN && ts->rxseq==ts->rxb.in)
    {
        if (rflags&TACK && ts->rxack==ts->txb.trial+1)
            new_state(ts, TCP_TWAIT);         /* If ACK+FIN, close */
        else if (!(rflags&TACK) || ts->rxack==ts->txb.trial)
            new_state(ts, TCP_CLING);         /* If FIN, wait for ACK */
        txlen = make_sock_tcp(gfp, ts, TACK, 0);
    }
    else if (rflags&TACK && ts->rxack==ts->txb.trial+1)
        new_state(ts, TCP_FINWT2);            /* If just ACK, half-close */
    break;
```

FIN Wait 2

FIN Wait 2 means you have half-Closed and are receiving data and waiting for a FIN.

```
case TCP_FINWT2:                             /* Half-closed: awaiting FIN */
    tsock_estab_rx(ts, gfp, dlen);           /* Fetch Rx data */
    if (rflags&TFIN && ts->rxseq==ts->rxb.in)
    {
        new_state(ts, TCP_TWAIT);            /* Got FIN, close */
        txlen = make_sock_tcp(gfp, ts, TACK, 0);
    }
    break;
```

Closing

In a simultaneous close, you are awaiting the final ACK.

```
    case TCP_CLING:                              /* Closing: need final ACK */
        if (rflags==TACK && ts->rxseq==ts->rxb.in+1)
            new_state(ts, TCP_CLOSED);
        break;
```

Timed Wait

The final state should theoretically wait to ensure that all segments in transit have been received, but such is the shortage of resources on an embedded system that it is better to close immediately and free the socket for the next transaction. There is a global handler that traps all unexpected segments and returns a Reset response.

```
    case TCP_TWAIT:                              /* Timed wait: just close! */
    default:
        new_state(ts, TCP_CLOSED);
        break;
    }
    return(txlen);
}
```

Incoming Segment Demultiplexer

So far, I have dealt with each TCP segment using a specific socket pointer, but how do you decide which incoming segment belongs to which socket? For this, you need a demultiplexer, which finds an appropriate socket for the incoming segment.

The demultiplexing steps are

1. search all the sockets for one with a matching IP address and port number;
2. if not found, get the first idle (closed) socket; and
3. if not found, send a Reset.

Segments for existing connections should be handled by step 1; new connection requests should be handled by step 2, or step 3 if no more sockets are available. Steps 1 and 2 call the TCP segment handler tsock_rx(). If it returns a non-zero length value, then it has prepared a segment for transmission and put_frame() is called to send it out.

```
/* Find socket for incoming segment; send TCP RESET if none found
** Receive length is non-zero if segment received (-1 if segment has no data) */
void tcp_receive(TSOCK tss[], int nsocks, GENFRAME *gfp, int dlen)
{
    int n, ok=0, txlen=0;
    TSOCK *ts;
    NODE loc, rem;
```

```
    if (gettcp_locdest(gfp, &loc))      /* Get local node */
    {
        gettcp_srce(gfp, &rem);         /* Get remote node */
        for (n=0; n<nsocks && !ok; n++) /* Try matching to existing socket */
        {
            ts = &tss[n];
            ok = loc.ip==ts->loc.ip && loc.port==ts->loc.port &&
                rem.ip==ts->rem.ip && rem.port==ts->rem.port;
        }
        for (n=0; n<nsocks && !ok; n++) /* If not, pick the first idle skt */
        {
            ts = &tss[n];
            if ((ok = ts->state==TCP_CLOSED)!=0)
            {
                ts->loc = loc;
                ts->rem = rem;
            }
        }
        if (ok)                         /* If found, socket gets the segment */
            txlen = tsock_rx(ts, gfp, dlen);
    }
    if (!ok)                            /* If not, send RESET */
    {
        if (statedebug)
            printf("    (?) sending reset\n");
        txlen = make_reset_resp(gfp, dlen);
    }
    if (txlen > 0)
        put_frame(gfp, txlen);
}
```

Retransmission

You may have been surprised at the omission of any retries in the preceding code; what happens if the transmitted segment goes astray? The easy way out would be to store a copy of the latest outgoing segment in each socket, so it can be retransmitted on demand. A more economical method is to store all the necessary information in the socket to permit the last transmission to be recreated. The circular buffer trial pointer is wound back to the out pointer, and the data, flags, or both are resent.

```
/* Remake the last transmission (TCP or ARP), return frame length */
int remake_tsock(TSOCK *ts, GENFRAME *gfp)
{
    int txlen=0;

    if (ts->state == TCP_ARPS)
        txlen = make_arp(gfp, &ts->loc, &ts->rem, ARPREQ);
    else if (ts->state==TCP_EST||ts->state==TCP_CLWAIT||ts->state==TCP_ACLOSE)
    {
        ts->txb.trial = ts->txb.out;
        txlen = tsock_estab_tx(ts, gfp, 1);
    }
    else if (ts->state)
        txlen = make_sock_tcp(gfp, ts, ts->txflags, 0);
    return(txlen);
}
```

To trigger this retransmission, each socket is polled to check for a time-out. Any socket that is functioning correctly will continually call new_state(), which keeps its timer refreshed. If the socket is stalled waiting for a response, a time-out occurs, and a retransmission is made, or if the maximum time-out count is exceeded, the connection is reset.

```
/* Poll all the sockets, checking for timeouts or Tx data to be sent */
void tcp_poll(TSOCK tss[], int nsocks, GENFRAME *gfp)
{
    TSOCK *ts;
    int n, txlen;

    for (n=0; n<nsocks; n++)
    {
        ts = &tss[n];
        txlen = 0;
        if (ts->state)
        {
            if (ts->state && ts->timeout && mstimeout(&ts->time, ts->timeout))
            {
                if (tcpdebug)
                    printf("    (%u) Timeout\n", ts->index);
                ts->timeout += ts->timeout;
                if (ts->retries-- >= 0)
                    txlen = remake_tsock(ts, gfp);
```

```
            else
            {
                new_state(ts, TCP_CLOSED);
                if (ts->state != TCP_ARPS)
                    txlen = make_sock_tcp(gfp, ts, TRST, 0);
            }
        }
        else if (ts->state)                  /* Send Tx data */
            txlen = tsock_rx(ts, gfp, 0);
    }
    if (txlen > 0)
        put_frame(gfp, txlen);
    }
}
```

After each time-out, the `timeout` value is doubled, so there is an exponential increase in the waiting time. This offers the best compromise between using a short time-out for local area networks and a long time-out for remote Internet connections.

TCP Application — Telnet

When looking for a sample application to exercise your TCP stack, there is one obvious candidate: Telnet. This was originally written to allow remote login to multi-user systems, but it can also be used to access a wide variety of other services. In case you don't have access to a multi-user system, this Telnet implementation will also offer limited capability as a server.

Network Virtual Terminal — NVT

At its simplest, Telnet is just a keyboard and screen interface to the TCP stack: whatever is typed on the keyboard is sent as TCP data (generally one character at a time), and whatever is received by the TCP stack is put on the display. The character set is restricted to seven bits, sent as a byte value with the most significant bit cleared. The end-of-line sequence is CR LF (carriage return, line feed). If a carriage return is to be sent on its own, it is followed by a null (byte value of zero) to avoid confusion with the end-of-line marker.

This imaginary terminal is known as a Network Virtual Terminal (NVT), and the character set is known as NVT-ASCII. A lot of TCP applications assume the existence of this baseline terminal; for example, the simple mail transfer protocol (SMTP) specifies its transactions using NVT-ASCII.

Although the NVT is sufficient to drive many protocols manually, it is inadequate for remote login to a multi-user system. The system and the NVT need to negotiate about a wide variety of options — for example, will Telnet echo all characters as they are typed, or should the remote system echo them back? To support this, an interpret as command (IAC) escape code (value FFh) can be sent by either side as a prefix to a command byte. The top few command bytes and their associated option-negotiation commands are listed in Table 6.3.

Table 6.3 **Command bytes and option-negotiation commands.**

Hex Value	Command Name
FEh	DON'T
FDh	DO
FCh	WON'T
FBh	WILL

Option Negotiation

When logging in to a remote system, Telnet generally receives a string of option requests. These are really a holdover from the days of serial terminals, when each terminal manufacturer implemented their own special characteristics, so lengthy negotiation was necessary to find common ground between the multi-user system and the terminal. The format of the negotiation is request-and-response, using the following requests and responses:

WILL   Sender wants to enable its option
DO   Sender wants receiver to enable its option
WON'T   Sender wants to disable its option
DON'T   Sender wants receiver to disable its option

Each step of the negotiation combines two commands, as shown in Table 6.4.

Table 6.4 **Requests and responses.**

Request	Response
DO (an option)	WILL (do an option)
DO (an option)	WON'T (do an option)
WILL (do an option)	DO (an option)
WILL (do an option)	DON'T (do an option)
WON'T (do an option)	DON'T (do an option)
DON'T (do an option)	WON'T (do an option)

Typical options are shown in Table 6.5.

Table 6.5 **Typical options.**

1	Echo
3	Suppress Go Ahead
5	Status
24	Terminal type
31	Window size
32	Terminal speed
34	Line mode

You don't really want to get bogged down in the minutiae of terminal emulation, so you need a simple way of disabling all the fancy features. Fortunately, if you respond WON'T to all requests, you end up with a sensible set of options:

Telnet Implementation

Your Telnet application needs to

- open a TCP connection to the remote host,
- send keyboard data and display received data, and
- close the connection.

The TCP stack does most of the work: you need to set the desired remote address and port number, and the TCP state machine will take over and set up the connection. You need some way of determining the current connection state: for example, how do you know when incoming data is available? You could just poll the incoming data buffer, but a cleaner method is to set up a callback function that is called every time a significant TCP event (opening, closing, data reception) occurs. It would be even more helpful if the callbacks were segregated into two types: those applicable to clients and those applicable to servers.

If you've been following the plot so far, you'll realize that, from a TCP perspective, there is very little difference between a client and a server. Both are capable of sending and receiving data and closing the connection. The only notable difference is that a client does an *Active Open* of a socket, using an *ephemeral* port number, whereas a server does a *Passive Open* of a *well-known* port number. However, from an application writer's perspective, there is a huge difference between a client and server: the client initiates the session, whereas the server merely responds to incoming requests. To make the application writer's life easier, you must draw a distinction between client and server callbacks, even though they are very similar in TCP terms.

TCP Callbacks

The TCP stack has two function pointers that are zero (inactive) by default.

```
extern int (*server_upcall)(TSOCK *ts, CONN_STATE conn);/* TCP server action */
extern int (*client_upcall)(TSOCK *ts, CONN_STATE conn);/* TCP client action */
```

These can be set to point to your server and client functions. When called, the application receives a pointer to the specific socket (since one application must be able to cover multiple sockets in different states) and a simplified *connection state* that indicates what is happening.

```
/* Simplified connection states for applications */
typedef enum {
    TCP_IDLE,           /* Idle */
    TCP_OPEN,           /* Opening connection */
    TCP_CONN,           /* Connected */
```

```
        TCP_DATA,               /* Connected, data received */
        TCP_NODATA,             /* Connected, no data received */
        TCP_CLOSE               /* Closing connection */
} CONN_STATE;
```

The application could deduce this connection state by looking at the socket data, but I'm making life easy by providing an *executive summary* of what is going on. During the life of a connection, the application will receive the following upcalls:

```
TCP_OPEN
TCP_CONN
TCP_NODATA interspersed with a few TCP_DATA
TCP_CLOSE
```

The application normally returns a non-zero value to indicate that the connection is OK. A zero value can be returned to

- deny a TCP_OPEN request (server),
- close an opening connection (server),
- close a connection (client or server), or
- keep a closing connection open (half-Closed server).

A useful extra feature is that all outgoing data will be sent before the closure request is acted upon. A very simple TCP server, such as the Daytime server described later, need only wait for TCP_CONN, put the outgoing data in the Transmit buffer, and return a zero value. The TCP stack will ensure the data is sent before the connection is closed.

Telnet Client Initialization

The program startup is similar to previous utilities, processing command-line arguments and the configuration file.

```
int main(int argc, char *argv[])
{
    int args=0, err=0, client=0, fail, n;
    char *p, temps[18], k=0, c;
    WORD dtype;
    GENFRAME *gfp;
    LWORD mstimer;
    TSOCK *ts;
    NODE rem;

    printf("TELNET v" VERSION "\n");          /* Sign on */
    get_locnode_n = locnode_n;                 /* Set upcall ptrs to funcs */
    server_upcall = server_action;
    client_upcall = client_action;
    signal(SIGINT, break_handler);             /* Trap ctrl-C */
```

```c
    ts = &tsocks[0];                         /* Pointer to client socket */
    memset(&rem, 0, sizeof(rem));            /* Preset remote port number */
    rem.port = DEFPORT;
    while (argc > ++args)                    /* Process command-line args */
    {
        if ((c=argv[args][0])=='-' || c=='/')
        {
            switch (toupper(argv[args][1]))
            {
            case 'C':                        /* -C: config filename */
                strncpy(cfgfile, argv[++args], MAXPATH);
                if ((p=strrchr(cfgfile, '.'))==0 || !isalpha(*(p+1)))
                    strcat(cfgfile, CFGEXT);
                break;
            case 'S':                        /* -S: display TCP states */
                statedebug = 1;
                break;
            case 'T':                        /* -T: display TCP segments */
                tcpdebug = 1;
                break;
            case 'V':                        /* -V: verbose packet disp */
                netdebug |= 1;
                break;
            case 'X':                        /* -X: hex packet display */
                netdebug |= 2;
                break;
            default:
                err = 1;
            }
        }
        else if (client==0 && isdigit(c))    /* If client mode.. */
        {
            rem.ip = atoip(argv[args]);      /* Get dest IP address */
            client++;
        }                                    /* ..then port num/name */
        else if (client==1 && (rem.port=str2service(argv[args]))!=0)
            client++;
        else if (client == 2)
        {
```

```
            clientcmd = argv[args];              /* ..then command string */
            client++;
        }
    }
    if (err || rem.port==0)                      /* Prompt user if error */
        disp_usage();
    else if (!(dtype=read_netconfig(cfgfile, &locnode)))
        printf("Invalid configuration '%s'\n", cfgfile);
    else
    {
        rem.dtype = genframe.g.dtype = dtype;   /* Set frame driver type */
        gfp = &genframe;                         /* Get pointer to frame */
        printf("IP %s", ipstr(locnode.ip, temps));
        printf(" mask %s", ipstr(locnode.mask, temps));
        if (locnode.gate)
            printf(" gate %s", ipstr(locnode.gate, temps));
        if (dtype & DTYPE_ETHER)
            printf(" Ethernet %s", ethstr(locnode.mac, temps));
        if (client && !on_subnet(rem.ip, &locnode) && !locnode.gate)
            printf("\nWARNING: no gateway specified!");
        printf("\nPress ESC or ctrl-C to exit\n\n");
        if (client)                              /* If client, open socket */
            open_tcp(ts, gfp, &locnode, &rem);
```

Main Loop

After initialization, the main loop

- receives incoming frames and extracts any TCP data into the circular buffer,
- polls the network drivers,
- polls the receive circular buffer and outputs any incoming TCP data, and
- polls the keyboard and checks for the escape or Ctrl-C characters.

```
    while (!breakflag && k!=0x1b)            /* Main loop.. */
    {
        do_receive(gfp);                     /* Receive frames */
        do_poll(gfp);                        /* Poll net drivers */
        if (client)
        {
            do_teldisp(ts);                  /* Telnet client display */
            if (!ts->state)                  /* If closed */
```

```
            breakflag = 1;                  /* ..exit from main loop */
        else if (k)                         /* If key pressed.. */
            buff_in(&ts->txb, (BYTE *)&k, 1);
        k = 0;                              /* ..send it */
    }
    if (kbhit())                            /* Get keypress */
        k = getch();
}                                           /* Close connection */
```

Closing Connections

If the user hits the escape key or Ctrl-C, the Telnet application must perform an orderly shutdown of any live connections. It is possible that an orderly shutdown may fail, in which case a forced shutdown (TCP Reset) is used instead. It is important to realize that a frame is not necessarily sent out on the network as soon as the put_frame() call is made, so the network drivers must be polled for a while afterwards, to give sufficient time for the frame to be sent.

```
    if (k)
        printf("Closing...\n");
    mstimeout(&mstimer, 0);                 /* Refresh timer */
    do
    {                                       /* Loop until conns closed */
        for (n=fail=0; n<NSOCKS; n++)
            fail += !close_tcp(&tsocks[n]);
        do_receive(gfp);
        do_poll(gfp);                       /* ..or timeout */
    } while (!mstimeout(&mstimer, 1000) && fail);
    if (fail)                               /* If still open.. */
    {
        printf("Resetting connections\n");
        for (n=0; n<NSOCKS; n++)            /* ..send reset */
            reset_tcp(&tsocks[n], gfp);
        while (!mstimeout(&mstimer, 1000))
        {                                   /* ..and wait until sent */
            do_receive(gfp);
            do_poll(gfp);
        }
    }
    close_net(dtype);                       /* Shut down net driver */
}
return(0);
```

```
}
/* Close a TCP socket. Return non-0 when closed */
int close_tcp(TSOCK *ts)
{
    if (ts->state==TCP_EST || ts->state==TCP_SYNR)
        new_state(ts, TCP_ACLOSE);
    return(ts->state == TCP_CLOSED);
}


/* Reset a TCP socket */
void reset_tcp(TSOCK *ts, GENFRAME *gfp)
{
    if (ts->state)
        put_frame(gfp, make_sock_tcp(gfp, ts, TRST, 0));
}
```

Polling

The TCP stack needs to be polled to check for time-outs, and the network drivers must be polled to keep them alive.

```
/* Poll the network interface to keep it alive */
void do_poll(GENFRAME *gfp)
{
    tcp_poll(tsocks, NSOCKS, gfp);
    poll_net(gfp->g.dtype);
}
```

Telnet Display

The Telnet client could use the callback to obtain incoming data, but it is simpler to just poll the Receive buffer and decode and display the incoming characters. You also need a miniature state machine to decode the Telnet options and send back suitably negative replies.

```
/* Telnet option byte values (subset) */
#define TEL_IAC    255         /* Interpret As Command */
#define TEL_DONT   254         /* Don't do it */
#define TEL_DO     253         /* Do it */
#define TEL_WONT   252         /* Won't do it */
#define TEL_WILL   251         /* Will do it */
#define TEL_ECHO   1           /* Echo option */
#define TEL_SGA    3           /* Suppress Go-Ahead option */
#define TEL_AUTH   37          /* Authentication option */
```

```
/* Telnet client display */
void do_teldisp(TSOCK *ts)
{
    static BYTE d[3]={0,0,0};

    while (buff_dlen(&ts->rxb))
    {
        if (d[0] == TEL_IAC)
        {
            if (!d[1] && buff_out(&ts->rxb, &d[1], 1))
            {
                if (d[1] == TEL_IAC)
                    d[0] = d[1] = 0;
                else if (!d[2] && buff_out(&ts->rxb, &d[2], 1))
                {
                    d[1] = TEL_WONT;
                    buff_in(&ts->txb, d, 3);
                    d[0] = d[1] = d[2] = 0;
                }


            }
        }
        else while (buff_out(&ts->rxb, d, 1) && d[0]!=TEL_IAC)
            putchar(d[0]);
    }
}
```

Telnet Client Callback

To short-circuit lengthy option negotiations, the Telnet client checks to see if it is being used on a so-called Telnet port (i.e., a port designated for remote login to a multi-user system) and, if so, sends a preemptive block of option negotiations. Failing to do this can introduce frustratingly long delays into the log-in process.

This Telnet utility also supports a command-line string option, allowing you to send the given message to a remote application before closing the connection. This, too, is loaded into the output buffer as soon as the connection is open.

```
/* Telnet options */
BYTE telopts[] =
{
    TEL_IAC, TEL_DO,    TEL_SGA,    /* Do suppress go-ahead */
```

```
        TEL_IAC, TEL_WONT,  TEL_ECHO,   /* Won't echo */
        TEL_IAC, TEL_WONT,  TEL_AUTH    /* Won't authenticate */
};

/* Upcall from TCP stack to client when opening, connecting, receiving data
** or closing. Return 0 to close if connected */
int client_action(TSOCK *ts, CONN_STATE conn)
{
    if (conn == TCP_OPEN)
    {
        if (ts->rem.port == TELPORT)          /* If login, send Telnet opts */
            buff_in(&ts->txb, telopts, sizeof(telopts));
        if (clientcmd)
        {
            buff_instr(&ts->txb, clientcmd);/* Send command-line string */
            buff_instr(&ts->txb, "\r\n");
        }
    }
    return(1);
}
```

Telnet Server Callback

A useful additional feature of this Telnet utility is the ability to act as a simple server. It implements the following server functions:

Echo   All input from the client is echoed back.

Daytime   On establishing a connection with the server, it reports the current data and time as a string.

HTTP   As a foretaste of things to come, the server produces a simple Web page in response to a browser request.

In the case of the Daytime and HTTP servers, the connection is closed as soon as the data is sent. The Echo server will wait until the client closes the connection.

```
/* Upcall from TCP stack to server when opening, connecting, receiving data
** or closing. Return 0 to prevent connection opening, or close if connected */
int server_action(TSOCK *ts, CONN_STATE conn)
{
    int ok=1;
    WORD port, len;
    BYTE temps[30];
    time_t t;
```

```
port = ts->loc.port;                        /* Connection being opened */
if (conn == TCP_OPEN)
{
    ok = port==ECHOPORT || port==DAYPORT || port==HTTPORT;
    if (port == DAYPORT)                     /* Daytime server? */
    {
        time(&t);                            /* ..send date & time string */
        buff_in(&ts->txb, (BYTE *)ctime(&t), 24);
        buff_in(&ts->txb, (BYTE *)"\r\n", 2);
    }
    else if (port == ECHOPORT)               /* Echo server? */
        ts->connflags = TPUSH;               /* ..use PUSH flag */
    else if (port == HTTPORT)                /* HTTP server ? */
    {                                        /* ..load my page */
        buff_in(&ts->txb, (BYTE *)MYPAGE, sizeof(MYPAGE));
    }
}
else if (conn == TCP_CONN)                   /* Connected */
{
    if (port==DAYPORT || port==HTTPORT)      /* If daytime or HTTP */
        ok = 0;                              /* ..close straight away */
}
else if (conn == TCP_DATA)                   /* Received data */
{
    if (port == ECHOPORT)                    /* If Echo */
    {                                        /* ..echo it back! */
        while ((len=buff_out(&ts->rxb, temps, sizeof(temps)))!=0)
            buff_in(&ts->txb, temps, len);
    }
}
return(ok);
}
```

Using Telnet

Daytime Protocol

In the previous chapter, I used the Datagram utility to fetch a date and time string from a UDP server. You can now use Telnet to fetch the same string over a TCP connection, since most Daytime servers support both UDP and TCP connections.

Assuming the server is at address 10.1.1.1, the command

```
telnet 10.1.1.1 daytime
```

may return the following response.

```
TELNET vx.xx
IP 10.1.1.11 mask 255.0.0.0 gate 10.1.1.111 Ethernet 00:c0:26:b0:0a:93
Press ESC or ctrl-C to exit

Tue May 23 09:48:12 2000
```

You can use the diagnostic capabilities of Telnet to show how the transfer works. The following lines show the same transfer and the response with the state display, TCP segment display, and verbose options enabled.

```
telnet -s -t -v 10.1.1.1 daytime

TELNET vx.xx
IP 10.1.1.11 mask 255.0.0.0 gate 10.1.1.111 Ethernet 00:c0:26:b0:0a:93
Press ESC or ctrl-C to exit

    (1) new state 'active open'
    (1) new state 'ARP sent'
Tx0 /len 42 ----BROADCAST---- ARP 10.1.1.11 -> 10.1.1.1
Rx0 \len 60 00:c0:f0:09:bd:c3 ARP 10.1.1.1 -> 10.1.1.11
    (1) new state 'ARP rcvd'
    (1) new state 'SYN sent'
    /ack 00000000 seq d34de133 port 1025->13 <SYN> MSS 1460 dlen 0h
Tx0 /len 58 00:c0:f0:09:bd:c3 IP 10.1.1.11 -> 10.1.1.1 TCP
Rx0 \len 60 00:c0:f0:09:bd:c3 IP 10.1.1.1 -> 10.1.1.11 TCP
    \seq 4edb0d1b ack d34de134 port 1025<-13 <SYN><ACK> MSS 1460 dlen 0h
    (1) new state 'established'
    /ack 4edb0d1c seq d34de134 port 1025->13 <ACK> dlen 0h
Tx0 /len 54 00:c0:f0:09:bd:c3 IP 10.1.1.11 -> 10.1.1.1 TCP
Rx0 \len 80 00:c0:f0:09:bd:c3 IP 10.1.1.1 -> 10.1.1.11 TCP
    \seq 4edb0d1c ack d34de134 port 1025<-13 <PSH><ACK> dlen 1Ah
    (1) Rx data 26 bytes
    /ack 4edb0d36 seq d34de134 port 1025->13 <ACK> dlen 0h
Tx0 /len 54 00:c0:f0:09:bd:c3 IP 10.1.1.11 -> 10.1.1.1 TCP
Tue May 23 09:50:15 2000
Rx0 \len 60 00:c0:f0:09:bd:c3 IP 10.1.1.1 -> 10.1.1.11 TCP
    \seq 4edb0d36 ack d34de134 port 1025<-13 <FIN><ACK> dlen 0h
    (1) new state 'last ACK'
    /ack 4edb0d37 seq d34de134 port 1025->13 <FIN><ACK> dlen 0h
```

```
Tx0 /len 54 00:c0:f0:09:bd:c3 IP 10.1.1.11 -> 10.1.1.1 TCP
Rx0 \len 60 00:c0:f0:09:bd:c3 IP 10.1.1.1 -> 10.1.1.11 TCP
      \seq 4edb0d37 ack d34de135 port 1025<-13 <ACK> dlen 0h
(63)   new state 'closed'
```

This display contains a lot of information, so it can be quite difficult to decode. Here is my analysis.

ARP Request

```
      (1) new state 'active open'
      (1) new state 'ARP sent'
Tx0 /len 42 ----BROADCAST---- ARP 10.1.1.11 -> 10.1.1.1
```

When transmitting, the state changes always appear *before* the packets to which they refer, so here you see a transition into the Active Open state, the creation of an ARP segment, a transition to the ARP Sent state, and finally the ARP frame emerging from the network buffer.

ARP Response

```
Rx0 \len 60 00:c0:f0:09:bd:c3 ARP 10.1.1.1 -> 10.1.1.11
      (1) new state 'ARP rcvd'
      (1) new state 'SYN sent'
      /ack 00000000 seq d34de133 port 1025->13 <SYN> MSS 1460 dlen 0h
Tx0 /len 58 00:c0:f0:09:bd:c3 IP 10.1.1.11 -> 10.1.1.1 TCP
```

As soon as the ARP response is received, a SYN can be sent to open the connection.

SYN ACK

```
Rx0 \len 60 00:c0:f0:09:bd:c3 IP 10.1.1.1 -> 10.1.1.11 TCP
      \seq 4edb0d1b ack d34de134 port 1025<-13 <SYN><ACK> MSS 1460 dlen 0h
      (1) new state 'established'
      /ack 4edb0d1c seq d34de134 port 1025->13 <ACK> dlen 0h
Tx0 /len 54 00:c0:f0:09:bd:c3 IP 10.1.1.11 -> 10.1.1.1 TCP
```

The server agrees to open the connection by returning a SYN ACK. With a response of ACK, the connection is established.

Data Transfer

```
Rx0 \len 80 00:c0:f0:09:bd:c3 IP 10.1.1.1 -> 10.1.1.11 TCP
      \seq 4edb0d1c ack d34de134 port 1025<-13 <PSH><ACK> dlen 1Ah
      (1) Rx data 26 bytes
      /ack 4edb0d36 seq d34de134 port 1025->13 <ACK> dlen 0h
Tx0 /len 54 00:c0:f0:09:bd:c3 IP 10.1.1.11 -> 10.1.1.1 TCP
Tue May 23 09:50:15 2000
```

In this case, you have no data to send; the Daytime server sends its data as soon as the connection is established. The display order is particularly confusing here, so it is important to check the data lengths, in order to establish that the date and time printout belongs to the incoming, not the outgoing, segment, which is purely an acknowledgment.

Closure

```
Rx0 \len 60 00:c0:f0:09:bd:c3 IP 10.1.1.1 -> 10.1.1.11 TCP
    \seq 4edb0d36 ack d34de134 port 1025<-13 <FIN><ACK> dlen 0h
  (1) new state 'last ACK'
    /ack 4edb0d37 seq d34de134 port 1025->13 <FIN><ACK> dlen 0h
Tx0 /len 54 00:c0:f0:09:bd:c3 IP 10.1.1.11 -> 10.1.1.1 TCP
```

When it receives your acknowledgment, the server starts closing the connection, because it has no more data. If you had some data to send, you could leave the connection half-Open by sending only an ACK, but there is little point, since a Daytime server will discard anything you send. Choose the quickest exit by acknowledging the server's FIN, and sending a FIN of your own.

Closed

```
Rx0 \len 60 00:c0:f0:09:bd:c3 IP 10.1.1.1 -> 10.1.1.11 TCP
    \seq 4edb0d37 ack d34de135 port 1025<-13 <ACK> dlen 0h
 (64)    new state 'closed'
```

Once the server acknowledges your FIN, the connection is closed.

Dropping Frames

To exercise your TCP state machine a bit more thoroughly, you need to introduce some errors into the system. At the top of the network driver file (NET.C), there are two compile-time definitions that allow you to deliberately introduce errors and see how the software responds.

```
/* Debug options to drop frames on transmit or receive */
#define TXDROP  0    /* Set non-zero to drop 1-in-N transmit frames */
#define RXDROP  0    /* Set non-zero to drop 1-in-N receive frames */
```

Altering the first value from 0 to 4 and rebuilding the software will cause one in four Transmit frames to be discarded. This is an extremely high error rate, so you wouldn't expect the TCP transfer to be as smooth, but you would expect it to go through correctly and not stall or crash. Here's the Daytime transaction again, using this setting.

```
TELNET vx.xx
IP 10.1.1.11 mask 255.0.0.0 gate 10.1.1.111 Ethernet 00:c0:26:b0:0a:93
Press ESC or ctrl-C to exit

    (1) new state 'active open'
    (1) new state 'ARP sent'
Tx0 /len 42 ----BROADCAST---- ARP 10.1.1.11 -> 10.1.1.1
Rx0 \len 60 00:c0:f0:09:bd:c3 ARP 10.1.1.1 -> 10.1.1.11
    (1) new state 'ARP rcvd'
    (1) new state 'SYN sent'
    /ack 00000000 seq e4e60b9b port 1025->13 <SYN> MSS 1460 dlen 0h
Tx0 /len 58 00:c0:f0:09:bd:c3 IP 10.1.1.11 -> 10.1.1.1 TCP
Rx0 \len 60 00:c0:f0:09:bd:c3 IP 10.1.1.1 -> 10.1.1.11 TCP
    \seq fe92c0e3 ack e4e60b9c port 1025<-13 <SYN><ACK> MSS 1460 dlen 0h
```

```
        (1) new state 'established'
           /ack fe92c0e4 seq e4e60b9c port 1025->13 <ACK> dlen 0h
Tx0 /len 54 00:c0:f0:09:bd:c3 IP 10.1.1.11 -> 10.1.1.1 TCP
Rx0 \len 80 00:c0:f0:09:bd:c3 IP 10.1.1.1 -> 10.1.1.11 TCP
           \seq fe92c0e4 ack e4e60b9c port 1025<-13 <PSH><ACK> dlen 1Ah
        (1) Rx data 26 bytes
           /ack fe92c0fe seq e4e60b9c port 1025->13 <ACK> dlen 0h
        Tx frame dropped for debug
Tue May 23 10:39:23 2000
Rx0 \len 60 00:c0:f0:09:bd:c3 IP 10.1.1.1 -> 10.1.1.11 TCP
           \seq fe92c0fe ack e4e60b9c port 1025<-13 <FIN><ACK> dlen 0h
        (1) new state 'last ACK'
           /ack fe92c0ff seq e4e60b9c port 1025->13 <FIN><ACK> dlen 0h
Tx0 /len 54 00:c0:f0:09:bd:c3 IP 10.1.1.11 -> 10.1.1.1 TCP
Rx0 \len 60 00:c0:f0:09:bd:c3 IP 10.1.1.1 -> 10.1.1.11 TCP
           \seq fe92c0ff ack e4e60b9d port 1025<-13 <ACK> dlen 0h
        (1) new state, Closed
```

Despite the deliberate error, this transaction looks the same as the previous one: What is going on? The answer is that the frame that was dropped happened to be the acknowledgment of the data, which is completely redundant. The server still sends its FIN, and your acknowledgment of that FIN includes, by implication, all the data that went before it. This clearly shows the hazards of TCP testing: introducing a one-in-four error rate may sound severe, but you must check that the errors are occurring where they can do the greatest potential damage and that your software handles the situation correctly.

At the risk of laboring this point, here is the same transaction, but with a one-in-three error rate.

```
TELNET vx.xx
IP 10.1.1.11 mask 255.0.0.0 gate 10.1.1.111 Ethernet 00:c0:26:b0:0a:93
Press ESC or ctrl-C to exit

        (1) new state 'active open'
        (1) new state 'ARP sent'
Tx0 /len 42 ----BROADCAST---- ARP 10.1.1.11 -> 10.1.1.1
Rx0 \len 60 00:c0:f0:09:bd:c3 ARP 10.1.1.1 -> 10.1.1.11
        (1) new state 'ARP rcvd'
        (1) new state 'SYN sent'
           /ack 00000000 seq e2e2c18b port 1025->13 <SYN> MSS 1460 dlen 0h
Tx0 /len 58 00:c0:f0:09:bd:c3 IP 10.1.1.11 -> 10.1.1.1 TCP
Rx0 \len 60 00:c0:f0:09:bd:c3 IP 10.1.1.1 -> 10.1.1.11 TCP
           \seq ea78cf01 ack e2e2c18c port 1025<-13 <SYN><ACK> MSS 1460 dlen 0h
        (1) new state 'established'
           /ack ea78cf02 seq e2e2c18c port 1025->13 <ACK> dlen 0h
        Tx frame dropped for debug
        (1) Timeout
           /ack ea78cf02 seq e2e2c18c port 1025->13 <ACK> dlen 0h
Tx0 /len 54 00:c0:f0:09:bd:c3 IP 10.1.1.11 -> 10.1.1.1 TCP
```

The above is a simple case of the Transmit frame being dropped, a time-out being triggered, and a duplicate being sent.

```
RxO \len 80 00:c0:f0:09:bd:c3 IP 10.1.1.1 -> 10.1.1.11 TCP
      \seq ea78cf02 ack e2e2c18c port 1025<-13 <PSH><ACK> dlen 1Ah
      (1) Rx data 26 bytes
      /ack ea78cf1c seq e2e2c18c port 1025->13 <ACK> dlen 0h
TxO /len 54 00:c0:f0:09:bd:c3 IP 10.1.1.11 -> 10.1.1.1 TCP
Tue May 23 10:43:47 2000
RxO \len 60 00:c0:f0:09:bd:c3 IP 10.1.1.1 -> 10.1.1.11 TCP
      \seq ea78cf1c ack e2e2c18c port 1025<-13 <FIN><ACK> dlen 0h
      (1) new state 'last ACK'
      /ack ea78cf1d seq e2e2c18c port 1025->13 <FIN><ACK> dlen 0h
      Tx frame dropped for debug
RxO \len 60 00:c0:f0:09:bd:c3 IP 10.1.1.1 -> 10.1.1.11 TCP
      \seq ea78cf1c ack e2e2c18c port 1025<-13 <FIN><ACK> dlen 0h
RxO \len 60 00:c0:f0:09:bd:c3 IP 10.1.1.1 -> 10.1.1.11 TCP
      \seq ea78cf1c ack e2e2c18c port 1025<-13 <FIN><ACK> dlen 0h
RxO \len 60 00:c0:f0:09:bd:c3 IP 10.1.1.1 -> 10.1.1.11 TCP
      \seq ea78cf1c ack e2e2c18c port 1025<-13 <FIN><ACK> dlen 0h
RxO \len 60 00:c0:f0:09:bd:c3 IP 10.1.1.1 -> 10.1.1.11 TCP
      \seq ea78cf1c ack e2e2c18c port 1025<-13 <FIN><ACK> dlen 0h
      (1) Timeout
      /ack ea78cf1d seq e2e2c18c port 1025->13 <FIN><ACK> dlen 0h
TxO /len 54 00:c0:f0:09:bd:c3 IP 10.1.1.11 -> 10.1.1.1 TCP
RxO \len 60 00:c0:f0:09:bd:c3 IP 10.1.1.1 -> 10.1.1.11 TCP
      \seq ea78cf1d ack e2e2c18d port 1025<-13 <ACK> dlen 0h
      (1) new state 'closed'
```

The above doesn't look so good. It got there in the end, but the error seems to have caused significant turmoil. The server sends its FIN, and your response is discarded. Not unnaturally, the server resends the FIN, and not unnaturally, you see this as a duplicate of something you've already received and so discard it. It's a little surprising that the server manages to transmit four repeats of the FIN before your time-out triggers, you resend the dropped frame, and all is well again.

Conclusion

The objective of TCP is simple: to create a reliable connection between two points on the network, such that data can flow bidirectionally in a timely fashion.

The concepts in TCP are simple. There is a *sequence space*, within which the data transfer fits with its start and end markers. A state machine keeps both participants in step and defines the negotiations between them to open and close the connections. Port numbers identify the logical endpoints of the connection, the service that is required, or the application that is to handle the data.

Implementing TCP is difficult. So much is expected of it that it takes an inordinately large amount of design effort and an even greater amount of testing.

In this chapter, I have worked through the concepts of TCP and implemented enough of the protocol for the purposes at hand. I have created a Telnet application, which not only provides a simple emulation of the standard utility but also has a server capability, so it can be used as a basis for further TCP client or server development.

Source Files

ether3c.c	3C509 Ethernet card driver
etherne.c	NE2000 Ethernet card driver
ip.c	Low-level TCP/IP functions
net.c	Network interface functions
netutil.c	Network utility functions
pktd.c	Packet driver (BC only)
serpc.c or serwin.c	Serial drivers (BC or VC)
tcp.c	TCP functions
telnet.c	Telnet utility
dosdef.h	MS-DOS definitions (BC only)
ether.h	Ethernet definitions
ip.h	TCP/IP definitions
net.h	Network driver definitions
netutil.h	Utility function and general frame definitions
serpc.h	Serial driver definitions (BC or VC)
tcp.c	TCP function definitions
win32def.h	Win32 definitions (VC only)

Telnet Utility

Utility	Simple emulation of standard Telnet with server capability
Usage	`telnet [options] [IP_address [port]]`
	Enters server mode (Echo, Daytime, and HTTP) if no IP address given
Options	`-c name` Configuration filename (default `tcplean.cfg`)
	`-s` State display mode
	`-t` TCP segment display mode
	`-v` Verbose display mode
	`-x` Hex packet display
Example	`telnet -v -c slip 172.16.1.1 daytime`
Keys	Ctrl-C or Esc to exit
	Keystrokes are sent on the network when connected to the telnet port
Config	`net` to identify network type
	`ip` to identify IP address
	`mask` optional subnet mask
	`gate` optional gateway IP address
Modes	Defaults to server mode (Echo, Daytime, HTTP) if no IP address
Notes	The port may be a number or one of the following:
	`echo`
	`daytime`
	`http`

Chapter 7

Hypertext Transfer Protocol: HTTP

Overview

Having worked your way through IP and TCP, you can finally create your first Web server. For this, you need to implement yet another protocol, hypertext transfer protocol (HTTP), which simply involves an exchange of text messages followed by the transfer of Web data down a TCP connection.

A Web server isn't much use without Web pages, so I'll look at the creation of pages that are particularly relevant to embedded systems and the display of real-world signals. The pages will be stored on disk and made available by the Web server on demand. In Chapter 8, I'll start looking at the insertion of live data into the pages.

HTTP GET Method

To fetch a Web document, the browser opens a TCP connection to server port 80 and then uses HTTP to send a request. Compared to TCP, HTTP is refreshingly simple: the request and response are one or more lines of text, each terminated by the new line (carriage return, line feed) characters. If the request is successful, the information (document text, graphical data) is then sent down the same connection, which is closed on completion.

HTTP commands are called *methods*; the one used to fetch documents is the get method.

Request

In its simplest form, a request consists of a text line containing the uppercase keyword `GET`, the filename, a protocol identifier, and the new line character.

```
GET /index.htm HTTP/1.0<CR><LF>
```

Following the `GET` method is an optional header containing further browser-specific details, such as its configuration and the document formats it can accept.

```
GET /index.htm HTTP/1.0
Connection: Keep-Alive
User-Agent: Mozilla/4.5 [en] (Win95; I)
Pragma: no-cache
Host: 10.1.1.11
Accept: image/gif, image/x-xbitmap, image/jpeg, image/pjpeg, image/png,
*/*
Accept-Encoding: gzip
Accept-Language: en
Accept-Charset: iso-8859-1,*,utf-8
```

The end of the header is marked by a blank line (just carriage return, line feed).

The header entries consist of a (case-insensitive) header name followed by a colon, a space, and a header value. The above example gives a good idea of the kind of information provided.

Response

The server replies with a *response line* containing the HTTP version, status code, and description, such as

```
HTTP/1.0 200 OK
```

or an error code as shown below.

```
HTTP/1.0 404 Not Found
```

The server may then send optional header information in a similar format to the request headers. Here is a response from the widely used Apache server.

```
HTTP/1.0 200 OK
Last-modified: Mon, 10 Apr 2000 09:01:53 GMT
Server: Apache/1.1.1
Content-type: text/html
Content-length: 1783
```

A blank line terminates the header field, and the *entity body* (text or graphic data) follows. Even when reporting a failure, the server may well include some text data, to give the browser something to display.

```
HTTP/1.0 404 Not found
Date: Mon, 10 Apr 2000 09:56:24 GMT
Server: Apache/1.1.1
Content-type: text/html
```

```
<HEAD><TITLE>File not found</TITLE></HEAD>
<BODY><H1>File not found</H1>
The requested URL /abc.htm was not found on this server
<P></BODY>
```

In this case, a text (HTML) message is sent to the browser, but binary data, most notably in the form of graphics files, can be sent instead.

```
HTTP/1.0 200 OK
Date: Mon, 10 Apr 2000 10:12:58 GMT
Server: Apache/1.1.1
Content-type: image/gif
Content-length: 8264
Last-modified: Mon, 10 Apr 2000 9:15:03 GMT
```

After a blank line, the binary image data starts. There is no special encoding for binary files; the disk file is sent over the network connection without any modifications.

Content-Types

The *content-type header* can have a wide range of values, encompassing all the well-known multimedia data standards. The notation used is type/subtype, such as `text/html`, where `text` is a general description and `html` is a specific kind of text. I'll concentrate on the following few listed in Table 7.1.

Table 7.1   Content-types.

Type	Subtype	Extension	Description
text	plain	txt	Simple unformatted text
text	html	htm	HTML-formatted text
image	gif	gif	GIF bitmap image
image	x-xbitmap	xbm	X-bitmap image

The original extension for HTML files was `.html`, but this has to be truncated to `.htm` for DOS systems that don't support long filenames. I will use the shorter extension throughout this book, since your software must be able to run on these legacy systems.

Plain Text

Not surprisingly, this is straightforward text with no special formatting. Line ends are signaled by a carriage return and line feed pair (CR, LF). Browsers usually display this text using a monospaced font such as Courier.

HTML

HyperText Markup Language (HTML) is the standard way of adding special formatting to text pages. The plain text is augmented by tags enclosed in angle brackets. The tags give an indication as to the type of formatting to be applied, so <h1> indicates the following text is a level one header, but doesn't specify which font should be used for the purpose. Most tags are begin–end pairs, such as this, where the trailing tag name is prefixed with a slash character.

The text is made "hyper" by the insertion of *hyperlinks*, which, if clicked in the browser, cause it to jump within the current document or fetch another document from the same or a different server.

GIF

The graphics interchange format (GIF) is the standard method of encoding bitmap images. It uses a discrete color (palette) representation, so it is better suited to computer graphics, rather than continuous-tone images (photographs).

X-Bitmap

The X-bitmap is an older monochrome standard that is relatively little-used but still supported by all browsers. Because it is text based, it can be quite useful for system testing and dynamic insertion of small images. The syntax is borrowed from C; here is a 12-by-12 pixel image of a plus sign (+).

```
#define myimage_height 12
#define myimage_width  12
static unsigned char myimage_bits[] = {
    0xf0, 0x00, 0xf0, 0x00, 0xf0, 0x00, 0xf0, 0x00,
    0xff, 0x0f, 0xff, 0x0f, 0xff, 0x0f, 0xff, 0x0f,
    0xf0, 0x00, 0xf0, 0x00, 0xf0, 0x00, 0xf0, 0x00,
};
```

The Browsers use simple parsing techniques, so are easily misled; if the top few lines are changed to the following, some browsers will fail to display the graphic, even though it is equally valid C code:

```
#define myimage_height 12
#define myimage_width  12
static unsigned char myimage_bits[] =
{
    0xf0, 0x00, 0xf0, 0x00, 0xf0, 0x00, 0xf0, 0x00,
```

File Extensions

There is a rather uneasy relationship between the file extension and the HTTP data type. When loading from disk, the browser has to decide the document type based purely on the file extension, the operating system's file associations, or both. When loading from the Web, there is normally both a file extension *and* a data type, with the latter taking precedence. However, there is no compulsion on the Web server to give an HTTP data type, even though its absence creates confusion, with some browsers using the file extension to derive a type, whereas others assume a default (HTML).

Extending this to its logical conclusion, it is possible to omit the HTTP header entirely, in which case most (but not all) browsers will Do The Right Thing. This protocol violation is not to be encouraged, but it can be useful for initial development and debugging.

Simple Web Server

In the previous chapter, I explained the upcalls (function callbacks) that are generated by the TCP/IP stack when a significant event occurs, such as a connection being opened or data arriving. Using these, you can create a simple Web server with the minimum amount of additional code.

A useful start is to create a Web server that emits a given file on request, with a minor refinement that it produces a file directory if no filename is given.

Webserve

You will recall that the TCP stack can maintain several simultaneous connections (i.e., file transfers). The TCP-specific information for each connection is stored in a socket structure, and there is an array of these, dimensioned to reflect the maximum number of simultaneous connections.

```
/* Socket storage */
#define NSOCKS 2
TSOCK tsocks[NSOCKS] =              /* Initialize index num and buffer sizes */
{
    {1,{_CBUFFLEN_},{_CBUFFLEN_}},
    {2,{_CBUFFLEN_},{_CBUFFLEN_}}
};
```

As a starting point, I've chosen to implement only two sockets, but this number can be increased by simply adjusting the equate.

Any connection-specific information (such as the handle of the file being transferred) must be stored on a similar basis of one item per socket. Rather than cluttering up the socket structure with application-specific data, I've opted to use a separate array of application-specific structures and then link them to the sockets.

```
/* Application-specific storage */
typedef struct {
    FILE *in;                      /* File I/P pointer */
} APPDATA;
APPDATA appdata[NSOCKS];
```

The linkage between socket and application data is done in the main program just after sign on.

```
    for (n=0; n<NSOCKS; n++)
        tsocks[n].app = &appdata[n];
```

Now you can use the app pointer in the socket structure to fetch the handle of the file being transferred.

The server code is entirely dependent on the TCP upcalls (function callbacks), which are generated when connected to a remote node.

```c
/* Upcall from TCP stack to server when opening, connecting, receiving data
** or closing. Return 0 to prevent connection opening, or close if connected */
int server_action(TSOCK *ts, CONN_STATE conn)
{
    int ok=1, len;
    WORD port;
    char *s, *name;
    APPDATA *adp;

    port = ts->loc.port;
    adp = (APPDATA *)ts->app;
    if (port != HTTPORT)
        ok = 0;
    else if (conn == TCP_OPEN)
    {
        adp->in = 0;
    }
    else if (conn == TCP_DATA)
    {
        if ((len = buff_chrlen(&ts->rxb, '\n'))!=0) /* Got request? */
        {
            len = mini(len+1, HTTP_MAXLEN);         /* Truncate length */
            buff_out(&ts->rxb, (BYTE *)httpreq, (WORD)len);
            httpreq[len] = 0;
            if (webdebug)                           /* Display if debugging */
                printf("%s", httpreq);
            s = strtok(httpreq, " ");               /* Chop into tokens */
            if (!strcmp(s, "GET"))                  /* 1st token is 'GET'? */
            {
                name = strtok(0, " ?\r\n");         /* 2nd token is filename */
                http_get(ts, name);                 /* Process filename */
            }
        }
    }
    http_data(ts);
    return(ok);
}
```

The call to http_data() keeps the Transmit buffer topped up. It copies as much as possible of the file into the remaining space.

```c
/* If there is space in the transmit buffer, send HTTP data */
void http_data(TSOCK *ts)
{
    APPDATA *adp;
    int len;

    adp = (APPDATA *)ts->app;              /* If space in buffer.. */
    if (adp->in && (len = buff_freelen(&ts->txb)) > 0)
    {                                      /* ..put out as much as possible */
        if ((len = buff_infile(&ts->txb, adp->in, (WORD)len)) == 0)
        {                                  /* If end of file, close it.. */
            fclose(adp->in);
            adp->in = 0;
            close_tcp(ts);                 /* ..and start closing connection */
        }
    }
}
```

A return value of zero from buff_infile() indicates the whole file has been put in the buffer (though not necessarily transmitted yet), so the connection closure can be started.

To help with formatted printing to the TCP transmission buffer, it is worthwhile creating your own version of printf().

```c
/* Version of printf() to write a string into a circular buffer.
** Return string length, or zero if insufficient room in buffer */
int buff_inprintf(CBUFF *bp, char *str, ...)
{
    char temps[200];
    int len;

    va_list argptr;
    va_start(argptr, str);
    len = vsprintf(temps, str, argptr);
    va_end(argptr);
    if (len<=0 || len>buff_freelen(bp))
        len = 0;
    else
        buff_in(bp, (BYTE *)temps, (WORD)len);
    return(len);
}
```

All that is needed to complete your Web server is a function to parse the file path and get the file handle.

```c
#define HTTP_OK      "HTTP/1.0 200 OK\r\n"
#define HTTP_NOFILE "HTTP/1.0 404 Not found\r\n"
#define HTTP_HTM     "Content-type: text/html\r\n"
#define HTTP_TXT     "Content-type: text/plain\r\n"
#define HTTP_GIF     "Content-type: image/gif\r\n"
#define HTTP_XBM     "Content-type: image/x-xbitmap\r\n"
#define HTTP_BLANK  "\r\n"

/* Process the filepath from an HTTP 'get' method */
void http_get(TSOCK *ts, char *fname)
{
    APPDATA *adp;
    char *s=0;

    adp = (APPDATA *)ts->app;
    strcpy(filepath, filedir);          /* Add on base directory */
    if (*fname)                         /* Copy filename without leading '/' */
        strcat(filepath, fname+1);
    strlwr(filepath);                   /* Force to lower-case */
    if (strlen(fname) <= 1)
    {                                   /* If name is only a '/'.. */
        if (webdebug)                   /* Display if debugging */
            printf("Sending directory\n", filepath);
        strcpy(filepath, filedir);
        strcat(filepath, "*.*");        /* ..display directory */
        buff_instr(&ts->txb, HTTP_OK HTTP_TXT HTTP_BLANK);
        buff_instr(&ts->txb, "Directory\r\n");
        if ((s = find_first(filepath)) != 0) do {
            buff_instr(&ts->txb, s);    /* One file per line */
            buff_instr(&ts->txb, "\r\n");
        } while ((s = find_next()) != 0);
        close_tcp(ts);
    }                                   /* If file not found.. */
    else if ((adp->in = fopen(filepath, "rb")) == 0)
```

```
    {                                        /* ..send message */
        if (webdebug)                        /* Display if debugging */
            printf("File '%s' not found\n", filepath);
        buff_instr(&ts->txb, HTTP_NOFILE HTTP_TXT HTTP_BLANK);
        buff_inprintf(&ts->txb, "Webserve v%s\r\n", VERSION);
        buff_inprintf(&ts->txb, "Can't find '%s'\r\n", filepath);
        close_tcp(ts);
    }
    else                                     /* File found OK */
    {
#if HTTP_HEAD == 0
                                             /* No HTTP header or.. */
#elif HTTP_HEAD == 1                         /* Simple OK header or.. */
        buff_instr(&ts->txb, HTTP_OK HTTP_BLANK);
#else                                        /* Content-type header */
        buff_instr(&ts->txb, HTTP_OK);
        s = strstr(filepath, ".htm") ? HTTP_HTM :
            strstr(filepath, ".txt") ? HTTP_TXT :
            strstr(filepath, ".gif") ? HTTP_GIF :
            strstr(filepath, ".xbm") ? HTTP_XBM : "";
        buff_instr(&ts->txb, s);
        buff_instr(&ts->txb, HTTP_BLANK);
#endif      if (webdebug)                    /* Display if debugging */
            printf("File '%s' %s", filepath, *s ? s : "\n");
    }
}
```

If the file can't be found, a message is displayed instead, using plain text so you don't have to worry about HTML tags.

```
HTTP/1.0 404 Not found
Content-type: text/plain

Webserve Vx.xx
Can't find '.\webdocs\abc.htm'
```

If no file is specified, a directory is displayed (Figure 7.1). The resulting browser display would win few prizes for elegance, but it serves its purpose.

Figure 7.1   Browser display of plain text directory.

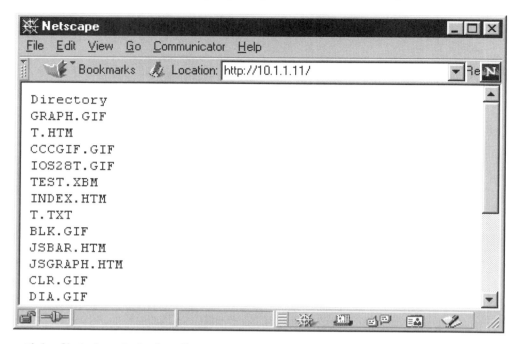

If the file is found, the handle is put in the application storage area, and conditional compilation is used to decide what to do next. If HTTP_HEAD is 0, no HTTP file headers are emitted. If it is 1, a simple HTTP OK header is displayed.

```
HTTP/1.0 200 OK
```

If HTTP_HEAD is 2, the full content-type is specified, as shown below.

```
HTTP/1.0 200 OK
Content-type: text/html
```

HTTP Header Tests

It is instructive to run variants of the Web server using different browsers to see which are able to accept the headless files. For these tests, I've used three browsers widely available at the time of writing: Internet Explorer v5.0, Netscape Communicator v4.5, and Pocket Internet Explorer v3.0 (for Windows CE v3).

The differences between browsers, as shown in Table 7.2, are commonplace. When creating an embedded Web server, it may be tempting to take short cuts in its implementation, but there is a significant risk of incompatibility with some browsers.

Table 7.2   Browser tests on headless files.

Header	Internet Explorer	Netscape Communicator	Pocket Internet Explorer
None	Text and graphics OK	Text and graphics OK	Plain text only; no graphics
HTTP OK	Text and graphics OK	Text OK; no graphics	Text and graphics OK
HTTP content-type	Text and graphics OK	Text and graphics OK	Text and graphics OK

Introducing HTML

The Web server, as presented so far, is rather dull. True, it can serve up all kinds of exciting files, but as a next step, it would be nice if the file directory was interactive so that you could click on a filename, rather than having to retype it in the browser.

To do this, you need to introduce some very basic HTML, sufficient for your purposes. You may already have created HTML documents using your word processor, but I'll be looking at HTML from a programmer's perspective, since the server has to create HTML pages from scratch. As far as possible, I'll stick to generic HTML to avoid the browser-specific problems I've already demonstrated.

HTML Tags

By default, all text in an HTML document is intended for display on the browser screen. HTML tags contain information that affects the text presentation and allows the insertion of extra information into the text, such as graphic images and navigational links. The tag consists of left and right angle brackets enclosing text that is generally insensitive to case.

```
<html>
<HTML>
```

As with all such escape-character schemes, the left angle brackets must be eliminated from normal text; otherwise, it will be treated as a tag delimiter.

```
This <tag> isn't a tag
```

Instead, it should be encoded as follows.

```
This &lt;tag> isn't a tag
```

In its turn, the ampersand needs an escape sequence.

```
Nuts &amp; bolts
```

Most tags are paired; the closing tag name is prefixed by a slash character. Care must be taken when nesting tags to avoid mismatches.

```
This is <b>bold <i>bold italic</i></b> text
```

There are a few tags that don't need pairing.

```
Line 1<br>Line 2
```

Browsers aren't rigorous in their enforcement of the HTML syntax, so there are some tags that should be paired but often aren't.

```
Paragraph 1<p>Paragraph 2<p>Paragraph 3
```

Tags can contain numeric or string attribute values. The attribute name is insensitive to case, but the attribute value may be sensitive to case. Quotation marks need not be used for a single-word value, but strings with embedded punctuation or spaces must be in single or double quotes (up to a maximum of 1,024 characters), as shown below.

```
<FONT FACE='Arial,Helvetica' SIZE=-2>
```

Some sample tags and attributes are shown in Table 7.3.

Table 7.3   HTML tags.

Tags	Attributes
`<!--This is a comment-->`	Comment — not displayed
`<a href="test.htm">Test page</a>`	Hyperlink
`<a name="help">Help info.</a>`	Anchor name
`<b>Bold text</b>`	Bold typeface
`<body>Sample text</body>`	Document body
` Start of new line`	Line break
`<em>Bold text</em>`	Bold text
`<font face='arial' size=-2>Smaller text</font>`	Set font
`<form action=resp.htm>...</form>`	Create form
`<h2>Heading<h2>`	Header level
`<head><title>My doc</tile></head>`	Document header
`<hr>New section`	Horizontal rule
`<i>Italic</i>`	Italic text
`<img src='graph.gif' alt='Test results'>`	Graphical image
`<input type=submit name=send>`	Input elements for forms
`<meta http-equiv=refresh content="5;default.htm">`	Additional document info
`<p>New paragraph`	Paragraph delimiter
`<pre>Program text</pre>`	Preformatted text
`<table><tr><td>Dat1</td><td>Dat2</td></tr></table>`	Tabular data
`<title>Test document</title>`	Document title

HTML Document

The fundamental HTML structure is a document, the extent of which is marked by opening and closing HTML tags. Within the document is a header section and a body.

```
<html>
<head>
...
</head>
<body>
...
</body>
</html>
```

Despite its usefulness in detecting tag mismatches, indentation is rarely used in HTML documents, because it can introduce spurious spaces into the document text. Don't forget that, by default, all nontagged text is intended to be displayed.

The header normally contains a title and document definitions, whereas the body contains the document's text.

```
<html>
<head>
<title>Skeletal HTML document</title>
</head>
<body>
This is a minimal HTML document
</body>
</html>
```

Links to other documents or locations within the current document are achieved using the hyperlink <a> tag.

```
<html>
<head>
<title>Test document</title>
</head>
<body>
Click <a href="test2.htm">here</a> for second document<br>
Click <a href="test.htm#help">here</a> for help<br>
<p>
<a name="help"></a>
This is the help text
</body>
</html>
```

The text bracketed within the <a> ... </a> tags is specially highlighted by the browser, to indicate a hyperlink that is activated when clicked.

Clickable File Directory

Returning to the clickable file directory, you'd expect the server-generated HTML text to look something like this.

```
<html>
<head>
<title>Directory</title>
</head>
<body>
<h2>Directory of \temp\webdocs</h2><br>
<a href='file1.htm'>file1.htm</a><br>
<a href='file2.gif'>file2.gif</a><br>
...
</body>
</html>
```

Each filename needs to be quoted twice: once as the hyperlink target and once as the clickable display text.

Implementation

The http_get() function needs some minor restructuring so that you can tinker with the individual tags for the directory header, directory entries, and the text after the directory.

```
/* Process the filepath from an HTTP 'get' method */
void http_get(TSOCK *ts, char *fname)
{
    APPDATA *adp;
    char *s=0;

    adp = (APPDATA *)ts->app;
    strcpy(filepath, filedir);          /* Add on base directory */
    if (*fname)                         /* Copy filename without leading '/' */
        strcat(filepath, fname+1);
    strlwr(filepath);                   /* Force to lower-case */
    if (strlen(fname) <= 1)
    {                                   /* If name is only a '/'.. */
        if (webdebug)                   /* Display if debugging */
            printf("Sending directory\n", filepath);
        strcpy(filepath, filedir);
```

```
        strcat(filepath, "*.*");        /* ..display directory */
        dir_head(&ts->txb, filedir);    /* Send out directory */
        if ((s = find_first(filepath)) != 0) do {
            dir_entry(&ts->txb, s);
        } while ((s = find_next()) != 0);
        dir_tail(&ts->txb);
        close_tcp(ts);
    }                                     /* If file not found.. */
    else if ((adp->in = fopen(filepath, "rb")) == 0)
    {                                     /* ..send message */
        if (webdebug)                     /* Display if debugging */
            printf("File '%s' not found\n", filepath);
        buff_instr(&ts->txb, HTTP_NOFILE HTTP_TXT HTTP_BLANK);
        buff_inprintf(&ts->txb, "Webserve v%s\r\n", VERSION);
        buff_inprintf(&ts->txb, "Can't find '%s'\r\n", filepath);
        close_tcp(ts);
    }
    else                                  /* File found OK */
    {
        buff_instr(&ts->txb, HTTP_OK);
        s = strstr(filepath, ".htm") ? HTTP_HTM :
            strstr(filepath, ".txt") ? HTTP_TXT :
            strstr(filepath, ".gif") ? HTTP_GIF :
            strstr(filepath, ".xbm") ? HTTP_XBM : "";
        buff_instr(&ts->txb, s);
        buff_instr(&ts->txb, HTTP_BLANK);
        if (webdebug)                         /* Display if debugging */
            printf("File '%s' %s", filepath, *s ? s : "\n");
    }
}
```

The directory _head, _entry, and _tail functions determine what HTML code is generated.

```
/* Write out head of HTML file dir, given filepath. Return 0 if error */
int dir_head(CBUFF *bp, char *path)
{
    return(buff_instr(bp, HTTP_OK HTTP_HTM HTTP_BLANK) &&
           buff_instr(bp, "<html><head><title>Directory</title></head>\r\n") &&
           buff_inprintf(bp, "<body><h2>Directory of %s</h2>\r\n", path));
}
```

```
/* Write out HTML file dir entry, given name. Return length, 0 if error */
int dir_entry(CBUFF *bp, char *name)
{
    return(buff_inprintf(bp, "<a href='%s'>%s</a><br>\r\n", name, name));
}
/* Write out tail of HTML file dir. Return length, 0 if error */
int dir_tail(CBUFF *bp)
{
    return(buff_instr(bp, "</body></html>\r\n"));
}
```

The carriage return and line feed characters are not strictly necessary, but they add to the readability of the HTML code generated.

Sample Display

The resulting browser display is a significant improvement over the plain-text original, as shown in Figure 7.2.

Figure 7.2   Browser display of HTML directory.

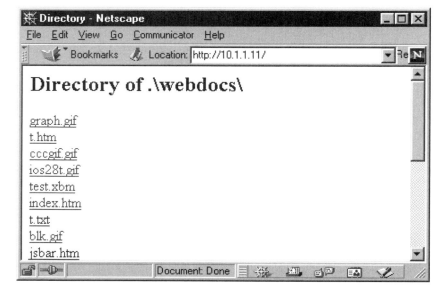

Tabular Format

It would be nice to report the size of each file in a neatly formatted display.

File	Size
test.gif	1234
x.htm	56789
... and so on ...	

The traditional way to line up the values in columns is to insert strategic padding, but HTML offers a much simpler method; let the browser do all the work. All you need to do is mark this as a table, indicate where each row starts and ends, and identify each cell (data item or heading).

HTML Table

To identify a table, use the `<table>` ... `</table>` tags. Each heading is delimited by `<th>` ... `</th>`, each row by `<tr>` ... `</tr>`, and each data item (cell) by `<td>` ... `<td>`. The resulting table with two dimensions is shown below.

```
<table>
              <th>Heading</th>    <th>Heading</th>
<tr>          <td>Data</td>       <td>Data</td>       </tr>
<tr>          <td>Data</td>       <td>Data</td>       </tr>
</table>
```

There is plenty of scope for adjusting the position, size, spacing, borders, and colors of the various elements, but I'll use the default values, so the HTML for the directory should look something like this.

```
<html><head><title>Directory</title></head>
<body><h2>Directory of .\webdocs\</h2>
<table><th>File</th><th>Size</th>
<tr><td><a href='test.gif'>test.gif</a></td><td>1234</td></tr>
<tr><td><a href='x.htm'>x.htm</a></td><td>56789</td></tr>
...
</table></body></html>
```

It is perfectly legal to nest the hyperlinks within the table data, so you don't have to sacrifice the ability to click on individual file names.

Implementation

Only a relatively minor adjustment to the directory _head, _entry, and _tail functions is required.

```
/* Write out head of HTML file dir, given filepath. Return 0 if error */
int dir_head(CBUFF *bp, char *path)
{
    return(buff_instr(bp, HTTP_OK HTTP_HTM HTTP_BLANK) &&
           buff_instr(bp, "<html><head><title>Directory</title></head>\r\n") &&
           buff_inprintf(bp, "<body><h2>Directory of %s</h2>\r\n", path) &&
           buff_instr(bp, "<table><th>File</th><th>Size</th>\r\n"));
}
/* Write out HTML file dir entry, given name. Return length, 0 if error */
int dir_entry(CBUFF *bp, char *name)
{
    return(buff_inprintf(bp, "<tr><td><a href='%s'>%s</a></td>", name, name) &&
           buff_inprintf(bp, "<td>%lu</td></tr>\r\n", find_filesize()));
}
/* Write out tail of HTML file dir. Return length, 0 if error */
int dir_tail(CBUFF *bp)
{
    return(buff_instr(bp, "</table></body></html>\r\n"));
}
```

Even using the browser's default table parameters, the result is quite passable (Figure 7.3).

Buffer Overflow

As you increase the amount of HTML ornamentation associated with each file, or if you happen to use a directory with a lot of files, you may encounter an unexpected problem: truncation of the directory display. This occurs because http_get() generates the complete directory in one "hit" and stores the result in the Transmit buffer, which is only 2Kb by default. If the actual document exceeds this length, the end is silently truncated, with unfortunate results.

The simplest solution is to increase the buffer size (by redefining _CBUFFLEN_), but, looking at the broader picture, it would be better to develop a general-purpose solution to accommodate an HTML display of arbitrary complexity in the future.

A clue to the answer comes from the way I've already handled the transmission of large files over the TCP connection. Instead of attempting to cram the whole file into the transmit buffer, http_data() continually checks the free space in the buffer and loads the next file fragment when there is room. When all the file is in the buffer (but not necessarily sent), the closure of the TCP connection is started.

```
if (adp->in && (len = buff_freelen(&ts->txb)) > 0)
{
    if ((len = buff_infile(&ts->txb, adp->in, (WORD)len)) == 0)
    {
        fclose(adp->in);
        adp->in = 0;
        close_tcp(ts);
    }
}
```

By analogy, you really should generate the directory one fragment at a time, until the Transmit buffer is full and then wait until it is empty before generating any more fragments. This implies the creation of a simple state machine to keep track of what has been generated. As ever, you must not forget the possibility of several simultaneous connections, each at a different point in the serving up of a directory, so you need to store the state information on a per-socket basis.

Figure 7.3   Browser display of tabular directory.

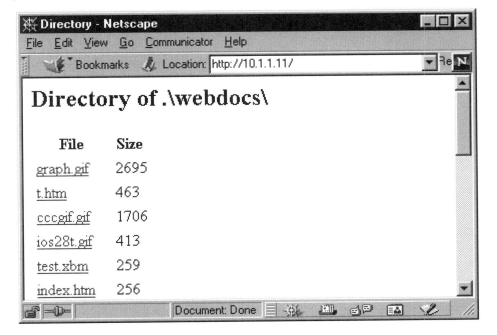

State Machine Implementation

A simple method of implementing the state machine is to keep a single state variable in the application-specific socket data, which will track the number of directory entries sent to the Transmit buffer.

0	nothing sent
1	HTTP, HTML, and table headers sent
2	first directory entry sent
...	
$n + 1$	all n directory entries sent
$n + 2$	table and HTML footers sent; start closure

It is important to remember that these states refer to the process of loading the Transmit buffer, not the process of sending TCP data segments (which is handled by the TCP stack). If there is sufficient room in the Transmit buffer, you can load the next data item and completely forget about it. The TCP stack handles the task of (reliably) getting the Transmit buffer data sent across to the other node, performing any retries as necessary.

The new state variable is included in the application-specific data structure.

```
/* Application-specific storage */
typedef struct {
    FILE *in;                        /* File I/P pointer */
    long count;                      /* State variable/counter */
} APPDATA;
APPDATA appdata[NSOCKS];
```

On receipt of an HTTP request, `http_get()` is called to send out the HTTP headers and prepare the file or directory for transmission.

```
/* Process the filepath from an HTTP 'get' method */
void http_get(TSOCK *ts, char *fname)
{
    APPDATA *adp;
    char *s=0;

    adp = (APPDATA *)ts->app;
    strcpy(filepath, filedir);       /* Add on base directory */
    if (*fname)                      /* Copy filename without leading '/' */
        strcat(filepath, fname+1);
    strlwr(filepath);                /* Force to lower-case */
    if (strlen(fname) <= 1)
    {                                /* If name is only a '/'.. */
        if (webdebug)                /* Display if debugging */
            printf("Sending directory\n", filepath);
```

```
            strcpy(filepath, filedir);
            strcat(filepath, "*.*");        /* ..display directory */
            dir_head(&ts->txb, filedir);    /* Send out directory */
            adp->count = 1;
        }                                   /* If file not found.. */
        else if ((adp->in = fopen(filepath, "rb")) == 0)
        {                                   /* ..send message */
            if (webdebug)                   /* Display if debugging */
                printf("File '%s' not found\n", filepath);
            buff_instr(&ts->txb, HTTP_NOFILE HTTP_TXT HTTP_BLANK);
            buff_inprintf(&ts->txb, "Webserve v%s\r\n", VERSION);
            buff_inprintf(&ts->txb, "Can't find '%s'\r\n", filepath);
            close_tcp(ts);
        }
        else                                /* File found OK */
        {
            buff_instr(&ts->txb, HTTP_OK);
            s = strstr(filepath, ".htm") ? HTTP_HTM :
                strstr(filepath, ".txt") ? HTTP_TXT :
                strstr(filepath, ".gif") ? HTTP_GIF :
                strstr(filepath, ".xbm") ? HTTP_XBM : "";
            buff_instr(&ts->txb, s);
            buff_instr(&ts->txb, HTTP_BLANK);
            if (webdebug)                   /* Display if debugging */
                printf("File '%s' %s", filepath, *s ? s : "\n");
        }
    }
}
```

Once this is done, http_data() takes over the routine task of checking for any space in the transmit buffer and sending the next file fragment or the next few directory entries.

```
/* If there is space in the transmit buffer, send HTTP data */
void http_data(TSOCK *ts)
{
    APPDATA *adp;
    int count=0, ok, len;
    char *s;

    adp = (APPDATA *)ts->app;
    if (adp->count)                         /* If sending a directory.. */
    {                                       /* Skip filenames already sent */
```

```
        ok = (s = find_first(filepath)) != 0;
        while (ok && ++count<adp->count)
            ok = (s = find_next()) != 0;
        strlwr(s);                      /* Force lower-case filename */
        while (ok && dir_entry(&ts->txb, s))
        {                               /* Send as many entries as will fit */
            adp->count++;
            ok = (s = find_next()) != 0;
            strlwr(s);
        }
        if (!ok && dir_tail(&ts->txb))  /* If no more entries.. */
        {
            adp->count = 0;             /* ..start closing connection */
            close_tcp(ts);
        }
    }                                   /* If sending a file.. */
    if (adp->in && (len = buff_freelen(&ts->txb)) > 0)
    {                                   /* ..put out as much as possible */
        if ((len = buff_infile(&ts->txb, adp->in, (WORD)len)) == 0)
        {
            fclose(adp->in);
            adp->in = 0;
            close_tcp(ts);
        }
    }
}
}
```

These modifications don't change the appearance of the directory but do make it able to cope with a very large number of files. The rather simplistic technique of skipping filenames that were already sent results in a significant CPU workload when handling large directories; it would be better to store the directory search data block in the application-specific data area, instead of the file count. However, my objective is to show the kinds of techniques that can be used to translate server information into HTML text and the importance of being able to fragment this process into bite-sized (or buffer-sized) chunks.

Tabular Graphics

HTML tables are a surprisingly versatile way of displaying data. I've already shown table elements that are text and hyperlinks, but graphical elements can be used as well.

HTML Images

To insert a graphic in a page, use the <img> tag, as in the following example (Figure 7.4).

```
<html>
<head><title>Test page</title></head>
<body><h2>Graphics test</h2>
<img SRC="ios28t.gif" height=28 width=66>
<img SRC="ios28t.gif" height=28 width=132><br>
<img SRC="blk.gif" height=2 width=200><br>
</body></html>
```

It is usual to specify the image height and width attributes, to speed up the browser's page rendering. In this example, the logo first is specified using its actual dimensions, then with a doubling of the horizontal dimension. Obligingly, the browser stretches the logo to suit.

Figure 7.4   Browser display of graphics.

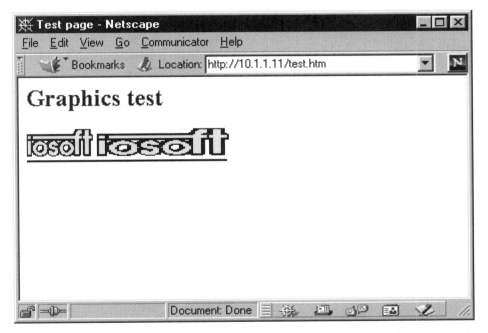

The black line under the logos is created by taking this image-stretching method to its logical extreme; blk.gif is actually a single black pixel, which the browser has stretched to 200 pixels wide and two pixels high.

Table Autosizing

You can place randomly sized graphic blocks in a table, and it will be resized automatically to accommodate them all. To create a simple column-type bar graph, it is only necessary to put the appropriately resized image blocks within a table (Figure 7.5). The columns are labeled using a second row of data.

```html
<html>
<head><title>Test page</title></head>
<body><h2>Image blocks in table</h2>
<table>
<tr valign=bottom>
<td><img SRC="yel.gif" height=110 width=20></td>
<td><img SRC="yel.gif" height=125 width=20></td>
<td><img SRC="red.gif" height=150 width=20></td>
<td><img SRC="red.gif" height=165 width=20></td>
<td><img SRC="yel.gif" height=140 width=20></td>
</tr>
<tr align=center><td>1</td><td>2</td><td>3</td><td>4</td><td>5</td></tr>
</tr></table>
</body></html>
```

Figure 7.5   Browser display of HTML image blocks in table.

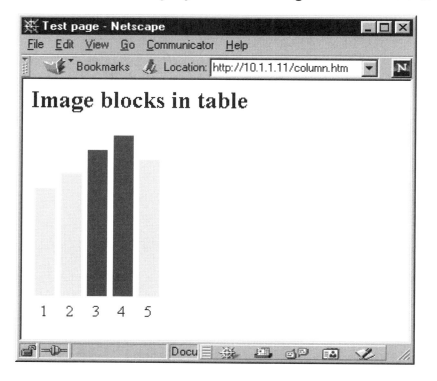

It is possible to create a *y*-axis using elaborately stacked columnar data, but it is much easier to use a predrawn GIF (Figure 7.6).

Figure 7.6   Browser display of HTML column graph.

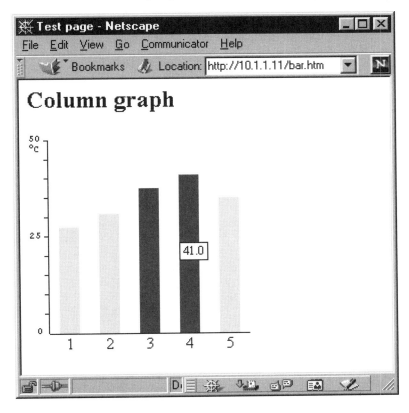

```
<html>
<head><title>Test page</title></head>
<body><h2>Column graph</h2>
<table border=0 cellpadding=0 cellspacing=0>
<tr align=center valign=bottom>
<td><img src="v50_200.gif"></td>
<td><img src="yel.gif" height=110 width=20 alt='27.5'></td>
<td><img src="yel.gif" height=124 width=20 alt='31.0'></td>
<td><img src="red.gif" height=150 width=20 alt='37.5'></td>
<td><img src="red.gif" height=164 width=20 alt='41.0'></td>
<td><img src="yel.gif" height=140 width=20 alt='35.0'></td>
</tr>
```

```
<tr><td></td><td colspan=5><img src="blk.gif" height=1 width=200></td></tr>
<tr align=center><td></td><td>1</td><td>2</td><td>3</td><td>4</td><td>5</td></tr>
</tr></table>
</body></html>
```

The addition of alternative text for each bar doesn't just help those with text-only browsers, it also allows the user of a modern browser to check the actual data value by positioning their cursor on the bar; the alternative text is displayed automatically.

The browser's table-autosizing algorithms are generally pretty good. In the graph in Figure 7.6, the total width is actually determined by the width of the scale plus the width of the *x*-axis.

```
<tr><td></td><td colspan=5><img src="blk.gif" height=1
width=200></td></tr>
```

This line spans all five data columns. Changing its width automatically changes the spacing between the bars and the positioning of the text and keeps the whole graph in proportion.

Stacked Graphical Data

So far, I have used only one data item per cell, but it is possible to stack several items. This is an obvious way to generate stacked bar charts, but it also can be used to generate *x–y* graphs. If you create a suitably small symbol (e.g., a diamond shape) and stack it on top of a transparent column, it will seem to float at the given height in the middle of the graph. This does sound rather bizarre, so it is best to give an example.

```
<html>
<head><title>Test page</title></head>
<body><h2>HTML table graph</h2>
<table border=0 cellpadding=0 cellspacing=0>
<tr align=center valign=bottom>
<td><img src="v50_200.gif"></td>
<td><img src="dia.gif"><br><img src="clr.gif" height=110 width=30></td>
<td><img src="dia.gif"><br><img src="clr.gif" height=123 width=30></td>
<td><img src="dia.gif"><br><img src="clr.gif" height=137 width=30></td>
<td><img src="dia.gif"><br><img src="clr.gif" height=142 width=30></td>
<td><img src="dia.gif"><br><img src="clr.gif" height=124 width=30></td>
<td><img src="dia.gif"><br><img src="clr.gif" height=102 width=30></td>
<td><img src="dia.gif"><br><img src="clr.gif" height=87  width=30></td>
<td><img src="dia.gif"><br><img src="clr.gif" height=72  width=30></td>
<td><img src="dia.gif"><br><img src="clr.gif" height=63  width=30></td>
<td><img src="dia.gif"><br><img src="clr.gif" height=70  width=30></td>
</tr>
<tr><td></td><td colspan=10><img src="blk.gif" height=1 width=300></td></tr>
<tr align=center>
```

```
<td></td><td><font FACE='Arial,Helvetica' SIZE=1>12:30</font></td>
<td></td><td></td><td><font FACE='Arial,Helvetica' SIZE=-2>13:00</font></td>
<td></td><td></td><td><font FACE='Arial,Helvetica' SIZE=-2>13:30</font></td>
<td></td><td></td><td><font FACE='Arial,Helvetica' SIZE=-2>14:00</font></td>
</tr></table>
</body></html>
```

Each diamond, `dia.gif`, is placed above a transparent column generated from a single-pixel image, `clr.gif`, of a specific height (Figure 7.7). The line break, `<br>`, is necessary to ensure the two are placed one above another.

Figure 7.7   Browser display of HTML *x*–*y* graph.

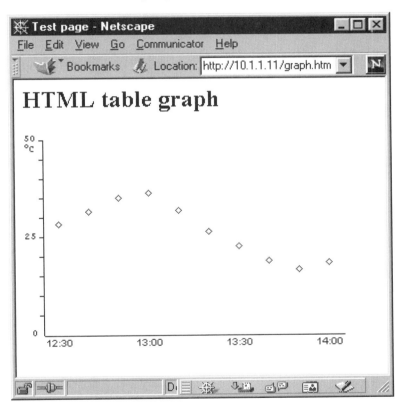

These graphs are certainly not the most elegant way of presenting the data, but they are achievable with a small amount of HTML code, which means they can be generated by a Web server using a small amount of C code. This makes it possible to add some simple, on-the-fly graphical reporting to a Web server with a minimum of programming effort.

Graphical Indicators

It is remarkable how many different displays can be created within the constraints of table construction.

- Elements are in a horizontal and vertical grid.
- Elements may span two or more rows or columns.
- Elements may not overlap.

The following HTML code generates column-type indicators (Figure 7.8). Alternate columns containing blanks, scales, and markers are used. The scales have to be propped up on small, 10-pixel-high blank blocks, to ensure the markers can indicate a zero value.

```
<html>
<head><title>Test page</title></head>
<body><h2>Column indicators</h2>
<table border=0 cellpadding=0 cellspacing=0>
<tr align=center valign=bottom>
<td><img src="clr.gif" height=1 width=20></td>
<td><img src="v40_200f.gif"><br><img src="clr.gif" height=10 width=10></td>
<td><img src="lptr.gif"><br><img src="clr.gif" height=5 width=10></td>
<td><img src="clr.gif" height=1 width=20></td>
<td><img src="v40_200f.gif"><br><img src="clr.gif" height=10 width=20></td>
<td><img src="lptr.gif"><br><img src="clr.gif" height=85 width=10></td>
<td><img src="clr.gif" height=1 width=20></td>
<td><img src="v40_200f.gif"><br><img src="clr.gif" height=10 width=10></td>
<td><img src="lptr.gif"><br><img src="clr.gif" height=170 width=10></td>
</tr>
</table>
</body></html>
```

With a small amount of artwork, these could become pretty bulb-thermometer indicators.

You might think it would be easier for the server to draw graphics on the fly from scratch; however, it is worth considering the advantages of using predrawn elements.

- CPU overhead is much lower. Drawing graphics from scratch can demand a large amount of CPU time, especially if there is a significant amount of detail. Fetching predrawn elements from files is always much faster.
- Browsers can cache the elements. Significant network bandwidth and display time can be saved if the Browser can cache the items to be displayed. This is possible only with predrawn graphical elements.
- Changing the page layout involves minimal programming. The graphical elements can be prepared by an artist, using a commercial painting package, with no software knowledge.

The Web client–server partnership was created on the understanding that servers are relatively dumb and just serve up files, while clients do all the hard work of rendering the displays. Although you will be compromising somewhat by adding some intelligence to the server, it is important not to burden it with too many creative tasks, lest it become unable to fulfil its primary function.

Figure 7.8   Browser display of column indicators.

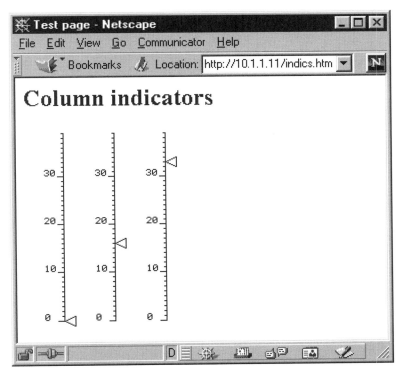

Summary

In this chapter, I looked at the HTTP protocol that allows the TCP/IP stack to serve up Web pages, and I implemented a simple Web server that can be used to browse files in a given directory.

Equally important is the content of the Web pages, and I have explored display techniques such as graphs and indicators that are particularly applicable to embedded systems. Although I used the creation of a clickable file index as an example of how Web pages can be created on the fly, the insertion of dynamic data into your Web pages has to wait until the next chapter.

Source Files

`ether3c.c`	3C509 Ethernet card driver
`etherne.c`	NE2000 Ethernet card driver
`ip.c`	Low-level TCP/IP functions
`net.c`	Network interface functions
`netutil.c`	Network utility functions
`pktd.c`	Packet driver (BC only)
`router.c`	Router utility
`serpc.c` or `serwin.c`	Serial drivers (BC or VC)
`tcp.c`	TCP functions
`webserve.c`	Web server utility
`dosdef.h`	MS-DOS definitions (BC only)
`ether.h`	Ethernet definitions
`ip.h`	TCP/IP definitions
`net.h`	Network driver definitions
`netutil.h`	Utility function and general frame definitions
`serpc.h`	Serial driver definitions (BC or VC)
`tcp.c`	TCP function definitions
`win32def.h`	Win32 definitions (VC only)

Webserve Utility

Utility	Web server with file directory capability	
Usage	`webserve [`*`options`*`] [`*`directory`*`]`	
	Uses directory `.\webdocs` if none given	
Options	`-c` *name*	Configuration filename (default `tcplean.cfg`)
	`-s`	State display mode
	`-t`	TCP segment display mode
	`-v`	Verbose display mode
	`-w`	Web (HTTP) display mode
	`-x`	Hex packet display
Example	`webserve c:\temp\webdocs`	
Keys	Ctrl-C or Esc to exit	
Config	`net`	to identify network type
	`ip`	to identify IP address
	`mask`	optional subnet mask
	`gate`	optional gateway IP address

Chapter 8

Embedded Gateway Interface: EGI

Overview

A classic Web server simply provides Web pages without alteration and is essentially just a file server. Modern Web servers can alter Web pages on the fly or create the pages from scratch each time they are requested. The umbrella term for this facility is common gateway interface (CGI), which, because of the large amount of string manipulation involved, is usually implemented on a powerful multi-user system running a language, such as Perl, that is well-suited for the job.

I need the ability to insert live data into Web pages, but I won't be using a powerful multi-user system or the Perl language, so I need to look carefully at what CGI offers the embedded-systems developer and whether there are ways of achieving dynamic content on a relatively modest system. To do this, I'll look at sample pages that could be used to control real-world signals and how they could be linked to real devices.

Interactive Displays

So far, the Web displays have been static: you click on a hyperlink and get another page. Now I'll start to introduce displays with greater interactive capability.

Switch and Lamp

To start simply, I'll create a page with a switch and lamp: the switch toggles when clicked and turns the lamp on or off. The easiest way to do this is to create two HTML pages, both with the button and lamp graphics in the same place: one with the lamp and switch on and the other with them off (Figure 8.1). By making the switch a hyperlink and cross-linking between the pages, the desired effect is achieved without any special Web server programming.

```
File switch1.htm
<html>
<head><title>Test page</title></head>
<body><h4>Click switch</h4>
<table><tr valign=middle>
<td><a href="switch2.htm"><img src="switchu.gif"></a></td>
<td><img src="ledoff.gif"></td>
</tr></table>
</body></html>

File switch2.htm
<html>
<head><title>Test page</title></head>
<body><h4>Click switch</h4>
<table><tr valign=middle>
<td><a href="switch1.htm"><img src="switchd.gif"></a></td>
<td><img src="ledon.gif"></td>
</tr></table>
</body></html>
```

When communicating over a relatively fast local area network (LAN), there is a short delay the first time the images are fetched; thereafter, the switch toggles with no noticeable delay — only the HTML page is refetched because the images are retained in the browser's cache.

Figure 8.1   Browser display of switch and lamp.

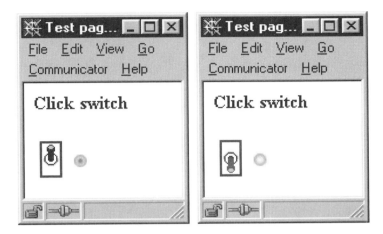

Array of Switches and Lamps

Unfortunately, this technique becomes unusable with several controls and indicators, such as an array of switches and lamps (Figure 8.2).

Figure 8.2   Browser display of switches and lamps.

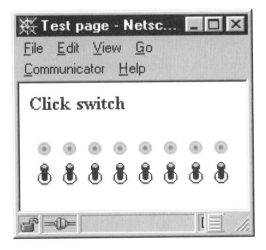

Drawing the controls and indicators is easy — just put them in a table.

```
<html>
<head><title>Test page</title></head>
<body><h4>Click switch</h4>
<table>
<tr align=center>
<td><img src="ledoff.gif"></td><td><img src="ledoff.gif"></td>
<td><img src="ledoff.gif"></td><td><img src="ledoff.gif"></td>
<td><img src="ledoff.gif"></td><td><img src="ledoff.gif"></td>
<td><img src="ledoff.gif"></td><td><img src="ledoff.gif"></td>
</tr><tr align=center>
<td><img src="switchu.gif"></td><td><img src="switchu.gif"></td>
<td><img src="switchu.gif"></td><td><img src="switchu.gif"></td>
<td><img src="switchu.gif"></td><td><img src="switchu.gif"></td>
<td><img src="switchu.gif"></td><td><img src="switchu.gif"></td>
</tr></table>
</body></html>
```

The problem is how to make the switches clickable. A simple method would involve creating one hyperlink per switch, but which page should the hyperlink point to? You could create nine pages, one with all the switches off and each of the others with one switch on, but what happens if the user wants two or more switches on? You could create 256 pages, one for each permutation of switches, but this is clearly becoming unworkable.

The problem is you are constrained by the following assumptions:

1. Each control is identified by a unique hyperlink destination, a URL.
2. Each URL refers to a unique file on the server.
3. Each server file can display only one unique state.

Of course, you can program your Web server to do what you like, so you need not conform to these assumptions.

1. Controls can share a common URL by being grouped together in an HTML "form," which is processed by a CGI.
2. The server can generate and interpret hyperlink filenames in any way it sees fit; they need not correspond to any real files.
3. A file can be modified dynamically as it is sent out, through the use of Server-Side Includes (SSIs).

The first and third methods are in common usage and will be examined in more detail later. The second technique is worthy of a brief examination, since it paves the way for the others.

Dynamic URL Generation

If the HTML file for the switches and lamps is generated completely on the fly, the hyperlinks for the switches can contain dummy filenames, adjusted to reflect the state change when that control is clicked, which need not have any relation to actual files on the system.

To take the first four switches as an illustration, imagine they are all in the "off" state and have been defined as follows.

```
<td><a href="sw1.htm"><img src="switchu.gif"></a></td>
<td><a href="sw2.htm"><img src="switchu.gif"></a></td>
<td><a href="sw4.htm"><img src="switchu.gif"></a></td>
<td><a href="sw8.htm"><img src="switchu.gif"></a></td>
```

Each switch has a binary weight corresponding to its position in the line.

Switch	1	2	3	4
Binary Value	0001	0010	0100	1000

If the user clicks switch 1, the server sees a request for sw1.htm, in which case it sends out the same fundamental page but with the URLs tweaked.

```
<td><a href="sw0.htm"><img src="switchd.gif"></a></td>
<td><a href="sw3.htm"><img src="switchu.gif"></a></td>
<td><a href="sw5.htm"><img src="switchu.gif"></a></td>
<td><a href="sw9.htm"><img src="switchu.gif"></a></td>
```

Switch 1 now has the correct graphic for "on," but the weights also have changed.

Switch	1	2	3	4
Binary Value	0000	0011	0101	1001

The binary pattern of the URL reflects the new switch (and lamp) states when that switch is clicked. In this way, you end up with 256 URLs for eight switches — one for each possible state — though this doesn't mean 256 files on a hard disk, since they are purely created on the fly.

Why hasn't this technique found much usage? Only because HTML forms are a neater and more flexible way of doing the same thing.

HTML Forms

An HTML form is a method of aggregating a set of controls under one URL (i.e., one file name on one host) without each control losing its individual identity.

The original purpose of forms was, as the name suggests, to allow the emulation of a fill-in-the-blanks paper form, with text areas, tick boxes, and the like. When the form is filled in, the user clicks a button, like Submit, to send it to the server. The browser then takes the information from all the form's fields, including the button pressed, and sends them to the server in one HTTP request. The server then picks its way through this information to decide what to do next, that is, what page to send in return.

Although I'll adapt the form technique to suit my own ends, it is worth keeping this fundamental model in mind, since it does impose some limitations on what I can do. I'll start with a simple form as its inventors intended.

Figure 8.3   Browser display of form.

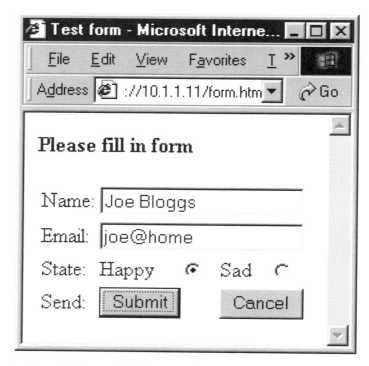

To create the form in Figure 8.3, I've used yet another table, because it lines the elements up with a minimum of effort.

```
<html><head><title>Test form</title></head>
<body><h4>Please fill in form</h4>
<form action="thank.htm">
<table>
<tr><td>Name:</td><td colspan=4><input type=text name=name></td></tr>
<tr><td>Email:</td><td colspan=4><input type=text name=addr></td></tr>
<tr><td>State:</td>
<td>Happy</td><td><input type=radio name=state value=happy checked></td>
<td>Sad</td><td><input type=radio name=state value=sad></td></tr>
<tr><td>Send:</td>
```

```
<td colspan=2><input type=submit name=send value=Submit></td>
<td colspan=2><input type=submit name=send value=Cancel></td></tr>
</table>
</body></html>
```

I can enter text in the boxes and click the radio buttons, and nothing is sent to the server. When I click the Submit button, all the state information is sent in an HTTP request to the server.

```
GET /thank.htm?name=Joe+Bloggs&addr=joe@home&state=happy&send=Submit HTTP/1.0
User-Agent: Mozilla/4.5 [en] (Win95; I)
Pragma: no-cache
Host: 10.1.1.11
```
... and so on ...

The file it has requested is the one named in the `form action` attribute, and it has appended the state information in what is called "URL encoding," strictly speaking, `applica-tion/x-www-form-urlencoded`. The information is separated from the filename by a question mark, the individual items are delimited by ampersands, and any spaces are replaced by plus signs. My simple Web server doesn't understand any of the state information, so it just returns the file requested (Figure 8.4), `thank.htm`.

```
<html><head><title>Form response</title></head>
<body><h3>Thank you</h3>
</body></html>
```

Figure 8.4   Browser display of form response.

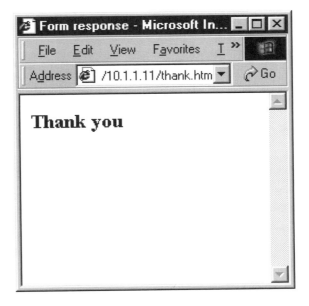

Clearly, it is necessary to add some software to the Web server to process these requests and make more appropriate responses. That software comes under the CGI umbrella.

Standard CGI Interface

So what is CGI? It is a standard method of enhancing a Web server using application-specific software, which can be written in any convenient language. Before creating your own CGI interface, it is worth examining the standard way it is done in a real-world implementation. I'll use the Apache Web server running under Linux for this model, but most mainstream Web servers operate in a similar fashion.

Linux CGI

The common gateway interface is a software module built into the Web server that can execute programs stored on the server's hard disk. These programs are often called scripts because they can be just a string of operating system commands written as a "shell script" (roughly equivalent to a DOS batch file).

Execution

A CGI application remains dormant until the Web server receives a request for a Web page that includes it in the URL, as shown below.

```
GET /cgi-bin/cgitest HTTP/1.0 ...
```

The Web server looks for an executable file called cgitest in its cgi-bin directory. If there are any additional arguments (perhaps because the user has clicked a form on the browser), then these are converted into environment variables. The executable file is then run, with its output being redirected into the TCP connection. Here's a simple CGI program written in C.

```
/* Simple CGI test in 'C' */
#include <stdio.h>
#include <stdlib.h>

void main(int argc, char *argv[])
{
    int i;

    printf("Content-type: text/html\n\n");
    printf("<HTML><HEAD><TITLE>CGI Test</TITLE></HEAD>\n");
    printf("<BODY><H2>CGI test</H2><PRE>\n");
    printf("Arguments:      %u\n", argc);
    for (i=0; i<argc; i++)
        printf("Argument %u:      %s\n", i, argv[i]);
    printf("Query string:   %s\n", getenv("QUERY_STRING"));
```

```
    printf("Remote addr:     %s\n", getenv("REMOTE_ADDR"));
    printf("</PRE></BODY></HTML>\n");
    exit(0);
}
```

You can see how this program creates an HTML page (Figure 8.5) using `printf()` calls that go straight into the TCP connection. I also had it fetch the values of a few environment variables just to show the kind of additional information that is available.

If this program (CGITEST) is compiled and run under a standard Web server (e.g., Apache running on Linux), it can be accessed from a Web client.

Figure 8.5   Browser display of CGI test.

The URL information after the filename is chopped up and fed to the application to simulate a succession of command-line arguments, so the HTTP command

```
GET /cgi-bin/cgitest?Hello+there HTTP/1.0
```

has effectively launched an application on the host with the following command line:

```
cgitest Hello there
```

The output from this program is sent back to the client.

This is really quite a neat remote-control mechanism; typing a suitable URL into the browser not only executes a program on the server, but it also passes additional arguments and displays the response on the browser. It does have its quirks, however. The logic behind some of its argument-parsing decisions does seem rather strange. For example, if you change the client request to look a bit more like the real output from a form,

```
GET /cgi-bin/cgitest?name=Jeremy&addr=At+home
```

the server's CGI front end realizes that this is form output, so it doesn't break it up into arguments. For forms processing, it is assumed that the CGI program will want to parse the string itself (Figure 8.6).

Figure 8.6   Browser display of form response.

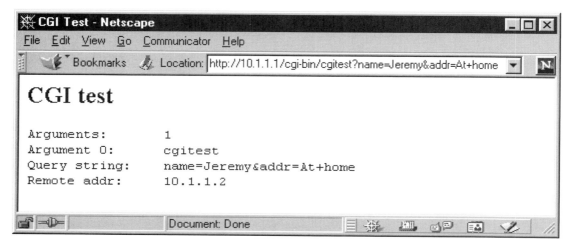

DOS Embedded Gateway Interface (EGI)

The most logical way to create a CGI interface for your Web server is to copy the standard method to run under DOS, Win32, and so on. You could provide the capability to run any DOS executable (including batch files) and emulate the same strange parsing decisions.

In practice, this approach is fraught with problems.

Reentrancy.   The Web server can run several connections simultaneously, each with its own CGI application. Unlike Linux, DOS is not designed to run several applications concurrently. Attempting to do so causes resource conflicts and system lockups.

Multitasking.   A single DOS application will hog the CPU until complete, starving the Web server and the other connections of CPU time.

Speed.   Because of its less sophisticated disk cache, repeatedly executing DOS programs is a slow process.

For these reasons, my CGI interface won't emulate the standard. Instead, I'll create an application-programming interface (API) that allows additional functionality to be integrated into the Web server by the use of callback functions triggered by certain network events. I'll call this an embedded gateway interface (EGI), to distinguish it from a "real" CGI.

What facilities do you require from the embedded gateway interface? As a minimum, it should call an application-specific function when the page is fetched from the server, but it should also allow you to call application-specific functions with the use of embedded tags while a file is being sent out.

Invoking the EGI

The EGI will be invoked every time a client requests a document with an `.egi` extension. The server will then

- look up the document name to find a matching EGI function; if not found, an error message is returned;
- call the EGI function so that it can initialize its workspace;
- set up local variables to match the tags in the document; and
- open a file containing the response to be sent.

Every time an appropriate tag is encountered in the response document, a variable value is substituted; for example, the response to a form could be in the following file:

```
<html><head>
<title>Form response</title></head>
<body>
<h3>To: <!--#$addr--></h3>
Hello, <!--#$name-->.<br>
<!--#$message-->
</body></html>
```

There is a distinct similarity with the standard mail-merge technique used to customize a standard letter for mass mailings. The customer's name and address are indicated by tags, which are replaced by values from a database when the letter is printed. In this case, you're replacing the tags with values from the form fields, so

```
Hello, <!--#$name-->.<br>
```

becomes

```
Hello, Joe Bloggs.<br>
```

Connection Variables

Because the Web server is capable of supporting several transactions (several connections) at the same time, conflicting variable values are possible. Perhaps both Joe Bloggs and Fred Smith are filling in copies of the form at the same time and happen to press the Submit button simultaneously. Which of these two names should the server use in response? The answer is, of course, that the server must respond to each client using the values from that client's form. The variable values are not just specific to the form and the server, they are also specific to the connection.

To emphasize this point, I will refer to the form variables as *connection* variables. They are created when the user submits a completed form to the server and destroyed when the client–server connection is lost (when the complete response has been sent to the client). Each connection must be equipped with a list of variables and their values, which are created from the form's fields.

Connection variables need not necessarily be directly tied into the form's field values. For example, the response file may contain a generic variable, such as message.

```
<h3>To: <!--#$addr--></h3>
Hello, <!--#$name-->.<br>
<!--#$message-->
```

This variable may contain an arbitrary response string, loosely derived from the "happy" or "sad" state variable.

```
I'm glad you are happy.
```

or

```
I'm sorry you are sad.
```

Put in all these variable substitutions, and the original response file

```
<html><head>
<title>Form response</title></head>
<body>
<h3>To: <!--#$addr--></h3>
Hello, <!--#$name-->.<br>
<!--#$message-->
</body></html>
```

is converted into this file (Figure 8.7).

```
<html><head><title>Form response</title></head>
<body><h3>To: joe@home</h3>
Hello, Joe Bloggs.<br>
I'm glad you are happy.
</body></html>
```

Figure 8.7 Browser display of form and response.

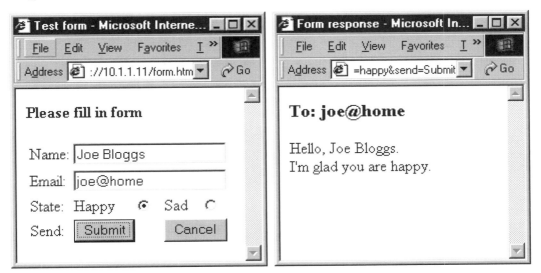

EGI Implementation

The main parts of the embedded gateway interface are the application-specific code and the server support functions.

Application-Specific Code

The code to support the form-handling example is very simple, consisting of only one function.

```
#define RESPFILE     "resp.htm"
#define CANCELFILE   "cancel.htm"
#define HAPPYMSG     "I'm glad you are happy."
#define SADMSG       "I'm sorry you are sad."

/* Function to handle the user's response to the form */
void form_resp(TSOCK *ts, char *str)
{
    APPDATA *adp;
    char *s;

    adp = (APPDATA *)ts->app;
    if (!str)
    {
        adp->egi = form_resp;
        s = !stricmp(get_connvar(ts, "send"), "submit") ? RESPFILE:CANCELFILE;
```

```
        adp->in = url_fopen(s);
        if (!adp->in)
        {
            buff_instr(&ts->txb, HTTP_OK HTTP_TXT HTTP_BLANK);
            buff_inprintf(&ts->txb, "File '%s' not found\n", s);
            close_tcp(ts);
        }
        else
        {
            buff_instr(&ts->txb, HTTP_OK HTTP_HTM HTTP_BLANK);
            s = !stricmp(get_connvar(ts, "state"), "happy") ? HAPPYMSG:SADMSG;
            put_connvar(ts, "message", s);
        }
    }
    else
    {
        printf("EGI Tag '%s'\n", str);
        str += EGI_STARTLEN;
        if (*str == '$')
        {
            s = strtok(str+1, "-");
            buff_instr(&ts->txb, get_connvar(ts, s));
        }
    }
}
```

When the connection is established, this function is called with a null argument. It must then

- set a function callback pointer (in the socket data) to point to itself, ensuring that the function will be called whenever an EGI-specific tag is found in the response file;
- open the response file, return a message to the browser if not found;
- set any other connection variables that are needed for the response file; in this case, check the happy/sad state and set a variable to contain the appropriate response text.

Control is then returned to the server. The handler function regains control only when an EGI-specific tag is encountered in the response file.

```
<!--#tag text-->
```

The handler is then called with the tag as an argument, and it can do one of several things before returning.

Do nothing.   This effectively deletes the tag from the response file.

Copy the tag to the output buffer. This serves little purpose. The tag looks like an HTML comment, so, although it would appear in the HTML code, it would have no effect on the browser display.

Look up a connection variable corresponding to the tag's text, and put its value in the output buffer. This is the most common course of action, so a dollar $ prefix is used to indicate a variable name that should be replaced with its string value.

Any other connection-specific action. For example, another response file could be opened in place of the current one (an include file). The only limits are the available space in the output (Transmit) buffer and the need to return soon-ish to keep the Web server alive.

In this example, the code needs to do only a variable lookup.

Connection-Variable Implementation

The application-specific storage area in each socket needs to be expanded to accommodate a pointer to the EGI handler function and the connection variable storage.

```c
/* Application-specific storage */
#define VARSPACE 1000
typedef struct {
    FILE *in;                     /* File I/P pointer */
    long count;                   /* State variable/counter */
    void (*egi)(TSOCK *ts, char *s);/* Pointer to EGI handler routine */
    int vlen;                     /* Total used length of variable space */
    char vars[VARSPACE];          /* Space for storing connection vars */
} APPDATA;
```

Within this space, I have to maintain a miniature database of variables and their (string) values. This could be done in a variety of ways: I have adopted an extremely simple "flat" technique, where the variable name and value are stored as two consecutive strings. To distinguish between the two, the name is prefixed by a length byte, with the most significant bit set.

```c
/* Add a variable to the connection variable space, return 0 if no room */
int put_connvar(TSOCK *ts, char *name, char *val)
{
    return(put_connvarlen(ts, name, strlen(name), val, strlen(val)));
}


/* Add a variable to the connection variable space, return 0 if no room
** String aren't necessarily null-terminated; they are pointers & lengths */
int put_connvarlen(TSOCK *ts, char *name, int namlen, char *val, int valen)
```

```
{
    int ok=0;
    APPDATA *adp;

    adp = (APPDATA *)ts->app;
    if (adp->vlen+namlen+valen+3<VARSPACE &&
        namlen<MAX_EGINAME && valen<MAX_EGIVAL)
    {
        adp->vars[adp->vlen++] = (char)(namlen | 0x80);
        strncpy(&adp->vars[adp->vlen], name, namlen);
        adp->vlen += namlen;
        adp->vars[adp->vlen++] = 0;
        strncpy(&adp->vars[adp->vlen], val, valen);
        adp->vlen += valen;
        adp->vars[adp->vlen++] = 0;
    }
    return(ok);
}
```

To retrieve the value of a given variable, search for the appropriate-length byte; then check the string after it. If they match, the next string is the variable's value.

```
/* Get variable from the connection space, return null string if not found */
char *get_connvar(TSOCK *ts, char *name)
{
    int n, len;
    APPDATA *adp;
    char *s=0, *end;

    adp = (APPDATA *)ts->app;
    end = &adp->vars[adp->vlen];
    n = strlen(name);
    if (n < MAX_EGINAME)
    {
        s = memchr(adp->vars, (char)(n | 0x80), adp->vlen-3);
        while (s && strncmp(s+1, name, n) && (len=(int)(end-s))>3)
            s = memchr(s+1, (char)(n | 0x80), len);
    }
    return(s ? s+n+2 : "");
}
```

URL Decoding

The URL obtained from the browser contains a large amount of form state information in a compact format. To eliminate the possibility of it being corrupted in transit between browser and server, it has escape sequences for various characters.

The following characters are reserved in URL encoding and do not need to be preceded by the escape code (escaped) if they have their conventional meaning.

   ; / ? : @ = &

Other characters are deemed unsafe (prone to alteration), so they need to be escaped.

   < > " # % [ ] { } | \ ^ ~ `

The escape code is the percent % character, followed by a two-digit hexadecimal value. So the original string

```
[test string]
```

becomes the following string.

```
%5Btest+string%5D
```

The + character is used in place of a space to avoid any unsafe gaps in the text.

The software has to decode these escape characters and extract the state variable values.

```c
/* Get connection variable values from URL */
void url_connvars(TSOCK *ts, char *str)
{
    int nlen, vlen, n;

    if (*str == '/')
        str++;
    vlen = strcspn(str, " .?");
    put_connvarlen(ts, "fname", 5, str, vlen);
    str += vlen;
    if (*str == '.')
    {
        str++;
        vlen = strcspn(str, " ?");
        put_connvarlen(ts, "fext", 4, str, vlen);
        str += vlen;
    }
    if (*str++ == '?')
    {
        while ((nlen=strcspn(str, "="))!=0 && (vlen=strcspn(str+nlen, "&"))!=0)
        {
            n = url_decode(str+nlen+1, vlen-1);
```

```
            put_connvarlen(ts, str, nlen, str+nlen+1, n);
            str += nlen + vlen;
            if (*str == '&')
                str++;
        }
    }
}


/* Decode a URL-encoded string (with length), return new length */
int url_decode(char *str, int len)
{
    int n=0, d=0;
    char c, c2, *dest;

    dest = str;
    while (n < len)
    {
        if ((c = str[n++]) == '+')
            c = ' ';
        else if (c=='%' && n+2<=len &&
                isxdigit(c=str[n++]) && isxdigit(c2=str[n++]))
        {
            c = c<='9' ? (c-'0')*16 : (toupper(c)-'A'+10)*16;
            c += c2<='9' ? c2-'0' : toupper(c2)-'A'+10;
        }
        dest[d++] = c;
    }
    return(d);
}
```

As an example, the URL

```
GET /form_resp.egi?name=Joe+Bloggs&addr=joe@home&state=happy&send=Submit
```

is converted into the variables in Table 8.1.

Table 8.1   URL decoding.

Variable	Value
fname	form_resp
fext	egi
name	Joe Bloggs
addr	joe@home
state	happy
send	submit

Finding the EGI Handler

Having decoded the URL, the Web server needs to find the correct handler function. Again, there are various ways of achieving this. I have opted for a simple table lookup.

```
void form_resp(TSOCK *ts, char *str);
#define EGI_FORM_FUNCS {form_resp, "form_resp"}

/* Structure to store an EGI handler function */
typedef struct
{
    void (*func)(TSOCK *ts, char *str);
    char name[MAX_EGINAME];
} EGI_FUNC;

EGI_FUNC egifuncs[] = {EGI_FORM_FUNCS, ..., {0}};

/* Execute the EGI function corresponding to a string, return 0 if not found */
int egi_execstr(TSOCK *ts, char *str)
{
    int ok=0, n=0;

    while (egifuncs[n].func && !(ok=!stricmp(str, egifuncs[n].name)))
        n++;
    if (ok)
        egifuncs[n].func(ts, 0);
    return(ok);
}
```

Each handler function has a table entry with the function pointer and the corresponding string. A simple linear search is performed to match the EGI filename to one of these names, and an error message is returned to the browser if none match.

```
/* Process the filepath from an HTTP 'get' method */
void http_get(TSOCK *ts, char *fname)
{
    APPDATA *adp;
    char *s=0;

    adp = (APPDATA *)ts->app;
    if (strstr(fname, EGI_EXT))              /* Find EGI handler */
    {
        url_connvars(ts, fname);
        if (!egi_execstr(ts, s=get_connvar(ts, "fname")))
        {
            buff_instr(&ts->txb, HTTP_NOFILE HTTP_TXT HTTP_BLANK);
            buff_inprintf(&ts->txb, "Can't find EGI function '%s'\r\n", s);
            close_tcp(ts);
        }
    }                                        /* If file not found.. */
    else if ((adp->in = url_fopen(fname)) == 0)
    {                                        /* ..send message */
        if (webdebug)                        /* Display if debugging */
            printf("File '%s' not found\n", fname);
        buff_instr(&ts->txb, HTTP_NOFILE HTTP_TXT HTTP_BLANK);
        buff_inprintf(&ts->txb, "Can't find '%s'\r\n", fname);
        close_tcp(ts);
    }
    else                                     /* File found OK */
    {
        strlwr(fname);
        buff_instr(&ts->txb, HTTP_OK);
        s = strstr(fname, ".htm") ? HTTP_HTM :
            strstr(fname, EGI_EXT)? HTTP_HTM :
            strstr(fname, ".txt") ? HTTP_TXT :
            strstr(fname, ".gif") ? HTTP_GIF :
            strstr(fname, ".xbm") ? HTTP_XBM : "";
        buff_instr(&ts->txb, s);
        buff_instr(&ts->txb, HTTP_BLANK);
```

```
            if (webdebug)                       /* Display if debugging */
                printf("File '%s' %s", fname, *s ? s : "\n");
        }
    }
```

The http_get() function from the old Web server has been modified to include the ability to process EGI URLs. The http_data() function also needs to be modified to call the handler when an EGI tag is encountered.

Trapping EGI Tags

When sending out an EGI response file, http_data() must isolate all EGI tags of the following form.

```
<!--#tag text-->
```

The tags are not to be transmitted but must be passed to the application-specific handler function for processing. This involves some slightly tricky coding, since you can't guarantee that the tag will come at a convenient point in the file. If you're sending the file out in blocks, the tag text, or the tag identifier, may straddle the boundary between two such blocks. Also, the file pointer must be positioned so that the next character to be sent is the one immediately after the tag. Finally, you must also ensure that there is a reasonable amount of space available in the Transmit buffer at the time the handler is called. There is no point in calling the handler if there is no room in the Transmit buffer for outgoing text.

```c
#define EGI_BUFFLEN 1024
char egibuff[EGI_BUFFLEN+1];

/* If there is space in the transmit buffer, send HTTP data */
void http_data(TSOCK *ts)
{
    APPDATA *adp;
    int len, n;
    char *start, *end;

    adp = (APPDATA *)ts->app;
    if (adp->in && (len = buff_freelen(&ts->txb)) >= EGI_BUFFLEN)
    {
        if (adp->egi)                        /* If EGI is active.. */
        {
            len = fread(egibuff, 1, EGI_BUFFLEN, adp->in);
            egibuff[len] = 0;
            if (len <= 0)
            {                                /* If end of file, close it.. */
                fclose(adp->in);
```

```
                    adp->in = 0;
                    close_tcp(ts);              /* ..and start closing connection */
            }
        else
        {                                        /* Check for start of EGI tag */
            if ((start = strstr(egibuff, EGI_STARTS)) == 0)
                start = strchr(&egibuff[len-EGI_STARTLEN], '<');
            if (start==egibuff && (end=strstr(egibuff, EGI_ENDS))!=0)
            {                                    /* If tag is at start of buffer.. */
                n = (int)(end - start) + sizeof(EGI_ENDS) - 1;
                fseek(adp->in, n-len, SEEK_CUR);
                egibuff[n] = 0;        /* ..call handler */
                adp->egi(ts, egibuff);
                len = 0;
            }                                    /* If tag not at start of buffer.. */
            else if (start)
            {                                    /* ..send file up to tag */
                n = (int)(start - egibuff);
                fseek(adp->in, n-len, SEEK_CUR);
                len = n;
            }
            if (len > 0)                /* Send next chunk of file */
                buff_in(&ts->txb, (BYTE *)egibuff, (WORD)len);
        }
    }
    else if (buff_infile(&ts->txb, adp->in, (WORD)len) == 0)
    {                                    /* If end of file, close it.. */
        fclose(adp->in);
        adp->in = 0;
        close_tcp(ts);                   /* ..and start closing connection */
    }
  }
}
```

The code has a 1Kb buffer for storing the next block of the response file. If there is at least 1Kb of space in the Transmit buffer, the next block is loaded and, if EGI is active on this transfer, checked for EGI tags. If the tag is in the middle of the buffer, the preceding text is put out, so that next the time http_data() is called, it will be conveniently at the front of the temporary buffer. The operating system file pointer is moved (using seek) to eliminate the tag and carry out the file transfer immediately after it.

Interactive Switches and Lamps

Having created a framework for processing HTML forms, I can return to the switch-and-lamp example mentioned earlier. Having created an array of eight switches and eight lamps, I need to encapsulate them in an HTML form so that clicking on a switch activates the corresponding lamp.

HTML Page

Each lamp and switch will be animated (i.e., created when the HTML page is requested) using EGI variable substitution, so the HTML page looks pretty simple.

```
<html>
<head><title>Switches and LEDs</title></head>
<body><h4>Click switch</h4>
<form action="switchfm.egi">
<table>
<tr align=center>
<td><!--#$led1--></td><td><!--#$led2--></td>
<td><!--#$led3--></td><td><!--#$led4--></td>
<td><!--#$led5--></td><td><!--#$led6--></td>
<td><!--#$led7--></td><td><!--#$led8--></td>
</tr><tr align=center>
<td><!--#$switch1--></td><td><!--#$switch2--></td>
<td><!--#$switch3--></td><td><!--#$switch4--></td>
<td><!--#$switch5--></td><td><!--#$switch6--></td>
<td><!--#$switch7--></td><td><!--#$switch8--></td>
</tr></table>
</form>
</body></html>
```

The EGI server will change each of the LED variables into an "off" or "on" LED image.

```
<img src='ledoff.gif'>
<img src='ledon.gif'>
```

The switches will also be changed into an "up" or "down" graphic, so you might think this would be of the following form:

```
<img src='switchu.gif'>
<img src='switchd.gif'>
```

However, this wouldn't make the switch clickable, so I must make the switch into an "image input element" (clickable graphic) instead. The `name` must be unique for each switch so that it can be identified in the form response.

```
<input type=image name='switch1' src='switchu.gif'>
<input type=image name='switch1' src='switchd.gif'>
```

EGI Code

The switch-handler function is similar in concept to my previous form handler. The main difference is that instead of having separate pages for the form and its response, the one page serves both purposes. This means that clicking a switch will cause the current page to be redisplayed, rather than a new page being fetched.

When first called, the EGI code has to

- open the response file and send an error message if not found and
- check for any switch that has been clicked and toggle the appropriate bit.

When an EGI tag is found, it must
- (if a switch) output the HTML code for that switch's state or
- (if an LED) output the HTML code for that LED's state.

To simplify processing the on/off values, the eight switch bits have been aggregated into a single BYTE value, `ledstate`.

```
#define LEDON     "<img src='ledon.gif'>"
#define LEDOFF    "<img src='ledoff.gif'>"
#define SWOFF     "<input type=image name='switch%u' src='switchu.gif'>"
#define SWON      "<input type=image name='switch%u' src='switchd.gif'>"

#define RESPFILE  "switchfm.htm"

BYTE ledstate;

/* Function to handle the user's response to switch clicks */
void switchfm_resp(TSOCK *ts, char *str)
{
    APPDATA *adp;
    char *s;
    int bit, mask;

    adp = (APPDATA *)ts->app;
    if (!str)
    {
        adp->egi = switchfm_resp;
```

```
        adp->in = url_fopen(RESPFILE);
        if (!adp->in)
        {
            buff_instr(&ts->txb, HTTP_OK HTTP_TXT HTTP_BLANK);
            buff_inprintf(&ts->txb, "File '%s' not found\n", RESPFILE);
            close_tcp(ts);
        }
        else if (*(s = find_connvar(ts, "switch"))!=0)
        {
            bit = *(s+6) - '1';
            mask = 1 << bit;
            ledstate ^= mask;
        }
    }
    else
    {
        printf("EGI Tag '%s'\n", str);
        str += EGI_STARTLEN;
        if (!strncmp(str, "$led", 4) && isdigit(*(str+4)))
        {
            bit = *(str+4) - '1';
            mask = 1 << bit;
            buff_instr(&ts->txb, (ledstate & mask) ? LEDON : LEDOFF);
        }
        if (!strncmp(str, "$switch", 7) && isdigit(*(str+7)))
        {
            bit = *(str+7) - '1';
            mask = 1 << bit;
            s = (ledstate & mask) ? SWON : SWOFF;
            buff_inprintf(&ts->txb, s, bit+1);
        }
    }
}
```

The switch-handler function must be added to the main function table so that the server knows which function to call when the user requests switchfm.egi.

```
#define EGI_SWIT_FUNCS {switchfm_resp, "switchfm"}
EGI_FUNC egifuncs[] = {EGI_FORM_FUNCS, EGI_SWIT_FUNCS, ..., {0}};
```

Client Display

The resulting display is usable for remote control of outputs (Figure 8.8). The first time the graphics are fetched, there is a noticeable delay, but after that, the speed of operation is limited only by the network transit time of the (very small) page, since the graphics are held in the browser's cache.

Figure 8.8 **Browser switch and LED display.**

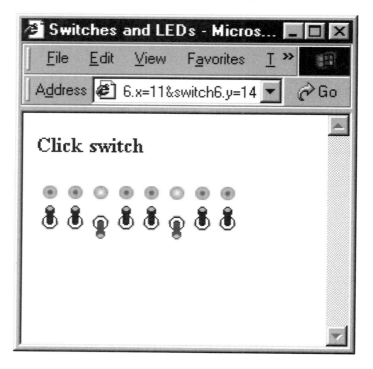

Switch State Storage

In creating the above code, I decided to store the LED and switch state in a global variable, on the assumption that this would be linked to some real-world I/O pins.

Consider what would happen if two or more users were to access the switch–LED page simultaneously and start clicking the switches. The result would be very confusing. For each switch clicked, more than one would appear to change. It might seem better to have one copy of the switch state variable for each user, but how do you map two or more switch state variables onto one set of hardware?

In a real application, the best solution is generally to allow multiple users to browse the server's diagnostic data, but allow only one user at a time to control the I/O. This can conveniently be achieved by asking the user to log in to the system via a form interface, with a password check to keep out unauthorized users. If a previous user is still logged in, access is denied. A wise precaution is to add a timed automatic logout in case the user forgets to do so.

Interactive Analog Controls

The EGI form interface handles strings, and the switch–LED interface handles Boolean on/off values, so now you need to look at the representation of analog values.

Studying the diagnostic messages from the previous example gives a clue as to how an analog interface might be implemented. Clicking switch 1 results in the following browser request.

```
GET /switchfm.egi?switch1.x=7&switch1.y=14 HTTP ...
```

The client hasn't identified just the graphical item that has been clicked, it has also identified the exact position of the cursor when it was clicked. In this case, the cursor was seven pixels in from the left-hand edge of the graphic and 14 pixels down from its top. You can use that information to add interaction to the analog indicators developed in the previous chapter (Figure 8.9).

Figure 8.9   Browser display of analog sliders.

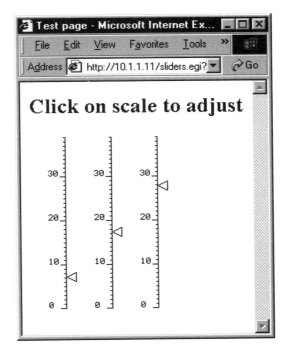

HTML Page

The HTML page to create these sliders is more complicated than in the previous examples, but it still relies on variable value substitution when the file is sent out.

```
<html>
<head><title>Test page</title></head>
<body><h2>Click on scale to adjust</h2>
<form action="sliders.egi">
```

```
<table border=0 cellpadding=0 cellspacing=0>
<tr align=center valign=bottom>

<td><img src="clr.gif" height=1 width=20></td>
<td><input type=image src="v40_200f.gif" name=scale1><br>
<img src="clr.gif" height=10 width=10></td>
<td><img src="lptr.gif"><br><!--#$ptr1--></td>

<td><img src="clr.gif" height=1 width=20></td>
<td><input type=image src="v40_200f.gif" name=scale2><br>
<img src="clr.gif" height=10 width=20></td>
<td><img src="lptr.gif"><br><!--#$ptr2--></td>

<td><img src="clr.gif" height=1 width=20></td>
<td><input type=image src="v40_200f.gif" name=scale3><br>
<img src="clr.gif" height=10 width=10></td>
<td><img src="lptr.gif"><br><!--#$ptr3--></td>


</tr></table>
</form>
</body></html>
```

Each indicator occupies three columns of the table.

1. A blank column for spacing (needs to be only one pixel high).
2. The scale, put on top of a transparent block (so that the pointer can reach zero).
3. The pointer, on top of a variable-height transparent block.

The variables ptr1 to ptr3, which control the height of the transparent block beneath each arrow pointer, are substituted with the following reference, for example:

```
<img src='clr.gif' height=41 width=10>
```

The value of 41 pixels corresponds to 8.2 scale units, which is the left-hand pointer in the previous browser display.

You will note that the clickable item is the scale and not the pointer.

```
<input type=image src="v40_200f.gif" name=scale1>
```

This is somewhat counter-intuitive. Not only can't you slide this slider (HTML doesn't support drag-and-drop), but also you have to click the scale adjacent to the pointer, rather than the pointer itself. It is possible to stack another transparent block on top of the pointer, make the blocks and pointer clickable, and compute the cursor offsets relative to each block and, hence, the desired value. That has been left, as they irritatingly say in textbooks, as an exercise for the reader.

EGI Code

The only additional complication is the determination of the cursor y-position when the scale is clicked. For this, you need to search the local variables for a variable `scalen`, using a partial matching scheme.

```
/* Find variable in the connection space, by matching first few chars
** Return full name string, null string if not found */
char *find_connvar(TSOCK *ts, char *name)
{
    int n;
    APPDATA *adp;
    char *s=0, *end;

    adp = (APPDATA *)ts->app;
    end = &adp->vars[adp->vlen];
    n = strlen(name);
    if (n < MAX_EGINAME)
    {
        s = adp->vars;
        while (*s && strncmp(s+1, name, n) && (int)(end-s)>2)
        {
            do {
                s++;
            } while (s<end-2 && !(*s & 0x80));
        }
    }
    return(*s & 0x80 ? s+1 : "");
}
```

Once found, the variable y-value can be determined, not forgetting that the origin is the top left corner of the image, so the bottom of the image has the maximum value.

```
#define PTR_STR     "<img src='clr.gif' height=%u width=10>"
#define PTR_OSET    5
#define PTR_MAX     200
#define NSLIDERS    3

#define RESPFILE    "sliders.htm"

int slidervals[NSLIDERS];
```

```c
/* Function to handle the user's response to slider clicks */
void sliders_resp(TSOCK *ts, char *str)
{
    APPDATA *adp;
    char *s, temps[10];
    int idx, val;

    adp = (APPDATA *)ts->app;
    if (!str)
    {
        adp->egi = sliders_resp;
        adp->in = url_fopen(RESPFILE);
        if (!adp->in)
        {
            buff_instr(&ts->txb, HTTP_OK HTTP_TXT HTTP_BLANK);
            buff_inprintf(&ts->txb, "File '%s' not found\n", RESPFILE);
            close_tcp(ts);
        }
        else if (*(s = find_connvar(ts, "scale"))!=0)
        {
            idx = *(s+5) - '1';
            sprintf(temps, "scale%u.y", idx+1);
            s = get_connvar(ts, temps);
            printf("Var %s val %s\n", temps, s);
            val = PTR_MAX - atoi(s);
            if (idx>=0 && idx<=3)
                slidervals[idx] = val;
        }
    }
    else
    {
        printf("EGI Tag '%s'\n", str);
        str += EGI_STARTLEN;
        if (!strncmp(str, "$ptr", 4) && *(str+4)>='1' && *(str+4)<='3')
        {
            idx = *(str+4) - '1';
            buff_inprintf(&ts->txb, PTR_STR, slidervals[idx]+PTR_OSET);
        }
    }
}
```

Again, the current slider values are stored in global variables on the assumption they will be mapped onto physical analog outputs.

The slider-function callback is added to the global list of handlers.

```
void sliders_resp(TSOCK *ts, char *str);
#define EGI_SLIDER_FUNCS {sliders_resp, "sliders"}


EGI_FUNC egifuncs[] = {EGI_FORM_FUNCS, EGI_SWIT_FUNCS, EGI_SLIDER_FUNCS, {0}};
```

Summary

I have looked at a variety of Web pages that are applicable to embedded systems and explored the techniques for inserting live data into them. The standard technique, common gateway interface (CGI), is not best suited for the task because the programmer is forced to create the entire page using a succession of print statements, rather than by updating the values in an existing page to incorporate the desired data.

I have named my alternative the embedded gateway interface (EGI), which simplifies the decoding of HTML form data and allows the insertion of dynamic data into existing pages or the creation of new pages from scratch. I gave various examples of the EGI technique, and I created a Web server incorporating these concepts.

Source Files

ether3c.c	3C509 Ethernet card driver
etherne.c	NE2000 Ethernet card driver
ip.c	Low-level TCP/IP functions
net.c	Network interface functions
netutil.c	Network utility functions
pktd.c	Packet driver (BC only)
router.c	Router utility
serpc.c or serwin.c	Serial drivers (BC or VC)
tcp.c	TCP functions
web_egi.c	EGI Web server utility
dosdef.h	MS-DOS definitions (BC only)
ether.h	Ethernet definitions
ip.h	TCP/IP definitions
net.h	Network driver definitions
netutil.h	Utility function and general frame definitions
serpc.h	Serial driver definitions (BC or VC)
tcp.c	TCP function definitions
win32def.h	Win32 definitions (VC only)

```
egi_form.c                      EGI function:form response
egi_form.h
egi_slid.c                                    slider response
egi_slid.h
egi_swit.c                                    switch response
egi_swit.h
```

WEB_EGI **Utility**

Utility:	Web server with embedded gateway interface	
Usage:	`web_egi [options] [directory]`	
	Uses directory `.\webdocs` if none given	
Options:	`-c name`	Configuration filename (default `tcplean.cfg`)
	`-s`	State display mode
	`-t`	TCP segment display mode
	`-v`	Verbose display mode
	`-w`	Web (HTTP) display mode
	`-x`	Hex packet display
Example:	`web_egi -v -w c:\temp\webdocs`	
Keys:	Ctrl-C or Esc to exit	
Config:	`net`	to identify network type
	`ip`	to identify IP address
	`mask`	optional subnet mask
	`gate`	optional gateway IP address

9

Chapter 9

Miniature Web Server Design

Overview

The following three chapters are an exercise in miniaturization — how small can a Web server get? I already know the answer: there is a server on the Web that claims to use only 256 bytes of read-only memory (ROM) for its TCP stack, but I can't help thinking that this must be a highly optimized chunk of assembly language, which is very difficult to adapt for any practical use. My objective is to create a miniature Web server in C that is still potentially useful, in that it can monitor and control real-world devices connected to the system's I/O lines.

There is little point in performing this exercise on the PC platform I've used in the previous chapters. I will have to switch to a smaller and cheaper system based on a single-chip computer (a microcontroller). The minimal resources of a microcontroller (ROM and RAM) will force you to look very carefully at the protocol-handling techniques already discussed and will challenge a lot of the comfortable assumptions I've been making over the past few chapters.

To do justice to the subject, it will take three chapters. In this chapter, I discuss the characteristics of microcontrollers and the techniques that can be used to miniaturize the TCP stack. In Chapter 10, I look at the hardware and TCP implementation, and in Chapter 11, I describe the Web server and the techniques used to insert dynamic data into its pages.

I'm also using these chapters as an opportunity to revisit the fundamental concepts of a Web server and TCP/IP protocols. A lot of ground has been covered since I looked at the IP datagram format, for example, so a little revision won't be amiss.

Microcontroller Software Development

A microcontroller is a computer on a chip and is designed for high-volume, low-cost applications. Microcontroller implementations and their associated software-development tools attempt to squeeze the maximum capability from the simplest of hardware. To do this, certain compromises have to be made, which makes the creation of *standard* software tools very difficult or impossible. The C compiler writer faces two choices: either stick to the standards, regardless, or adapt the C programming environment to reflect the hardware constraints. The former approach, although superficially attractive, can stretch the humble resources of the microcontroller to the limit. A few lines of standard code can use up a disproportionately large amount of the microcontroller's meagre resources. For this reason, most development environments are highly specific to the microcontroller being used.

To fit TCP into a microcontroller, you'll have to adapt your thinking to reflect the quirks of the development tools and the limitations of the hardware. Regrettably, the generic source-code approach must be sacrificed for the greater good of miniaturization. The software in the next few chapters must be tied firmly to a specific microcontroller and its associated development environment. That's not to say that the techniques aren't equally applicable to similar devices; it's just that there would be a significant amount of work to adapt, or port, the code to another device.

Hardware

Which microcontroller to choose? There is a bewildering array of possibilities, but I have adopted the following criteria:

1. Low cost and widespread availability
2. Reasonable ROM and RAM size
3. Small package size (footprint); low component count
4. Easily (re)programmed
5. Low-cost development tools with high-quality C compiler

Item 1 is easy to achieve. A few dollars buys a capable microcontroller nowadays, but how much ROM and RAM is considered reasonable? There seems to be a direct conflict between items 2 and 3. To get a few kilobytes of on-chip RAM, you must buy a device with at least forty legs, most of which are I/O pins you won't need. Item 5 is also difficult. Some low-cost and public domain compilers don't support structures and unions (which would make life difficult) or are somewhat erratic in their code generation (which would make life extremely difficult). There are good-quality cross-compilers for a thousand dollars or more, but this is an unrealistic price for an experimental project.

It's at this stage that personal preference comes in. It happens that I have done a large amount of work with the PICmicro® MCUs produced by Microchip Technology Inc. The PIC16C76 seems to be the best match to my criteria.

1. Available for less than $10.00 in one-off quantities
2. Has 8K words of ROM (program memory), and 368 bytes of RAM
3. Is in a narrow 28-pin package
4. Is available with in-circuit (flash) programmable memory
5. Has a low-cost in-circuit debugger and C compiler

The ROM size sounds reasonable, but the RAM size is rather too small for comfort. To make matters worse, it is segmented, so the maximum size of any one structure or array is 96 bytes. I'll need to investigate new coding techniques to fit my software into this minimal environment.

PIC16C76/16F876

The PIC16C76 has the following on-chip hardware:

- 8K × 14-bit words of EPROM program memory
- 368 bytes of RAM data memory
- eight-level hardware stack
- interrupt capability
- eight-bit analog-to-digital converter (ADC) with input multiplexer
- one 16-bit and two eight-bit timers
- two capture/compare/PWM (pulse-width modulation) modules
- synchronous serial port (SSP)
- universal synchronous/asynchronous receiver–transmitter (USART)
- 22 input/output pins (parallel I/O shared with the above functions)

The PIC16F876 is essentially the same as the PIC16C76, but with flash memory in place of the EPROM. This is more convenient because it can be programmed and erased in-circuit.

The memory architecture (Figure 9.1) will seem fairly strange to a PC programmer. There are three completely separate memory spaces:

- read-only program memory
- read/write data memory
- dedicated hardware stack

It must be stressed that these spaces are absolutely separate and are accessed using completely different addressing schemes.

Figure 9.1   Block diagram of PICmicro.

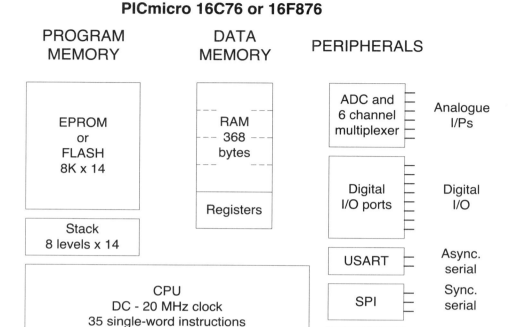

PICmicro 16C76 or 16F876

Program Memory

Program memory is 14 bits wide so that every instruction can fit in a single program word and take a single CPU cycle (four external clock cycles) to execute, with the exception of branches, which take two CPU cycles. The program memory contains only program instructions — no data, not even constant data because there is no mechanism to access it. The program memory is segmented into four banks of 2Kb words each. There are specific bit-manipulation operations to switch between banks, which are automatically inserted by the compiler with the limitation that functions may not straddle two banks.

The inability to put constant data in the program space (or, more specifically, the inability of the CPU to read any such data) makes it difficult to store large amounts of constant data such as constant strings. There is a "return with a byte value" instruction, which has to be used repeatedly to form the string from a series of character-value instructions. Such strings have awkward properties: they can't be accessed by pointers and must be copied into RAM before use. Although the compiler provides some support for this, it is very easy for mistakes to go undetected, producing garbage instead of the desired string. It is best to avoid string constants wherever possible.

Data Memory

The data-memory space is eight bits wide and is shared between the I/O registers and the workspace RAM. Like the program memory, it is segmented into four banks, but the bank-switching uses a separate set of control bits. The workspace RAM occupies the memory locations that aren't taken up by I/O registers (of which there are a large number), so the 368 bytes of RAM is fragmented into the following address ranges.

Bank 0: 96 bytes 20h–7Fh
Bank 1: 80 bytes A0h–EFh
Bank 2: 96 bytes 110h–16Fh
Bank 3: 96 bytes 190h–1EFh

In addition, the following areas are common to all banks. Data written in one bank can be read in all others.

70h–7Fh
F0h–FFh
170h–17Fh
1F0h–1FFh

It is fortunate that you don't have to work your way around this strange map. You can off-load the job onto the compiler; however, you can confuse the compiler into generating wrong code, so it is best to be cautious in the use of data pointers and avoid them wherever possible.

Hardware Stack

There is a hardware stack for machine code calls and returns. It is 13 bits wide (to accommodate the full 8Kb address range) and eight levels deep. There is no provision for extending the stack into RAM or detecting overruns. Nesting function calls more than eight deep will have unforeseen consequences, though this should be detected at compile time.

Data values cannot be stored in the stack, so how are function arguments and local variables handled? The compiler assigns fixed locations in RAM for these variables, having carefully assessed the function nesting to ensure that one function won't destroy another's variables.

External Memory

A Web server needs ample storage for Web pages, and the on-chip ROM is clearly inadequate for this. In Chapter 10, I'll look at how to add external memory for Web page storage.

Network Interface

It is possible to use the spare I/O lines of a PICmicro as an eight-bit data bus to drive an Ethernet controller, but this is too complicated to include in a book that is supposedly about software. Mindful of the large numbers of laptop and Palmtop computers that have an infrared interface and the fragility of subminiature serial connectors, I decided to implement an infrared interface using the IrDA (infrared data association) standards for low-level communications.

After considerable work, it became clear that the IrDA protocols weren't as low-level as I had thought, and the simple task of sending IP datagrams over an infrared link demanded a very significant additional programming effort, which threatened to eclipse the rest of the project. In view of this, I decided to revert to an RS232 SLIP interface, which may be an interface of the lowest common denominator, but it is still very useful for a wide range of applications.

The PICmicro has an on-chip asynchronous serial interface (USART), so the only extra network components are the level shifters for the RS232 voltage levels. To allow modem interfacing, a three-wire interface is implemented using a general-purpose output line for the output handshake.

Development Environment

Compiler

When I referred to the existence of a sensibly priced C compiler for the PICmicro, I was referring to the PCM compiler from Custom Computer Services. As part of its key features, it

* handles the vagaries of PICmicro program and data memory addressing;
* supports structures and unions;
* provides simple interface for the I^{2}C (inter-IC) bus I'll be using in Chapter 10;
* supports serial port I/O, including formatted printing;
* has straightforward DOS command-line and IDE interfaces (the PCW product has a Windows interface); and
* can be integrated into an emulator-based development environment.

It is not without its peculiarities, which I'll explain when I implement the code. Possibly most surprising is the absence of a linker. The whole project must be contained within one `main()` file. The only way to cope with large projects is to move some of the source code into one or more `#include` files. This isn't too much of a problem in practice because the use of a microcontroller imposes severe limits on the project size anyway.

Emulator or Debugger

Until recently, there has been an implicit assumption that if you are developing microcontroller software, it must be for a mass market, so your company can afford to spend a lot of money on a processor emulator, which replaces the microcontroller with a more sophisticated version (called a "bond-out" chip) that allows an external computer to control and monitor its execution, browse or modify its memory, and even trace its operation while it runs at full speed.

It is true that an emulator makes the software developer's life a lot easier, and I confess to using the excellent RICE17A emulator from Advanced Transdata Corporation to develop the code for this book. For initial experimentation, it is worth investigating the lower-cost alternatives, such as the PIC-ICD in-circuit debugger from the same company. By accepting a simpler debug capability and a longer download time, there is a very significant cost saving.

Whatever the development hardware, it is very important that it integrates well with the chosen compiler. My only experience is with the Advanced Transdata products, which work very well with the Custom Computer Services compiler. An integrated Windows environment provides a text editor, single-key compile and download, and complete control and monitoring of the target CPU.

Software Techniques

Having discussed the limitations of microcontrollers, you need to work out how you can squeeze the complete TCP/IP stack into one of them.

RAM Limitation

The most acute problem, by far, is the lack of RAM. All the code so far has made the assumption that the incoming and outgoing frames are stored in RAM, and structures are overlaid onto this RAM so that specific values can be checked, read, or modified.

```
...
ip = getframe_datap(gfp);            /* Get pointer to IP frame */
ver = ip->i.vhl >> 4;                /* Get IP version & hdr len */
hlen = (ip->i.vhl & 0xf) << 2;
sum = ~csum(&ip->i, (WORD)hlen);     /* Do checksum */
...
```

The only way to cram these structures into 368 bytes of RAM is to severely restrict the maximum frame size you can send or receive. At best, you might be able to accommodate a 128-byte frame, albeit fragmented because you have a maximum of 96 bytes of contiguous memory. Unfortunately, the TCP and IP headers will occupy at least 40 bytes of this frame, so the net data throughput will be poor. I know because I've tried it.

To achieve a good response time, the incoming frame must be decoded on the fly as it is received, and the outgoing frame must be prepared on the fly as it is transmitted.

Creating Protocols on the Fly

Consider the following code fragment.

```
/* ***** IP (Internet Protocol) header ***** */
typedef struct
{
    BYTE  vhl,             /* Version and header len */
          service;         /* Quality of IP service */
    WORD  len,             /* Total len of IP datagram */
          ident,           /* Identification value */
          frags;           /* Flags & fragment offset */
    BYTE  ttl,             /* Time to live */
          pcol;            /* Protocol used in data area */
    WORD  check;           /* Header checksum */
    LWORD sip,             /* IP source addr */
          dip;             /* IP dest addr */
} IPHDR;
...
```

```
ip->i.vhl = 0x40+(sizeof(IPHDR)>>2);/* Version 4, header len 5 LWORDs */
ip->i.service = 0;                   /* Routine message */
ip->i.len = len + sizeof(IPHDR);     /* Data length */
...
```

It isn't too hard to imagine the same data created on the fly as it is transmitted using, for example, the following code.

```
put_byte(0x40+(sizeof(IPHDR)>>2));  /* Version 4, header len 5 LWORDs */
put_byte(0);                         /* Routine message */
put_word(len + sizeof(IPHDR));       /* Data length */
```

The put_ functions insert the necessary SLIP escape sequences and send the values to the serial port. The abolition of the structure reduces RAM consumption significantly and also reduces the code size slightly. The complicated indexed-addressing assignment operation is replaced by a single-word function call. Unfortunately, the new code is harder to debug because there isn't a convenient memory image to browse when you want to check the last frame sent. To compensate, you may have to employ the services of a protocol analyzer to monitor the external communications when debugging.

Checksums

Another problem occurs with the calculation of IP and TCP checksums. In the past, I've scanned the memory image to compute these. Now I don't have a memory image to scan, so I'll have to get the put_ functions to do the job.

```
/* Send a byte out to the SLIP link, then add to checksum */
void put_byte(BYTE b)
{
    putchar(b);
    check_byte(b);
}
```

It would be really helpful if the checksum were the last word transmitted, because then you could just send the calculated value as part of the end-of-frame sequence.

```
put_byte(0x40+(sizeof(IPHDR)>>2));  /* Version 4, header len 5 LWORDs */
put_byte(0);                         /* Routine message */
put_word(len + sizeof(IPHDR));       /* Data length */
...
put_word(sum);                       /* Checksum value */
```

Unfortunately, there are two checksum values to be computed (TCP and IP), and they are toward the end of their respective headers, where the data to be checked hasn't been transmitted yet. Then there is the question of how to include the IP pseudoheader in the checksum — clearly, a more sophisticated checksum technique is needed. I will return to this subject later.

Reception

Just as the transmit structures can be replaced by a string of function calls, so can those used for receive. Rather than store the complete frame then use structure references to pick out the elements of interest, the frame is scanned on input, storing only the important information, for example, the following code scans the start of the IP header.

```
    if (match_byte(0x45) && skip_byte() &&        // Version, service
        get_word(iplen) && skip_word() &&         // Len, ID
        skip_word() &&  skip_byte() &&            // Frags, TTL
        ...
```

I use three basic types of input function.

match_   ensures that the specified value is present

skip_   checks that the bytes are present and then discards them

get_   checks that the bytes are present and saves them for re-use later

These functions are surprisingly versatile in that they allow you to indicate those values that are unimportant, those that must be checked but needn't be retained, and those that must be saved for later use. If these values are separated out when the frame is received, you can free up a significant amount of storage space.

All these functions return a Boolean TRUE/FALSE value, so the surrounding if statement will return TRUE only if all the required data has been obtained and matched correctly. The obvious disadvantage with this method is that the processor will tend to lock up if communications fail in midmessage; it will wait forever for a byte that never comes. This can be avoided by including a timeout in the function that fetches the bytes from the serial link, which makes all subsequent input calls fail (return FALSE) until communication is restored.

Blocking

One of the more difficult design decisions is the extent to which it is sensible to make the CPU block on serial output. Blocking means that the CPU sits in a tight loop, waiting for each character to clear the Transmit buffer before sending the next one. In some applications, this might be acceptable, since it could be argued that there isn't much point in handling new input data if the CPU is still in the process of responding to the old. However, there are counter-arguments.

- If the processor misses a message from the host, there can be a significant delay (at least two seconds) before it is resent.
- The CPU cycle time would be dominated by the speed of the serial link. If it happened to be slow (e.g., through an old modem), then a proportionately greater amount of time would be spent waiting for characters to be sent, and a smaller time would be available for other tasks.
- It isn't a good idea for any one peripheral to dominate CPU usage. There may well be other housekeeping tasks (e.g., data collection) for the CPU to do that must be completed within a specific time frame.

The dilemma comes from the inability to buffer the output data (because of a lack of RAM), conflicting with a desire to buffer the data (so it can be output on interrupt or polling). A very similar logic applies to the handling of incoming frames: you'd like to receive them on interrupt, but you can't afford the buffer space.

To resolve these problems, you have to delve deeper into the underlying transaction the CPU will be called upon to perform. Perhaps now is a good time to review exactly how SLIP, IP, TCP, and HTTP work together to make a Web server:

Web Server Protocols

From the top down, the following protocols are needed for a Web server:

HTTP   Web page request/response
TCP   reliable communications
IP   low-level data transport
ICMP   diagnostics (Ping)
SLIP   serial interface
Modem emulation

I will now review each of these, with the aim of creating a small, yet fully functional, Web server implementation.

HTTP Request

The hypertext transfer protocol (HTTP) defines a request–response mechanism for obtaining documents from a Web server.

The Web browser sends a request to the server in the form of a multiline string, each line terminating with carriage return and line feed (CR, LF) characters. The first line specifies an uppercase method (i.e., command) followed by an argument string. The most common method is GET, followed by a filename to be fetched and a protocol version identifier. Subsequent lines contain additional information about the browser configuration.

```
GET /index.htm HTTP/1.0
User-Agent: Mozilla/4.5 [en] (Win95; I)
Pragma: no-cache
Host: 10.1.1.11
Accept: image/gif, image/x-xbitmap, image/jpeg, image/pjpeg, image/png,
*/*
Accept-Encoding: gzip
Accept-Language: en
Accept-Charset: iso-8859-1,*,utf-8
```

If you're keen to keep memory usage to a minimum, you must ask of what use the additional information is. You don't care about the type of browser; your server doesn't have sufficient resources to maintain a cache or an access log; and if the file you're sending has an unacceptable character set, there's nothing that can be done about it. Even the HTTP version number on the first line isn't needed, as I'll be using the simplest possible HTTP interface.

It would seem that you could chop off the remainder of the command line after the filename without losing functionality, but what is the maximum length of the filename? Surprisingly, this is largely under the control of the server.

When the user wants to access the server for the first time, an IP address is entered into a browser window.

```
http://172.16.1.2
```

The Web client (browser) locates the given address and submits the request to that server using a null filename.

```
GET / HTTP/1.0
```

By convention, most Web servers interpret this as a request for the default index file, which is index.htm or index.html, which, in turn, contains pointers to other files on the server. While the user clicks on pages you've provided, they will only be requesting filenames you've defined, so if you keep these short, you won't have to handle long filenames.

A possible exception to this rule would occur if you included HTML forms on your server. When a form is submitted, all the state information is appended to the filename, making a much longer string. There are various other difficulties associated with forms handling on a very small Web server, so, for the time being, assume that forms aren't being used.

HTTP Response

The response from the server to the client consists of an HTTP header and, if the request succeeded, the document itself. The header consists of several text lines, each terminated with a CR, LF delimiter. The header is separated from the document's contents by a single blank line.

As a minimum, the header must identify the HTTP protocol version, the success or failure status, and the content-type of the document (plain text, HTML, GIF graphic, for example).

```
HTTP/1.0 200 OK
Content-type: text/html

<html>
<head>Test page</head>
<body>This is a test page</body>
</html>
```

Unfortunately, it isn't possible to send out the same HTTP header for all files; it must be adapted to reflect the file's contents. The following list is the minimum number of file formats that must be supported:

```
text/plain
text/html
image/gif
```

However, it would be highly desirable to add other formats to the list.

If the client's request fails, an appropriate HTTP error message must be sent out, and it is also desirable to send out a document explaining the problem so that the browser has something to display. The simplest explanation could be in the form of a plain-text document, which has no fancy formatting.

```
HTTP/1.0 404 Not found
Content-type: text/plain

File 'abc.htm' not found
```

TCP

To convey the HTTP request and response between client and server, a reliable communications channel is required. This is provided by transmission control protocol (TCP), which provides a reliable-logical connection between two endpoints on the network, known as sockets. The objective of TCP is to make the network connection appear as transparent as possible. Regardless of the network type or the distances involved, data should be transferred between sockets in as timely and as error-free a fashion as possible.

TCP Sockets

A socket is an endpoint of a network connection that acts as a source and sink of connection data. Each active socket is implicitly linked to an application that sends and receives this data. In the case of a Web client, the application is a browser; in the case of a Web server, the application is an HTTP server, as described in the previous section.

Aside from the IP (Internet protocol) addresses of client and server, the other parameters that define a socket are the port numbers. In the case of servers, a port number defines the service being offered; for example, a Web server should respond only to incoming requests on port number 80.

At any one time, a server may support several simultaneous transactions, each of which involves a unique client–server socket pair. Clients will frequently open up several simultaneous connections to the same server in order to fetch several items in parallel, such as the graphical items on a Web page. To save on resources, a Web server can restrict the number of connections to just one, but this leads to very sluggish response, even when in use by only one client. If the client attempts to fetch, say, a page of text and three graphic images simultaneously, the server can do one of two things:

Ignore the request.   The TCP client will retry after about 1.5 seconds; the next retry time doubles to three seconds, the one after to six seconds, and so on. If several images are being requested, there can be an unacceptably long wait until the last one is successfully obtained.

Reject the request.   If a TCP "reset" is sent, the client will quickly retry the request; I have observed retry rates of around two per second for 40 seconds. This is a lot of extra traffic for the serial link to handle and will succeed only in slowing the data transfer further.

Ideally, you would respond to as many simultaneous network requests as the network bandwidth would permit. The problem is how to do this without requiring a large amount of socket storage.

Passive Open

Convention dictates that a server application passively open a TCP socket before exchanging data through it. This model is derived from standard implementations on multi-user systems, where there is a strong separation between the system's TCP code and the user's application code. To fit in the microcontroller, your application code will have to be tightly coupled to the TCP stack so that the distinction between the two becomes blurred.

There is no point in maintaining the fiction of Passive Open. If a network frame arrives and the server has the resources to handle it, then it should do so.

Sequence Space

To control and monitor the establishment of a connection, transfer of data, and closure of a connection, each TCP transmission (segment) is identified by a TCP sequence number, which refers to its position in an imaginary sequence space. The start and end of a transaction (known as SYN and FIN) can be seen as fixed points in this space. Using the sequence number, the recipient can place an incoming segment in its rightful location within that space and detect whether it forms a logical progression from the last segment received, is a duplicate of a previous segment, or is a future segment received out of sequence.

The sequencing process is symmetrical: both client and server use sequence numbers to place their transmitted segments in the outgoing sequence space and send acknowledgment numbers to confirm the point they have reached in the (completely separate) incoming sequence space. Figure 9.2 shows a sample data transfer of 120 hex bytes sent in two unequally sized blocks. In addition to the actual 32-bit sequence number, the relative sequence number is shown in parentheses. I've used the convention that the first data byte has a relative sequence number of zero, which means that the first synchronizing byte has a value of minus one.

Figure 9.2   Sequence space.

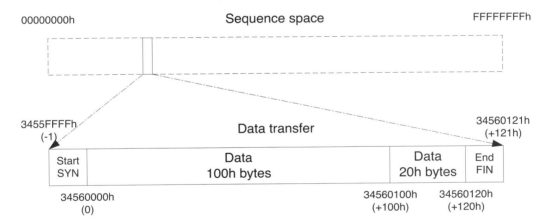

At the start of any transaction, the 32-bit sequence number must be set to a completely new value to avoid confusion with past transactions. The good news is that, within certain constraints, the server can choose any 32-bit starting sequence value it likes. This suggests it might be possible to use the sequence number as a kind of file pointer, indicating the current location of the file (in ROM) being sent. The bad news is that the sequence value must be chosen before the first segment of the new transaction is sent. At this time, the client hasn't yet revealed the filename to be accessed, so it isn't possible to choose a value that is convenient for accesses to that file.

A lesser option, which is still very useful, is to choose a sequence value that reflects the relative position within the file. This has already been done in Figure 9.2. The least significant word of the sequence number has the following hex values.

FFFFh   Initial SYN marker
0000h   First data byte of file
xxxxh   Offset of *xxxx* into the file

So long as the file size is less than 64Kb (a reasonable assumption, given the ROM size limitations of the miniature Web server), this technique will result in a useful simplification to your code.

Of course, you have no control over the client's choice of sequence number, so you cannot use a similar trick there. However, the client's request string is sufficiently short that it should fit within a single frame, so you have only the one incoming data frame to worry about. If any more arrive, they can be discarded.

Managing Connections

It is traditional to view the opening and closing of a TCP connection in terms of the state diagram transitions. This implies the need to maintain an individual state machine for each simultaneous connection (i.e., each socket), which will consume a lot of RAM.

To solve this problem, it is worth bearing in mind that the Web server always responds to client HTTP requests; it never takes the initiative. If you could also guarantee that it only ever responded to TCP segments, rather than initiating them, then a massive simplification in the TCP stack would result. You wouldn't have to store node addresses or port numbers, since these could just be copied from the incoming to the outgoing message. By careful choice of initial sequence number (as discussed in the previous section), you can deduce the location in the current file by examining the incoming acknowledgment value, so you would always know what action to take next. You'd create an implicit, rather than explicit, TCP state machine using the incoming acknowledgment value as a state variable.

This looks very promising, but there are two further problems to solve.

Current filename.   If the server doesn't keep a record of which client requested which file, how does it know what to send next? It may know the relative *position* within the file from the sequence number, but if it doesn't know the filename, this isn't a lot of use.

Retransmissions.   A normal TCP stack will retransmit a TCP segment if it doesn't receive an acknowledgment within a certain time. If your stack doesn't store any information about its clients, it won't ever be able to retransmit anything without being prompted.

My first attempt to solve the first problem was a fiendishly clever plan to use the least significant bits of the sequence number to indicate the filename (or more precisely, the index number in a file directory). So long as the file is sent in blocks of, say, 64 bytes and the front is padded with a variable-length HTTP message that depends on the filename, then ... well, work it out for yourself. The disadvantage of this technique is the inflexibility of having to send out fixed-length blocks, which is a real nuisance when generating Web pages dynamically.

To solve the second problem, it is tempting to rely on the client's retry mechanisms, but this doesn't work. The server can't rely on receiving an acknowledgment for every segment, since some may have been lost in transit. If the server stops sending data (because it failed to see an acknowledgment), it will be a long time (two hours) before the client's "keepalive" timer triggers it to send a keepalive probe, to see if the server has crashed. That is a long time to wait; the client application will have abandoned the connection long before.

A simple solution to both problems is to restrict the outgoing page to one TCP data block (segment) and to send it out as soon as the client's HTTP request has been received.

One-Segment Pages

This is a small Web server with extremely limited resources, so the idea of fitting a Web page into a single TCP segment isn't quite as daft as it sounds. True, this may force Web designers to be less lavish in their use of page embellishments, but is that necessarily a bad thing? A small Web server has a small amount of information to convey, and there is no point padding it out unnecessarily.

I have adopted the usual maximum SLIP size of 1,006 bytes. The IP and TCP headers are a minimum of 20 bytes each, so the maximum TCP data size is 966 bytes. It is remarkable how much can be achieved within this limitation; for example, all the HTML examples in the previous chapters are significantly smaller than this.

The key advantage of one-segment pages is that there is a one-to-one relationship between the actions of the client and the server (Figure 9.3). This relationship is reinforced if the closing FIN is piggybacked onto the page data.

Figure 9.3   One-segment TCP data transfer.

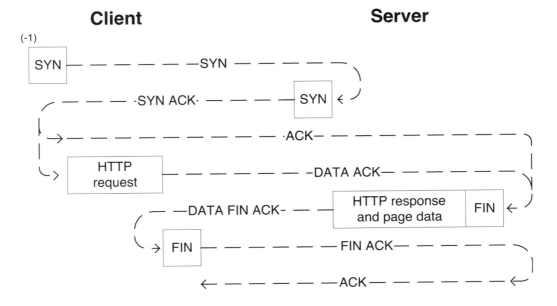

TCP Segment Format

A TCP header plus data block is known as a segment and has the format shown in Figure 9.4.

Figure 9.4 TCP segment format.

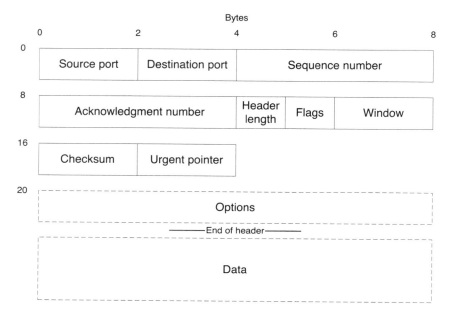

Destination port. You'll check the destination port field of every incoming TCP segment to see if it refers to a supported service. At present, there are only two of these.

Port 13: Daytime service
Port 80: HTTP (Web) server

The Daytime service returns a simple string giving the current date and time. It is by no means essential for a Web server to provide this, but it is a useful step nevertheless on the road to creating an HTTP server because the TCP transaction is simpler and easier to debug.

Sequence and acknowledgment numbers. The 32-bit sequence and acknowledgment numbers have already been discussed. Thirty-two-bit arithmetic is a problem on the PICmicro because the CPU supports only eight-bit operations directly. The chosen compiler provides no support for 32-bit data types, so you have to create all your own functions. If you assume that the incoming and outgoing data is less than 64Kb, then a useful simplification is to perform only 16-bit operations on the 32-bit values, propagating the carry value to the upper 16 bits.

Header length. The header length reflects the length of the standard header plus the options. There is no length value for the data because it can be deduced from the value in the lower protocol layer (IP).

Flags. These are one-bit Boolean option flags.

FIN	0x01
SYN	0x02
RESET	0x04
PUSH	0x08
ACK	0x10
URGENT	0x20

I have defined the header length and flags as two single-byte values for simplicity, but I should point out that the standard defines a four-bit header length, a six-bit reserved field, and then six "code bits."

Window size. The window size indicates the amount of Receive buffer space that is available and is used for flow control. It can be set to a fixed size on transmit and ignored on receive. You can safely assume that any client contacting your server has sufficient buffer space for your humble pages, and if they don't, it was pretty stupid of them to send the request in the first place!

Urgent pointer. This can be ignored because all data can be treated with equal priority.

Options. In addition to a variable-length data field, there is a variable-length header options field. Mercifully, you don't have to generate options, and you can safely discard any incoming options.

Checksum. The only awkward point to note about this header is that it must include a valid checksum value, which is computed across the whole TCP segment, plus a pseudo-header (Figure 9.5) containing parts of the IP header.

Figure 9.5   IP pseudoheader.

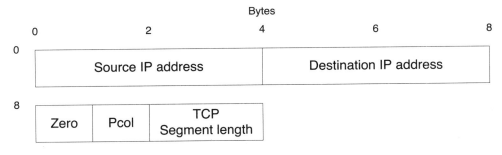

The usual checksum computation technique is to scan the TCP segment image in memory, but you don't have sufficient RAM to do this. If the checksum came at the end of the data, it would be easy to compute on the fly as the header and data are sent out and then appended to the data.

As you'll see in the next chapter, TCP checksum generation is a major issue in this small implementation, particularly when attempting to include dynamic data on the Web pages.

Long Segments

If you are to make any headway creating a Web server, you'll have to handle TCP segments that are larger than the available RAM. In the case of transmitted segments, the bulk of data will reside in external ROM, so as far as possible, it will be copied directly from there to the RS232 output. You'll also receive long HTTP requests, where the only items of interest are in the first few tens of bytes.

Transmit.   The IP and TCP headers must be created in RAM so that their checksums can be computed. For normal (short) segments, these images are then SLIP encoded (by inserting escape sequences) while they are being sent down the serial line. If the segment is long (i.e., it includes ROM data), then a flag is set such that the ROM-to-SLIP transmission takes over when the RAM-to-SLIP transmission stops. This critically depends on the TCP checksum being known in advance; that is, it is precomputed for the ROM image and added in when the TCP header and pseudoheader checksum is calculated.

Receive.   You're only interested in the start of an HTTP request and can happily discard the rest. TCP doesn't possess a mechanism for discarding data. If you don't acknowledge it, it will simply be resent until you do. There are two possible solutions: you could reduce the TCP window size so that the request is sent in two or more chunks and then discard all but the first chunk or, using a simpler method, you can store only the start of each segment data in RAM and discard the rest. I've chosen the latter method. It is tempting to ignore the checksum on the incoming TCP segment and assume that it is correct, but this is rightly frowned on in the TCP community. Instead, you could compute a checksum for the discarded portion of the segment and add it on after the complete segment is received. The approach I have adopted is a minor variation of this, whereby the checksum of all the incoming TCP data is computed separately, irrespective of how much is stored in RAM or discarded. This is added to the value computed from the TCP header and pseudoheader, which are always stored in RAM.

IP

To convey the TCP segments between hosts, use the Internet protocol (IP). After the difficulties of TCP, IP is relatively easy to implement.

Datagram Format

An IP header plus data block is known as a datagram (Figure 9.6).

Version and header length.   I'm using IPv4, and the default header size (measured in 32-bit words) is five, so I'll assume a value of 45h for both transmit and receive.

Service.   Prioritize datagrams by setting this field to zero, which is normal precedence.

Length.   The length is the total datagram length in bytes, including the IP header.

Figure 9.6   IP datagram.

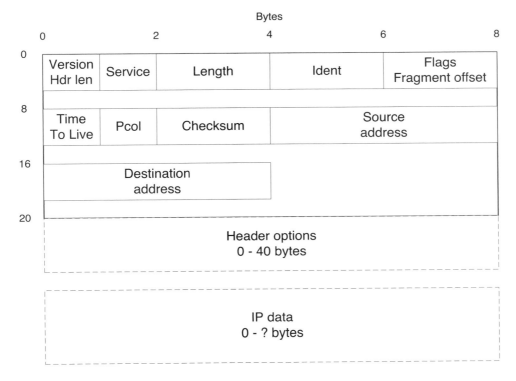

Ident.   The Ident value is incremented for each datagram sent.

Fragmentation.   IP allows a large datagram to be split into two or more smaller datagrams using a process called fragmentation. Considering the acute lack of RAM on the microcontroller, it is impossible to support fragmentation. This is unlikely to be a problem in practice, since it carries a very significant performance penalty and is generally avoided wherever possible.

Time to live.   Time to live is an expire time for the datagram, to prevent it from endlessly circulating the Internet. It is a constant value that is generated on transmit and ignored on receive.

Protocol.   This is an indicator of which protocol is used in the data area of the datagram. I'll only use the following values:

- ICMP (described later)
- TCP

Checksum.   This is a simple checksum of the IP header only.

Source and destination addresses. These are IP addresses expressed as 32-bit values. Important questions are (1) what IP address is assigned to your system and (2) how is it programmed with that address. This issue can be side-stepped by making the assumption that because you're using a point-to-point serial link, there can be only one intended recipient for all the network traffic — namely, your Web server. Hence, you can disregard the destination IP address value, but you must be careful to use this value in the source address field of your outgoing datagrams.

Options. Header options are very occasionally used to give tighter control over datagram routing. For simplicity, don't accept or transmit any options.

Long Datagrams

To accommodate long TCP segments, you have to accept IP datagrams that are longer than the available RAM. Unlike TCP, there are no checksum problems, since the IP checksum does not include the data area, so you can discard excessive input data or add extra output data without any checksum problems.

ICMP

Internet Control Message Protocol (ICMP) is very useful for performing network diagnostics. An ICMP message (Figure 9.7) is contained within the data field of an IP datagram.

Figure 9.7   ICMP message.

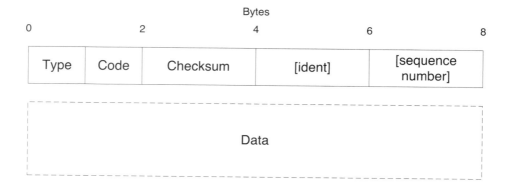

Ping

The most commonly used facility is the ICMP Echo request, or Ping. You don't have to implement this on your Web server, but it will be very useful to check the lower protocol layers prior to implementing the Web server itself. The Echo request is type 8 code 0, and the reply is type 0 code 0. The checksum covers the complete ICMP header and data area. The Ident and sequence numbers and all the data are echoed back to the sender as a check of network integrity.

Buffer Size

The default data size for a UNIX Ping is 64 bytes, which is too large for the available buffer RAM. Fortunately, the Ping utility has an argument to specify the data size, so it can be reduced to, say, 32 bytes, which is the default size for DOS systems. This requires a buffer size of 60 bytes (including a 20-byte IP header and an eight-byte ICMP header), which is more realistic.

SLIP

SLIP is a simple method of converting a stream of serial data characters into a defined block, which I'm calling a frame. It is easy to implement. A delimiter character is put at the end of each frame (and also, by convention, at the start). If the delimiter character is encountered in the data stream, a two-character escape sequence is substituted.

```
#define SLIP_END    0xc0    // SLIP escape codes
#define SLIP_ESC    0xdb
#define ESC_END     0xdc
#define ESC_ESC     0xdd
/* Start a transmission */
void tx_start(void)
{
    putchar(SLIP_END)
}


/* Encode and transmit a single SLIP byte */
void tx_byte(BYTE b)
{
    if (b==SLIP_END || b==SLIP_ESC)
    {
        putchar(SLIP_ESC);
        putchar(b==SLIP_END ? ESC_END : ESC_ESC);
    }
    else
        putchar(b);
}
```

Modem Driver

The implicit assumption in most PC communications software is that serial networking should be configured for access via a modem and telephone line. I would like to be able to link a PC directly to my PICmicro server's serial port, so the server will have to impersonate a modem to keep the PC happy.

Fortunately, this only involves accepting modem command strings, which are prefixed by "AT" and delimited by a carriage return character, and returning an "OK" string. This is usually sufficient, though it is wise to assert the data carrier detect (DCD) hardware handshake line to the PC as well, in case its software uses it to check that the (emulated) phone link is still functioning correctly.

A typical PC-to-modem interaction is shown in Table 9.1.

Table 9.1   PC-to-modem interaction.

PC	Modem	
AT<CR>	OK<CR><LF>	Check that modem is responding
ATE0V1<CR>	OK<CR><LF>	Disable command echo; enable text message responses
AT<CR>	OK<CR><LF>	Check that modem is responding
ATDT12345<CR>	OK<CR><LF>	Tone-dial telephone number 12345

Some modem scripts look for a CONNECT message after dialing, but this does not appear necessary when using the standard Windows modem types.

Disconnection follows a similar pattern (Table 9.2).

Table 9.2   Disconnect from modem.

PC	Modem	
ATH<CR>	OK<CR><LF>	Disconnect from line
ATZ<CR>	OK<CR><LF>	Reset modem

Summary

I have looked at the challenges associated with creating a miniature Web server, with particular reference to the PICmicro microcontroller I chose. The constraints imposed by the PICmicro are quite severe. In particular, there is an acute lack of RAM, which forces you to employ novel techniques to handle protocol and TCP state management. If I weren't writing a book, I'd be tempted to quit now, while I'm ahead.

Chapter 10

TCP/IP on a PICmicro® Microcontroller

Overview

Having chosen a microcontroller and looked at the techniques for miniaturizing the TCP/IP stack, you now have to implement these ideas. First I'll look at the additional hardware devices that will be needed, and I'll produce a complete circuit diagram. The circuit is sufficiently simple that a prototype can be hand-wired without much difficulty. Alternatively, a commercially produced board can be purchased.

I'll then look at the code for modem emulation, serial (SLIP) interface, IP, and TCP, ending up with a fully working system that will respond to ICMP Pings and accept TCP Daytime requests. This will make a firm foundation on which to build your miniature Web server, which is covered in Chapter 11.

Peripherals

External Memory

A Web server needs ample storage for Web pages, and the on-chip ROM is clearly inadequate for this. Conventionally, you would add memory to a microcontroller using an external address and data bus, but I've deliberately chosen a device with a small pin count, which has no provision for external memory addressing. A satisfactory alternative is to use an external device with a synchronous (clocked) serial interface (Figure 10.1). There are a wide range of devices available, generally using a four-wire or two-wire interface (plus power and ground).

Figure 10.1 External device interfaces.

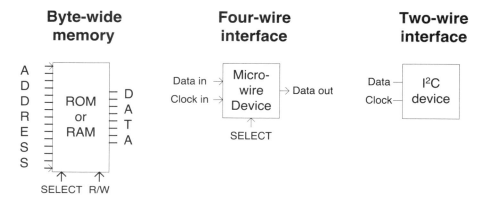

A four-wire device has a clock line, chip select, and data input and output. The microcontroller starts a transaction by asserting the chip select line and then toggles the clock line to mark each data bit as it is sent out. First an address value is sent bit-by-bit to the device; then either the data is sent as well (write cycle) or is read back over the other line (read cycle).

A two-wire device has a clock line and a single bidirectional data line. The inter-integrated circuit (I^2C) bus protocol defines the state transitions on these two lines so as to identify the start and end of a transfer cycle (Figure 10.2). At the start, an address byte is sent to identify a specific device on the I^2C bus, and then a further address value is sent to indicate the location of interest in the device. After this, the device's memory can be read or written one bit at a time over the serial interface.

Figure 10.2 Single-byte I^2C transfer.

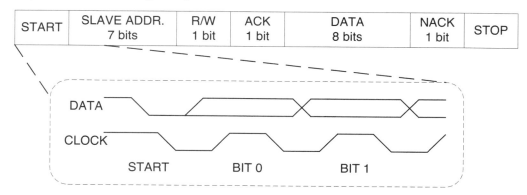

If the external memory is to store Web pages, it needs to have a sensible capacity. At the time of writing, the largest nonvolatile (E^2ROM) serial memory in common usage was a 32Kb I^2C device in an eight-pin package. By the time you read this, the 64Kb devices may be commonplace.

The PICmicro I^2C hardware is mainly targeted at supporting a slave device and not a bus controller, so the standard I^2C software supplied with the compiler has to do a significant amount of bit bashing. Of course, this carries a significant speed penalty, but in practice, a transfer rate of 8Kb per second or more can be achieved, which will be adequate for my needs.

Digital Outputs

To demonstrate the ability of the PICmicro to control external hardware, I need to be able to drive some parallel output lines, equipped with indicator light-emitting diodes (LEDs). This is ridiculously easy with the PICmicro because there are plenty of spare I/O lines with the ability to drive LEDs directly without any buffering. I will use port B, which is a convenient block of eight unused I/O pins, configurable as outputs.

Temperature Sensor

It is nice to include a real-world signal in the demonstration, particularly one that has a natural variability. Having used a temperature sensor in previous demonstrations, this seemed to be the obvious candidate, though PICmicro interfacing proved to be slightly more difficult than anticipated.

The LM35 series of devices are low-cost analog temperature sensors. When fed from a five-volt supply, they produce a calibrated output of ten millivolts per degree Celsius. The PICmicro has an on-chip analog-to-digital converter (ADC), so it should be a simple task to use the LM35, but unfortunately, there is a problem when the processor is being emulated.

The ADC works by measuring the ratio between the unknown voltage and a known reference voltage. It can take its reference from the five-volt supply, but this is much larger than the voltage being measured (250 millivolts when the temperature is 25C), so the accuracy is very poor. There is provision for using the ADC with an external reference, but this needs to source a significant amount of current during conversion, so it must be of low impedance. Unfortunately, the emulator pins have series resistors (to protect them from stray voltages), so it is impossible to use an external voltage reference.

Instead, I opted for an all-digital solution by using an I^2C device, the LM75 type. It, too, is fully calibrated, though its surface-mount package style is less convenient for my purposes. However, the temperature value is read using a method that is very similar to accessing the external ROM, so some software complexity is saved.

Digital Inputs

To allow the connection of simple on/off devices such as switches and sensors, I need some digital input lines. Since I'm not using the analog input capability of the PICmicro, the six port A lines can be configured as digital inputs.

Real-Time Clock

The final peripheral is a real-time clock (RTC), which I will use to demonstrate a simple HTML form interface. The PICmicro has sufficient resources to keep time itself, but, for simplicity, I decided to use an external chip on the I²C bus.

The Philips PCF8583 has time and date registers and 240 bytes of nonvolatile RAM that would be useful in a data-logging application. The registers and memory are linearly mapped using a simple eight-bit pointer value, so little extra software is needed.

Block Diagram

The main functional blocks are connected to the processor via three main interfaces: parallel I/O, asynchronous serial, and I²C bus (Figure 10.3).

Figure 10.3   PWEB hardware block diagram.

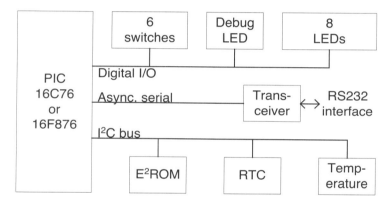

To simplify construction, a PICmicro evaluation or experimentation board may be used as the basis for any hardware construction. Most PICmicro evaluation boards are equipped with a voltage regulator, RS232 interface, and some LEDs, so it may be necessary only to add the I²C devices.

Circuit Diagram

This is supposed to be a software book, but the hardware is so simple that I have included a circuit diagram, with the positions of the devices shown to assist the creation of a prototype. If you don't want to construct your own, prebuilt boards (with additional features such as optically isolated digital I/O) are available from Io Ltd. (see Appendix B for details).

I've made some notes about the hardware design shown in Figure 10.4.

Figure 10.4   PWEB circuit diagram.

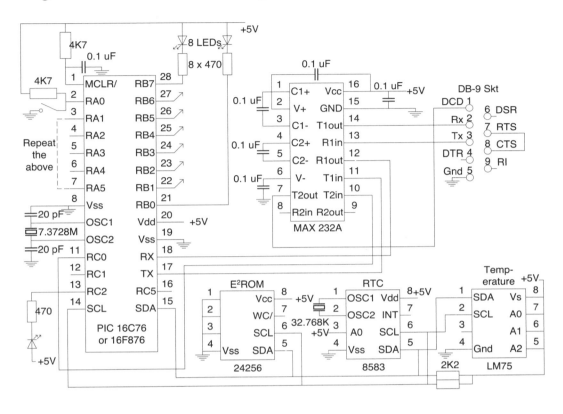

Power supply.   The circuit runs off a single five-volt supply. Normally, a voltage regulator would be included on-board to allow the use of an unstabilized power supply. With an appropriate choice of components, the same design can run off a battery supply as low as three volts.

CPU crystal.   A value of 7.3728MHz was chosen because it permits the use of a wide range of serial baud rates.

D-type connector.   I have followed the convention adopted by most PICmicro evaluation boards, and used a female D-type serial connector, which should link to a PC serial port using a straight-through male–female serial cable.

Battery backup.   The real-time clock chip is shown without a battery. To keep it running while the CPU is powered down, its supply voltage would normally have battery backup.

RTC address.   The E^2ROM and RTC share the same base I^2C address. Some E^2ROM devices don't have a means of changing their addresses, so the address line A0 of the RTC has been strapped high to avoid a conflict.

I²C pull-up resistors. It is often forgotten that both the I²C clock and the data lines are bidirectional, with open-collector drivers on the devices. Pull-up resistors are essential, and a value of 2K2 (2,200 ohms) is adequate for the relatively slow bus speeds. This could probably be increased to reduce current consumption in a battery-powered application.

Temperature sensor. Various types of I²C temperature sensors are available, with a range of accuracies. I used one with a two-degree accuracy and a half-degree resolution, which is sufficient for demonstration purposes.

CPU emulation. I found it very convenient to use a CPU emulator (Advanced Transdata Corp., RICE17A) when developing the software, though these can be quite expensive. There are lower-cost alternatives, such as the PIC-ICD in-circuit debugger from the same company, though it is important to carefully scrutinize the specifications, especially download and update times, before opting for a lower-cost option such as this.

CPU EPROM/Flash programming. The flash-based CPUs can be programmed in-circuit using an appropriate connector. I confess to having little experience with this, and I used conventional EPROM-based parts for the prototype. A wide variety of device programmers are available. It is beyond the scope of this book to look at these development methods in detail.

E²ROM programming. You will need some means of preprogramming the E²ROM device. Most EPROM programmers are capable of programming I²C devices using an appropriate adaptor. I use the excellent EMP20 programmer from Needham's Electronics, with personality module 18. You may need to experiment to customize it to the device you are using. I happened to use SGS M24256 devices in my prototype, which aren't on the device menu, though they can be programmed satisfactorily using the ATMEL 24C256 setting.

I/O ports. Output devices are generally better at sinking current than sourcing it (unusually, the maximum source and sink currents for the PICmicro are both 25mA). The common side of the LEDs goes to the supply, rather than ground, and it will be necessary to drive the outputs low in order to turn on the LEDs. For similar reasons, there is a resistor pulling the inputs high, and the switches drive them low, so the input is high when the switch is off.

Omitting devices. For initial experimentation, the eight parallel port LEDs and all the I²C devices can be omitted because they are only needed for the final Web server. Accessing a non-existent I²C device should simply return an all-ones data value for any read cycle. However, the eight LEDs are very useful when debugging because they can be programmed to indicate the state of various internal signals.

Low-Level Software

The various PICmicro C compilers provide differing levels of hardware support. One of the reasons for selecting the Custom Computer Services (CCS) compiler is its comprehensive support for serial and I²C communications. Regrettably, you may need to make significant modifications to my software if using an alternative-development environment.

CPU Identification

The CCS compiler can handle a wide range of processors within the PICmicro family. You need to tell it which one you're using. This is by means of a #include file.

```
#include <16c76.h>              // CPU definitions, must be followed by..
#device *=16                    // ..enable 16-bit (!) data pointers
#FUSES HS,NOWDT,NOPROTECT,PUT,BROWNOUT   // PICmicro fuse settings
#ID CHECKSUM                    // ID bytes are checksum
#ZERO_RAM                       // Wipe RAM for safety
```

Having identified the 16C76 CPU, you have to decide whether to handle the strangely segmented data memory yourself or let the compiler insert the appropriate bank-switching code. Since I want to create the clearest and simplest C code, and not necessarily the smallest or fastest, I have enabled the compiler bank-switching support. It is important to do this at the top of the file; otherwise, the resulting code mix leads inevitably to disaster.

It is also convenient to clear the RAM before use and preprogram the CPU configuration fuses that enable such facilities as the crystal type, watchdog timer, memory protection, and the like. The fuse settings can be overridden when the device is programmed.

Data Types

The CCS compiler's native data types are highly nonstandard, though not that unusual for a microcontroller.

1-bit	short, short int
8-bit unsigned	unsigned, unsigned int, int, char, byte
8-bit signed	signed, signed int
16-bit unsigned	long, long int
16-bit signed	signed long
32-bit float	float

The one-bit data type should be used for Boolean values wherever possible, since the compiler is able to perform useful code optimizations.

To avoid confusion, I'll use BOOL, WORD, and LWORD, defined as follows.

```
#define WORD unsigned long    // Data type definitions
#define BOOL short            // 'short' is very short (1 bit) in this compiler
typedef union                 // Longword definition (not a native type)
{
    BYTE b[4];                //  Byte array
    WORD w[2];                //  ..or word array..
    BYTE l;                   //  ..or dummy reference for call-by-reference
} LWORD;
```

Strangely, the compiler seems to be case insensitive, so you can use BYTE as usual without having to define it. You can also use int for all small-value variables, such as loop counters.

Serial Drivers

Before using the RS232 interface, the CPU clock speed and baud rate need to be defined.

```
#use DELAY(CLOCK=7372800)    // CPU clock frequency 7.3728 MHz
#use RS232 (BAUD=38400, XMIT=PIN_C6, RCV=PIN_C7, RESTART_WDT, ERRORS)
#bit TX_READY =      0x98.1 // Tx ready status bit
```

The 38,400 baud rate was chosen as a reasonable compromise between transfer speed and CPU loading. Once the serial port has been defined, `putchar()` and `printf()` can be used for serial output. You could use `getchar()` for serial input, but it is better to use serial interrupts. The compiler provides full support for serial interrupt handlers written in C, which are easy to use once the underlying interrupt structure is mastered.

```
enable_interrupts(INT_RDA);             // Enable serial Rx interrupts
enable_interrupts(GLOBAL);

...
/* Rx interrupt handler */
#INT_RDA
void rx_handler(void)
{
    BYTE b;

    b = getchar();
    ...
}
```

Don't forget that the PICmicro has an eight-level hardware stack, so you should avoid function calls within the interrupt handler. To help, the compiler can translate a function call into in-line code, sparing you the stack overhead. For example, `getchar()`, shown above, is actually inserted in-line.

I²C Drivers

Although an I²C slave is largely implemented in hardware, you need a master to drive the slave peripheral devices. This requires some bit-bashing serial I/O code, which is included in with the compiler. First you need to specify the I/O pins and data rate.

```
#use I2C (MASTER, SDA=PIN_C4, SCL=PIN_C3, RESTART_WDT, FAST)
#define SENSOR_ADDR  0x9e   // i2c addr for temperature sensor
#define EEROM_ADDR   0xa0   // i2c addr for eerom
#define RTC_ADDR     0xa2   // i2c addr for real-time clock
```

Each I²C device address has a seven-bit value, plus a least significant bit indicating that a read (1) or write (0) is to be performed. Some drivers keep the read/write bit separate from the address field, whereas others (including the CCS compiler) treat the whole address field as a single eight-bit number, with the read/write flag as the least-significant bit. Hence, the above RTC address of A2h is really a seven-bit address of 51h (as given in the device manual) with a write cycle. To perform a read cycle, the least-significant bit is set. This can best be illustrated by looking at the code for getting and setting the clock values.

```c
/* Get current time from Real-Time Clock */
void get_rtc_time(void)
{
    int i;

    i2c_start();
    i2c_write(RTC_ADDR);                    // i2c address for write cycle
    i2c_write(2);                           // RTC register offset
    i2c_start();
    i2c_write(RTC_ADDR | 1);                // i2c address for read cycle
    for (i=0; i<sizeof(rtc); i++)           // Read bytes
        rtc.b[i] = i2c_read(i<sizeof(rtc)-1);
    i2c_stop();
}

/* Set current time in Real-Time Clock */
void set_rtc_time(void)
{
    int i;

    i2c_start();
    i2c_write(RTC_ADDR);                    // i2c address for write cycle
    i2c_write(2);                           // RTC register offset
    for (i=0; i<sizeof(rtc); i++)           // Write bytes
        i2c_write(rtc.b[i]);
    i2c_stop();
}
```

To read the time values, a register offset value (2) is first written to the device; then the I^2C cycle is restarted in read mode. It is very important that all the bytes read are acknowledged, excluding the last one. To this end, i2c_read() takes an argument, which is non-zero if an acknowledgment is to be sent.

Reading the temperature sensor is a very similar exercise, except that the default value of the internal register pointer is the 16-bit temperature value, so no write cycle is necessary.

```
/* Get current temperature from i2c sensor */
void get_temperature(void)
{
    i2c_start();
    i2c_write(SENSOR_ADDR | 1);
    temphi = i2c_read(1);
    templo = i2c_read(0);
    i2c_stop();
}
```

A word of warning: the I²C interface appears deceptively simple, but it carries a few traps for the unwary. Read the device data sheets carefully; for example, misreading a 16-bit register on the LM75 using an eight-bit cycle can lead to the device jamming the data bus for an additional nine clock cycles. Furthermore, all transitions on the clock and data lines are significant, so any electrical interference on these lines will cause major problems — good-quality screening and grounding are essential if the temperature sensor is mounted away from the circuit board.

Having said that, the I²C bus is an extremely useful way of connecting additional devices to the PICmicro and is ideally suited to the application I'm developing.

Parallel I/O

The PICmicro has a variety of on-chip peripherals that may be connected internally to the device's pins. When not required for this purpose, the pins can be used as general-purpose inputs or outputs. The compiler has the ability to automatically configure each pin as an input or output at run time. I prefer to declare all inputs and outputs manually up-front.

```
#use fast_io(A)              // I'll set the direction bits on I/O ports
#use fast_io(B)
#use fast_io(C)


#byte   PORTA=5              // Main I/O ports
#byte   PORTB=6
#byte   PORTC=7
#define ALL_OUT    0         // Direction (TRIS) values
#define ALL_IN     0xff
#define DIAG_LED   PIN_C2 // Diagnostic LED Pin ident

setup_port_a(NO_ANALOGS);            // No analogue I/Ps
set_tris_a(ALL_IN);                  // Port A digital I/P
```

```
PORTB = 0xff;                    // Initialize ports
PORTC = 0;
set_tris_b(ALL_OUT);             // Port B LEDs
set_tris_c(0xf8);                // Port C mostly I/Ps
```

The set_tris_ statements define whether each pin of a given port is an input (1) or output (0). Once the port has been defined using the #byte directive, it can be accessed using assignments just like any other variable, though I would recommend you perform only simple assignments. The following code has a rather unfortunate side-effect in the current release of the compiler.

```
#byte    PORTB=6


BYTE val=0x55;
PORTB = ~val;
```

The compiler generates code that first sets the output port to 55h and *then* complements it to AAh. The resulting glitch can cause problems when driving fast I/O devices.

An alternative method for single-bit I/O, such as the diagnostic LED, is to use a simple definition and output_bit().

```
#define DIAG_LED      PIN_C2 // Diagnostic LED Pin ident


void setled(BOOL on)
{
    ...
    output_bit(DIAG_LED, !on);
    ...
}
```

When driving peripherals such as LEDs, it is traditional to connect the common side to the supply, so that a PICmicro output must be low in order to switch the LED on. I have adopted this scheme, but two evaluation boards (from Advanced Transdata and Microchip) both connect the common side of the port B LEDs to ground, so you will need to modify the code if using such boards or accept an inverted LED indication.

Timers

A polled timer scheme has been used throughout the book for event scheduling, and it is particularly useful here. I have three hardware timers in the PICmicro. I'll use timer 1 because it is a 16-bit timer.

```
// Timer 1 trigger value; tick time = (1024 x DIV) / CPU_CLK
// 100 ms ticks with 7.3728 MHz clock requires divisor 90 prescale 8
#define TIMER1_DIV    90
#define TIMER1_SET    (T1_INTERNAL | T1_DIV_BY_8)
```

```
...
setup_timer_1(TIMER1_SET);              // Initialize timer
...
/* Update the current tick count, return non-zero if changed */
BOOL geticks(void)
{
    static unsigned tc, lastc=0;
    BOOL changed=0;

    tc = TIMER_1_HIGH;
    if (tc - lastc >= TIMER1_DIV)
    {
        tickcount++;
        lastc += TIMER1_DIV;
        changed = 1;
    }
    return(changed);
}


/* Check for timeout using the given tick counter */
BOOL timeout(int &var, int tout)
{
    BOOL ret=0;

    if (!tout || tickcount-var>=tout)
    {
        var = tickcount;
        ret = 1;
    }
    return(ret);
}
```

The timeout function can be used to check whether the given delay time has
expired, for example

```
#define LEDONTIME    1        // Ticks for LED on
#define LEDOFFTIME  50        // Ticks for LED off
BOOL ledon;                   // Diagnostic LED state
int ledonticks, ledoffticks;// LED tick counts
int tickcount;                // Timer tick count
```

```
...
geticks();                      // Get timer ticks
if (timeout(ledoffticks, LEDOFFTIME))
    setled(1);
else if (ledon && timeout(ledonticks, LEDONTIME))
    setled(0);
```

This will cause the LED to flash briefly (100 milliseconds) every five seconds.

SLIP and IP Drivers

In Chapter 9, I discussed various techniques for sending and receiving datagrams on the fly using a minimum of RAM. It is now time to put the theory into practice.

SLIP Receive

The receive interrupt handler needs to

- Fetch the next character from the serial interface
- Decode the SLIP escape sequences
- Check for a modem command string
- Compute a data checksum if beyond the IP and TCP headers
- Store the character if there is room in the buffer

There are a few unusual features, which I explain below.

Buffer sizes. Because of the segmentation of the RAM area, the maximum contiguous data area that can be defined is 96 bytes. I'll use this as a Transmit buffer, so I have to make do with a smaller area for the Receive buffer. The ultimate constraint is that I must be able to fit the IP and TCP headers into it, with sufficient space left over for a reasonable length of HTTP request.

```
#define RXBUFFLEN   80      // Rx buffer: more than enough for a 32-byte Ping
#define RXHDR_LEN   40      // After header, data is checksummed on arrival

#define SLIP_END    0xc0    // SLIP escape codes
#define SLIP_ESC    0xdb
#define ESC_END     0xdc
#define ESC_ESC     0xdd

BYTE rxbuff[RXBUFFLEN];     // Receive buffer, I/O ptrs, and 'Rx full' flag
int rxin, rxout, rxcount;
BOOL rxflag;
BOOL modemflag;             // Flag to show modem command received
```

SLIP decoding.   To make the maximum use of the available RAM, I'll decode the SLIP escape sequences as they arrive.

Buffer method.   You would normally think in terms of a circular buffer for receive input, but there doesn't seem much point in introducing this additional complication when the buffer is too small to accommodate more than one incoming frame. To improve throughput, you can allow a new frame to be loaded into the buffer while you are analyzing the old one, though this presupposes that you can analyze the old frame faster than the new characters arrive.

Modem emulation.   If there is a modem command string, you need to send an "OK" response, but it would be unwise to do this within the interrupt handler. Instead, set a global flag so that a modem handler can be called from the main loop.

Checksum of data area.   The tangled problem of TCP checksums was tackled in the previous chapter. It was noted that the RAM is too small to hold all the incoming TCP segment, yet you need to verify the incoming TCP checksum. The solution is to checksum all incoming data after the TCP and IP headers. The resulting value is added on to the header checksum when the TCP segment is being verified.

Checksum algorithm.   Traditionally, the checksum is computed by adding 16-bit word values together, counting the carry flags out of the most significant byte and then adding them on at the end.

```
WORD csum(void *dp, WORD count)
{
    register LWORD total=0L;
    register WORD n, *p, carries;

    n = count / 2;
    p = (WORD *)dp;
    while (n--)
        total += *p++;
    if (count & 1)
        total += *(BYTE *)p;
    while ((carries=(WORD)(total>>16))!=0)
        total = (total & 0xffffL) + carries;
    return((WORD)total);
}
```

Sixteen-bit arithmetic is complicated enough on a PICmicro, so 32-bit `long`s are to be avoided at all costs. A cleaner method of calculating the checksum is to take advantage of the inherent symmetry of the algorithm (in addition to the usual lo-byte to hi-byte carry, there is a hi-byte to lo-byte carry) and the inherent eight-bit nature of the PICmicro and to split the calculation into two completely separate halves.

```c
/* Rx interrupt handler */
#INT_RDA
void rx_handler(void)
{
    BYTE b;
    BOOL escflag;

    if (kbhit())                        // Make sure serial character received
    {
        if ((b = getchar()) == SLIP_ESC)
            escflag = 1;                // SLIP escape character?
        else if (b == SLIP_END)
        {                               // SLIP end-of-frame character?
            if (rxin > 0)
            {                           // If non-zero frame length..
                rxflag = 1;             // ..set Rx frame flag
                rxcount = rxin;         // Reset counter for new frame
                rxin = 0;
            }
        }                               // Modem command 'AT...<CR>'?
        else if (b=='\r' && rxbuff[0]=='A')
        {
            modemflag = 1;              // Set flag (can't be handled on interrupt)
            rxin = 0;
        }
        else                            // Normal SLIP character..
        {
            if (escflag)                // ..following an escape char?
                b = b==ESC_END ? SLIP_END : b==ESC_ESC ? SLIP_ESC : b;
            escflag = 0;
            if (rxin == RXHDR_LEN-1)
            {                                   // If nearly at end of hdrs, reset checksum
                rdcheckhi = rdchecklo = 0;
                rdcheckflag = 0;
            }                           // If after headers, calculate checksum
            if (rxin >= RXHDR_LEN)
            {                                   // Alternate between checksum bytes..
```

```
            if (rdcheckflag)
            {                           // Update lo byte value
                if ((rdchecklo = b+rdchecklo) < b)
                {                       // If lo byte overflow, increment hi byte
                    if (++rdcheckhi == 0)
                        rdchecklo++;
                }                       // ..and maybe carry back to lo byte!
            }
            else
            {                           // Update hi byte value
                if ((rdcheckhi = b+rdcheckhi) < b)
                {                       // If hi byte overflow, increment lo byte
                    if (++rdchecklo == 0)
                        rdcheckhi++;
                }                       // ..and maybe carry back to hi byte!
            }
            rdcheckflag = !rdcheckflag;  // Next time, check other byte
        }
        if (rxin < RXBUFFLEN)  // Save char if room left in buffer
        {
            rxbuff[rxin++] = b;
        }
    }
  }
}
```

SLIP Transmit

Transmission is performed by polling, rather than on interrupt. The polling function must

- check to see if transmission is enabled;
- check to see if there is room in the USART for the next character;
- if dropping frames for debug, decide whether this character is to be dropped;
- fetch the next character from RAM — if all are sent, and an E^2ROM file is open, fetch the next character from there; and
- send the character using SLIP encoding — if none are left, send the SLIP end-of-frame character and disable transmission.

Buffer sizes.   I've decided to give the lion's share of the RAM (a whole 96 bytes) to the Transmit buffer. This allows me to demonstrate the generation of Web pages, albeit extremely small ones, on the fly.

```
#define TXBUFFLEN  96         // Tx buffer

BYTE txbuff[TXBUFFLEN];        // Transmit buffer
int txin, txout;               // Buffer I/P and O/P counters
WORD txi2c;                    // Count of i2c bytes to be sent
BOOL txflag;                   // Flag to start sending out Tx data
```

Frame dropping.   By setting an equate at the top of the file to a value n, the Transmit function will discard one-in-n frames. This is a useful debug test to simulate an unreliable network. I typically choose a value of 4, which results in rather slow and erratic Web page updates, but they should still display correctly. For the time being, I'll set the value to zero, so no frames will be dropped.

```
#define TXDROP   0      // Set to 4 to drop 1-in-4 Tx frames for test
```

Transmit ready flag.   You may recall an earlier definition associated with the RS232 initialization.

```
#bit TX_READY =       0x98.1 // Tx ready status bit
```

This allows you to check whether the USART is ready to receive another character using a very simple test.

```
    if (txflag && TX_READY)              // If something to transmit
```

As with a port defined using #BYTE, the Boolean value can be manipulated just like any other variable.

```
/* Tx poll: send next SLIP character if possible, adding escape sequences */
#separate
void tx_poll(void)
{
    static BOOL escflag=0;
    BYTE b;

    if (txflag && TX_READY)              // If something to transmit
    {
#if TXDROP
        if (txout==0 && txin>4)          // Check if dropping frames for test
```

```
            {
            dropcount++;
            if ((dropcount % TXDROP) == 0)
            {
                txflag = 0;
                txin = txout = 0;
            }
        }
    }
#endif
    if (txout == txin)                 // If all RAM headers sent..
    {
        if (txi2c)                     // ..and ROM file to be sent..
        {
            open_file();               // ..open it for O/P
            txout++;
        }
        else                           // All sent: terminate SLIP frame
        {
            putchar(SLIP_END);
            txin = txout = 0;
            txflag = 0;
        }
    }
    else if (txout > txin)             // If sending ROM file
    {
                                       // ..and all sent..
        if (txi2c && !tx_file_byte())
        {                              // ..terminate frame, close file
            putchar(SLIP_END);
            close_file();
            txin = txout = 0;
            txflag = txi2c = 0;
        }
    }
    else                               // If sending RAM header..
    {
        b = read_txbuff(txout);        // ..encode next byte and send it
        if (escflag)
        {                              // Escape char sent, now send value
            putchar(b==SLIP_END ? ESC_END : ESC_ESC);
```

```
                    txout++;
                    escflag = 0;
                }                               // Escape char required
            else if (b==SLIP_END || b==SLIP_ESC)
                {
                    putchar(SLIP_ESC);
                    escflag = 1;
                }
            else                        // Send byte unmodified
                {
                    putchar(b);
                    txout++;
                }
            }
        }
    }
```

Modem Emulation

The receive handler has already checked for the presence of a modem command and set a flag accordingly. If the flag is set, you need to send an "OK" response to the modem.

```
/* Rx poll: check for modem command, send response */
void rx_poll(void)
{
    if (modemflag && !txflag)
    {
        strcpy(txbuff, "OK\r\n");        // Send OK in response to modem cmd
        txin = 4;
        txout = 0;
        txflag = 1;
        modemflag = 0;
    }                                    // Diagnostic; send index if '?'
    else if (rxbuff[0] == '?' && !txflag)
    {
        rxbuff[0] = rxin = 0;
        romdir.f.name[0] = 0;
        find_file();
        txflag = txi2c = 1;
    }
}
```

I have used this opportunity to include some extra diagnostic code for the E^{2}ROM file system. If the first character of a frame is ?, then the default HTML file is transmitted out the serial port. I'll look at the E^{2}ROM files in a lot more detail later.

Protocol Parsing

I looked at protocol-parsing techniques in the previous chapter, with a view to minimizing RAM usage. Basically, I'll call three types of function:

get_*xx*()   Get value from buffer

skip_*xx*()   Skip over value in buffer

match_*xx*(val)   Match the given value to the buffer value

xx is a byte, word, or lword.

Each function returns a Boolean value, which is FALSE if the value could not be obtained (e.g., end-of-file condition) or the given value could not be matched.

These parsing functions work on a *memory image*, because it is assumed that anything that needs parsing (such as the TCP and IP headers) will have been stored in RAM first. I have used the technique to parse bytes as they arrive down the serial link, but this makes debugging very difficult: when parsing fails, there is no memory image to inspect.

Checksum.   As an additional sophistication, I automatically compute the checksum of any parsed bytes. If they are not needed, the resulting value can always be discarded. I'll use the single-byte method described earlier.

```
/* Add byte to checksum value */
void check_byte(BYTE b)
{
    if (checkflag)
    {
        if ((checklo = b+checklo) < b)
        {
            if (++checkhi == 0)
                checklo++;
        }
    }
    else
    {
        if ((checkhi = b+checkhi) < b)
        {
            if (++checklo == 0)
                checkhi++;
```

```
        }
    }
    checkflag = !checkflag;
}
```

The high and low values, and the high/low selector flag will have to be cleared before parsing is started.

The basic character I/P function, on which all the parsing functions are based, returns a single character from the buffer and sets a global flag if the end of the buffer has been reached.

```
/* Get a SLIP byte from buffer; if end, set flag */
BYTE getch_slip(void)
{
    BYTE b=0;

    slipend = rxout>=rxcount;
    if (!slipend)
    {
        b = rxbuff[rxout++];
        check_byte(b);
    }
    return(b);
}
```

Parsing functions use the byte value, and return a FALSE Boolean flag if the end has been reached.

```
/* Get an incoming byte value, return 0 if end of message */
BOOL get_byte(BYTE &b)
{
    b = getch_slip();
    return(!slipend);
}
```

Reference parameters.   I have taken advantage of the compiler's support for reference parameters, also known as call-by-reference, which allows the above function to be used in the following way:

```
BYTE val;

if (get_byte(val))
    ...
```

This is considerably more economical than the usual method of passing a pointer, although there are limitations in the parameters that can be passed.

```
/* Write a byte to the Tx buffer */
void write_txbuff(int &oset, BYTE &b)
{
    txbuff[oset] = b;
}
...
BYTE a=12, b=34;
write_txbuff(a, b);      // Compiles OK
write_txbuff(a, b+1);    // Won't compile: argument is not a variable
write_txbuff(a, 34);     // Won't compile: argument is a constant
```

get_ **functions.**   Having defined get_byte( ), the rest of the functions slide into place.

```
/* Get an incoming word value, return 0 if end of message */
BOOL get_word(WORD &w)
{
    BYTE hi, lo;

    hi = getch_slip();
    lo = getch_slip();
    w = ((WORD)hi<<8) | (WORD)lo;
    return(!slipend);
}
/* Get an incoming lword value, return 0 if end of message */
BOOL get_lword(LWORD &lw)
{
    lw.b[3] = getch_slip();
    lw.b[2] = getch_slip();
    lw.b[1] = getch_slip();
    lw.b[0] = getch_slip();
    return(!slipend);
}
```

The PICmicro processor has no 16-bit data types and completely separate code and data spaces, so it is not inherently big-endian or little-endian. The CCS compiler stores its 16-bit values with the least significant byte first, so you have to take care to convert the big-endian network byte order. As in previous software, you'll always do the conversions as soon as possible on input and as late as possible on output.

An oddity among the get_ functions is get_hexbyte(), which decodes a character string into a hex value, stopping after the last hex digit has been scanned.

```
/* Get an incoming byte value as 1 or 2 hex characters */
BOOL get_hexbyte(BYTE &val)
{
    BYTE b;
    BOOL ok=0;

    val = 0;
    while (isxdigit(rxbuff[rxout]) && get_byte(b))
    {
        ok = 1;
        val <<= 4;
        if (b <= '9')
            val += b - '0';
        else
            val += (b-'A'+10) & 0xf;
    }
    return(ok);
}
```

match_ **functions.**   These functions allow the given value to be matched in the input stream. In the case of a mismatch, the incorrect input character is discarded.

```
/* Match an incoming byte value, return 0 not matched, or end of message */
BOOL match_byte(BYTE b)
{
    return(b==getch_slip() && !slipend);
}
/* Match an incoming byte value, return 0 not matched, or end of message */
BOOL match_word(WORD w)
{
    WORD inw;

    return(get_word(inw) && inw==w);
}
```

There is also a string-matching function that works a little differently: in the case of a mismatch, the input pointer is unaltered so that another match can be tried.

```
/* Match a string value, return 0 (and don't move O/P ptr) if not matched */
BOOL match_str(char *s)
{
    BOOL ok=1;
    int rxo;

    rxo = rxout;
    while (ok && *s)
        ok = match_byte(*s++);
    if (!ok)
        rxout = rxo;
}
```

Unfortunately, this function only works on strings in RAM, so any constant strings must be copied before use.

```
char temps[6];

strcpy(temps, "hello");
if (match_str(temps))
    ...
```

The following code won't work and (worse still) won't generate a warning message.

```
if (match_str("hello"))     // Don't do this!!
    ...
```

skip_ **functions.** Some values in the input stream are of no interest and can be skipped, though they must still be included in the checksum.

```
/* Skip an incoming byte value, return 0 if end of message */
BOOL skip_byte()
{
    getch_slip();
    return(!slipend);
}
```

I won't insult your intelligence by printing out the code for skip_word() and skip_lword().

put_ **functions.** The code for the put_ functions, although important, is unlikely to win many awards for originality.

```
/* Send a byte out to the SLIP link, then add to checksum */
void put_byte(BYTE b)
{
    if (txin < TXBUFFLEN)
    {
        write_txbuff(txin, b);
        txin++;
    }
    check_byte(b);
}
/* Send a null out to the SLIP link */
void put_null(void)
{
    put_byte(0);
}

/* Send a word out to the SLIP link, then add to checksum */
void put_word(WORD w)
{
    put_byte(w >> 8);
    put_byte(w);
}

/* Send a null word out to the SLIP link */
void put_nullw(void)
{
    put_byte(0);
    put_byte(0);
}
void put_lword(LWORD &lw)
{
    put_byte(lw.b[3]);
    put_byte(lw.b[2]);
    put_byte(lw.b[1]);
    put_byte(lw.b[0]);
}
```

The existence of put_null() and put_nullw() derives from my early concerns about the small size of the program ROM and my belief that I had to make every effort to minimize the code size of the protocol handlers. My relaxed coding style in other areas may give you a clue that these fears were largely groundless, so these functions could safely be replaced by put_byte(0) and put_word(0). The use of write_txbuff() is also a luxury that can be dispensed with, although the ability of the compiler to insert functions in-line and the use of call-by-reference parameters mean that it isn't as expensive a luxury as you'd think.

IP Receive

The IP input function is called from the main program loop.

```
rx_poll();                          // Check for Rx modem commands
if (rxflag)                         // If frame received..
{
    rxflag = 0;                     // ..prepare for another
    rxout = 0;
    get_ip();                       // ..and process Rx frame
}
```

It assumes that it will call other protocol handlers as required, each handler being responsible for generating a response frame if necessary.

```
/* Get an IP datagram, send response */
BOOL get_ip(void)
{
    BYTE b, hi, lo;
    int n=0;
    BOOL ret=1;

    slipend = checkflag = 0;                        // Clear checksum
    checkhi = checklo = 0;
    if (match_byte(0x45) && skip_byte() &&          // Version, service
        get_word(iplen) && skip_word() &&           // Len, ID
        skip_word() &&  skip_byte() &&              // Frags, TTL
        get_byte(ipcol) && skip_word() &&           // Protocol, checksum
        get_lword(remote.l) && get_lword(local.l) &&// Addresses
        checkhi==0xff && checklo==0xff)             // Checksum OK?
    {
        if (ipcol==PCOL_ICMP && get_ping_req())     // Ping request?
        {
            tpdlen = rpdlen;                        // Tx length = Rx length
            if (!txflag)
```

```
            {
                put_ip(tpdlen+ICMPHDR_LEN);          // Send ping reply
                put_ping_rep();
            }
        }
        else if (ipcol==PCOL_TCP && get_tcp())       // TCP segment?
            tcp_rx();                                 // Call TCP handler
        else
            discard_data();                           // Unknown; discard it
    }
    else
        discard_data();
    return(ret);
}
```

To help you work through this code, here's the IP header structure I've used in other implementations.

```
/* ***** IP (Internet Protocol) header ***** */
typedef struct
{
    BYTE  vhl,                  /* Version and header len */
          service;              /* Quality of IP service */
    WORD  len,                  /* Total len of IP datagram */
          ident,                /* Identification value */
          frags;                /* Flags & fragment offset */
    BYTE  ttl,                  /* Time to live */
          pcol;                 /* Protocol used in data area */
    WORD  check;                /* Header checksum */
    LWORD sip,                  /* IP source addr */
          dip;                  /* IP dest addr */
} IPHDR;
```

The surprising discovery is that the new function-based parsing method looks so clean and can handle the IP header checksum in such a simple manner. In many ways, the new method is better at handling error conditions. Imagine receiving a few garbage bytes in place of the desired frame. As soon as the parser detects the corruption, it will stop parsing and fall through to discard the whole frame.

To get to the stage of responding to pings (ICMP Echo requests), you need only an IP transmission function and ICMP handlers.

IP Transmit

Normally, you'd first build a memory image of a TCP/IP datagram, checksum it, and send it out. The lack of RAM on the PICmicro makes this approach rather difficult, so in early experimentation, I decided to transmit each (SLIP-encoded) byte on the serial link as soon as it was created. The disadvantage with this approach is that the task of transmitting the bytes dominates the CPU and leaves very little time for anything else.

By careful juggling of the available RAM (and making the Receive buffer smaller than I'd like), I evolved a compromise where the TCP and IP header are buffered in RAM before they are sent out, but the ROM file data is transferred directly to the serial output (via the SLIP encoder). However, the software still bears the signs (scars?) of the earlier method; for example, the put_ functions include a checksum calculation, even though that could be done later on the RAM image. Thus the get_ and put_ functions are analogous: on one side they have a normal serial byte stream (not SLIP encoded), and on the other side they have a memory buffer; they checksum whatever passes between the two.

There is a minor inconvenience in — surprise, surprise — the checksum calculation, in that you must emit the checksum value before the end of the header, that is, before the checksum of the source and destination IP addresses has been computed. As a simple work-around, calculate the checksum of these addresses using specific function calls.

```
/* Send out an IP header, given data length */
void put_ip(WORD len)
{
    static BYTE id=0;

    checkhi = checklo = 0;              // Clear checksum
    checkflag = 0;
    tx_start();
    put_byte(0x45);                     // Version & hdr len */
    put_null();                         // Service
    len += IPHDR_LEN;
    put_byte(len>>8);                   // Length word
    put_byte(len);
    put_null();                         // Ident word
    put_byte(++id);
    put_nullw();                        // Flags & fragment offset
    put_byte(100);                      // Time To Live
    put_byte(ipcol);                    // Protocol
    check_lword(local.l);               // Include addresses in checksum
    check_lword(remote.l);
    put_byte(~checkhi);                 // Checksum
```

```
        put_byte(~checklo);
        put_lword(local.l);                    // Source & destination IP addrs
        put_lword(remote.l);
    }
```

A word of warning in case you're tinkering with the output drivers: don't forget that put_byte() calls that are emitting the checksum values are also changing them as well. If you get these steps in the wrong order, the checksum values will be wrong. It is important that put_byte() sends the character out first and *then* adds it to the checksum.

```
/* Send a byte out to the SLIP link, then add to checksum */
void put_byte(BYTE b)
{
    if (txin < TXBUFFLEN)
    {
        write_txbuff(txin, b);
        txin++;
    }
    check_byte(b);
}
```

ICMP

The only ICMP function I'll implement is the ability to respond to Echo requests (pings).

ICMP Receive

Assuming the IP header has been parsed, it is necessary only to check for the presence of an ICMP Echo request and that the checksum is correct. Here's a reminder of what's in the ICMP header.

```
/* ***** ICMP (Internet Control Message Protocol) header ***** */
typedef struct
{
    BYTE  type,              /* Message type */
          code;              /* Message code */
    WORD  check,             /* Checksum */
          ident,             /* Identifier (possibly unused) */
          seq;               /* Sequence number (possibly unused) */
} ICMPHDR;
```

I'll use two global variables, rpdlen and tpdlen, to record the lengths for incoming and outgoing data. I can make matters simpler by treating the ICMP identifier and sequence number as if they are data. After all, they have to be echoed back to the sender without modification.

```
/* Get an ICMP echo request message, return 0 if error */
BOOL get_ping_req(void)
{
    int i, n=0;
    BYTE b;
    BOOL ret=0;

    checkhi = checklo = 0;
    if (match_byte(8) && match_byte(0) && skip_word())
    {
        rpdlen = 0;
        while (skip_byte())
            rpdlen++;
        ret = (checkhi==0xff) && (checklo==0xff);
    }
    return(ret);
}
```

You may recall that skip_byte() computes a checksum on the byte it discards and returns FALSE when all the data is exhausted, hence, its use as shown above.

ICMP Transmit

The code fragment in get_ip() that sends the ICMP response is repeated below.

```
tpdlen = rpdlen;                        // Tx length = Rx length
if (!txflag)
{
    put_ip(tpdlen+ICMPHDR_LEN);         // Send ping reply
    put_ping_rep();
}
```

After the IP header has been sent out (or rather, has been sent to the Transmit buffer), put_ping_rep() must alter the ICMP header, calculate a new checksum, and echo the data. The following code is suboptimal, but it does the job.

```
/* Put out an ICMP echo response message */
void put_ping_rep(void)
{
    int i;

    put_nullw();                        // Type and code
    checkhi = checklo = 0;              // Clear checksum
```

```
    checkflag = 0;                          // Reset flag in case odd data len
    check_bytes(&rxbuff[IPHDR_LEN+4], tpdlen);   // Calculate checksum of data
    put_byte(~checkhi);                     // Checksum value
    put_byte(~checklo);
    rxout = IPHDR_LEN + 4;
    for (i=0; i<tpdlen; i++)                 // Copy data
        put_byte(rxbuff[rxout++]);
    tx_end();
}
```

Pinging PWEB

The code so far allows you to test PWEB using the Ping utility described in Chapter 4; for example, if the PICmicro is connected to COM2 and the configuration file slip.cfg contains the following lines:

```
net     slip pc com2:38400,n,8,1
ip      172.16.1.2
```

then the PICmicro can be pinged in the following way:

```
ping -c slip 172.16.1.1

PING Vx.xx
IP 172.16.1.2 mask 255.255.0.0 SLIP
Pinging 172.16.1.1 - ESC or ctrl-C to exit
Reply from 172.16.1.1 seq=1 len=32 OK
Reply from 172.16.1.1 seq=2 len=32 OK
Reply from 172.16.1.1 seq=3 len=32 OK
Reply from 172.16.1.1 seq=4 len=32 OK
^C
ICMP echo: 4 sent, 4 received, 0 errors
```

The -v (verbose) flag provides more detail on the transfers, and the -f (flood) flag causes the pings to run much faster.

TCP

Chapter 9 contains a comprehensive review of the methods I will employ to cram TCP onto the PICmicro microcontroller. In this chapter, I'll concentrate on the software itself. The code in get_ip() that calls out TCP handler is reviewed below.

```
else if (ipcol==PCOL_TCP && get_tcp())       // TCP segment?
    tcp_rx();                                // Call TCP handler
```

The get_tcp() function is responsible for parsing and checking the incoming TCP segment, whereas tcp_rx() decides whether to send a response or pass the segment on to an HTTP (Web) handler.

TCP Receive

The TCP parsing function must

- ensure that all of the TCP header is present,
- save the source and destination port numbers,
- save the sequence and acknowledgment numbers, and
- validate the TCP checksum.

The checksum applies to the TCP header and an IP pseudoheader, which was previously defined as two structures.

```
/* ***** TCP (Transmission Control Protocol) header ***** */
typedef struct tcph
{
    WORD   sport,              /* Source port */
           dport;             /* Destination port */
    LWORD  seq,               /* Sequence number */
           ack;              /* Ack number */
    BYTE   hlen,              /* TCP header len (num of bytes << 2) */
           flags;            /* Option flags */
    WORD   window,            /* Flow control credit (num of bytes) */
           check,            /* Checksum */
           urgent;           /* Urgent data pointer */
} TCPHDR;


/* ***** Pseudo-header for UDP or TCP checksum calculation ***** */
/* The integers must be in hi-lo byte order for checksum */
typedef struct                /* Pseudo-header... */
{
    LWORD srce,               /* Source IP address */
          dest;              /* Destination IP address */
    BYTE  z,                 /* Zero */
          pcol;              /* Protocol byte */
    WORD  len;               /* UDP length field */
} PHDR;
```

I'll use the usual mix of get_ and skip_ functions to achieve the same objective as these structures, but without necessarily storing all the values in RAM.

```
/* Get a TCP seqment, return 0 if error */
BOOL get_tcp(void)
{
    int hlen, n;
    BOOL ret=0;
```

```
    checkhi = checklo = 0;
    if (get_word(remport) && get_word(locport) &&    // Source & dest ports
        get_lword(rseq.l) && get_lword(rack.l) &&    // Seq & ack numbers
        get_byte(hlen) && get_byte(rflags) &&        // Header len & flags
        skip_word() && skip_lword())                 // Window, csum, urgent ptr
    {
        iplen -= IPHDR_LEN;                          // Get TCP segment length
        check_byte(iplen>>8);                        // Check pseudoheader
        check_byte(iplen);
        check_lword(local.l);
        check_lword(remote.l);
        check_byte(0);
        check_byte(PCOL_TCP);
        rxout = (hlen>>2) + IPHDR_LEN;
        rpdlen = iplen - rxout + IPHDR_LEN;
        checkhi += rdcheckhi;
        checklo += rdchecklo;
        ret = (checkhi==0xff) && (checklo==0xff);
    }
    return(ret);
}
```

The TCP header checksums are computed as the values are read out from the Receive buffer; then the pseudoheader values are added on, and, finally, the data checksum is computed by the Receive handler. The check_xx() functions are designed for use with data in the native (little endian) byte order of the compiler, so they implicitly swap the byte order before making their calculations.

```
/* Add word to checksum value */
void check_word(WORD w)
{
    check_byte(w>>8);
    check_byte(w);
}
/* Add longword to checksum value */
void check_lword(LWORD &lw)
{
    check_byte(lw.b[3]);
    check_byte(lw.b[2]);
    check_byte(lw.b[1]);
    check_byte(lw.b[0]);
}
```

TCP Transmit

Transmitting a TCP segment is a mirror image of the Receive code. There is no need to create the pseudoheader in RAM; it is only necessary to add the appropriate values using calls to the check_*xx*() functions.

```c
/* Put out a TCP segment. Checksum must be set to correct value for data */
void put_tcp(void)
{
    WORD len;

    checkflag = 0;                      // Ensure we're on an even byte
    put_word(locport);                  // Local and remote ports
    put_word(remport);
    put_lword(rack.l);                  // Seq & ack numbers
    put_lword(rseq.l);
    put_byte(TCPHDR_LEN*4);             // Header len (no options)
    put_byte(tflags);
    put_byte(0x0b);                     // Window size word
    put_byte(0xb8);
    check_lword(local.l);               // Add pseudoheader to checksum
    check_lword(remote.l);
    check_byte(0);
    check_byte(PCOL_TCP);
    len = tpdlen + TCPHDR_LEN;
    check_byte(len>>8);
    check_byte(len);
    checkflag = 0;
    put_byte(~checkhi);                 // Send checksum
    put_byte(~checklo);
    put_nullw();                        // Urgent ptr
    if (!txi2c)                         // If data in RAM (i.e. not in ROM)..
        txin += tpdlen;                 // ..update Tx data pointer
    tx_end();                           // Transmit the packet
}
```

The adjustment to the Transmit buffer input pointer txin reflects the two possible scenarios for sending TCP data: either it is already in RAM and its length is given by tpdlen, or it is in E^2ROM and the length is given by the ROM file directory entry. In the former case, you need to set the buffer pointer beyond the data in RAM so that the Transmit functions knows how much to send.

TCP Sequencer

The heart of the TCP stack is the code that decides what action to take based on the incoming segment. The stack provides two services (i.e., can handle requests to two ports).

Daytime port 13
HTTP port 80

The code off-loads generation of the data to two functions: daytime_rx() and http_rx().

```
/* Handle an incoming TCP segment */
void tcp_rx(void)
{
    BYTE *p, *q;
    BOOL tx=1;

    tpdlen = 0;                             // Assume no Tx data
    tflags = TACK;                          // ..and just sending an ack
    if (txflag || (rflags & TRST))          // RESET received, or busy?
        tx = 0;                             //..do nothing
    else if (rflags & TSYN)                 // SYN received?
    {
        inc_lword(rseq.l);                  // Adjust Tx ack for SYN
        if (locport==DAYPORT || locport==HTTPORT)
        {                                   // Recognized port?
            rack.w[0] = 0xffff;
            rack.w[1] = concount++;
            tflags = TSYN+TACK;             // Send SYN ACK
        }
        else                                // Unrecognized port?
            tflags = TRST+TACK;             // Send reset
    }
    else if (rflags & TFIN)                 // Received FIN?
        add_lword(rseq.l, rpdlen+1);        // Ack all incoming data + FIN
    else if (rflags & TACK)                 // ACK received?
    {
        if (rpdlen)                         // Adjust Tx ack for Rx data
            add_lword(rseq.l, rpdlen);
        else                                // If no data, don't send ack
            tx = 0;
        if (locport==DAYPORT && rack.w[0]==0)
        {                                   // Daytime request?
            daytime_rx();                   // Send daytime data
```

```
        tx = 0;
    }
    else if (locport==HTTPORT && rpdlen)
    {                                   // HTTP 'get' method?
        if (http_rx())                  // Send HTTP data & close
            tx = 0;
        else                            // ..or just close connection
            tflags = TFIN+TACK;
    }
}
if (tx)                                 // If ack to send..
{
    put_ip(TCPHDR_LEN);                 // ..send IP header
    checkhi = checklo = 0;              // ..reset checksum
    put_tcp();                          // ..send TCP header
}
}
```

I'll examine this code by looking at how it responds to the various TCP segments that can be received.

Reset. If you receive a reset, you mustn't send a response; otherwise, you might get another reset in response, and so on ....

Invalid port number. The client might request a service you don't support, in which case, send a TCP reset. The code does this check only for SYN. There is no reason why you should receive an ACK or FIN with an invalid port number (although it might be argued that you should still check for the possibility).

SYN. When the client wants to open a connection, you create a new sequence number using your connection count as the high word and a value of FFFFh as the low word and respond with an acknowledgment value one greater than the received sequence number (because SYN counts as one byte in the sequence space).

FIN. When the client wants to close the connection, most probably in response to your FIN, respond with an acknowledgment value reflecting all the data in the received segment (in the unlikely event there is any) plus one (for the FIN).

ACK. First bump up the received sequence number to reflect the data received — you may not actually be interested in the data, but if you don't do this, the client will only resend it. If there isn't any data, suppress any automatic response: the segment is most likely an ACK of your SYN or FIN. Now check the low word of the acknowledgment number. If it's zero, this must be an ACK of your SYN, possibly with some data, so it is the right time to call the Daytime or HTTP handler.

Daytime Service

The response to a Daytime request is a simple string giving the date and time. I'm going to return only the time because this is intended only as a demonstration of a simple TCP transaction. The sequence of events follows:

1. Client initiates a connection to port 13.
2. Client may send some data, but it is discarded.
3. Server generates a single-segment response containing the time string.
4. Server closes the connection.
5. Client acknowledges the closure.

The main work is done by the TCP sequencer, leaving only the Daytime TCP segment to be generated. The closure is signaled by including the FIN flag with the data.

```
/* Handle an incoming daytime request */
void daytime_rx()
{
    tpdlen = DAYTIME_LEN;                    // Data length of response
    get_rtc_time();                          // Read clock
    put_ip(TCPHDR_LEN+tpdlen);               // Send IP header
    checkhi = checklo = 0;                   // Reset checksum
    txin = IPHDR_LEN + TCPHDR_LEN;           // O/P data to buffer, calc checksum
    printf(put_byte, DAYTIME_STR, rtc.b[2], rtc.b[1], rtc.b[0]);
    txin = IPHDR_LEN;                        // Go back to end of IP header
    tflags = TFIN+TACK;                      // O/P TCP header
    put_tcp();
}
```

Testing the Daytime Service

Using the same SLIP configuration file as for Ping, the Telnet utility from the book can be used to check the Daytime service using the following command line:

```
telnet -c slip 172.16.1.1 daytime
```

This would generate something like the following response:

```
TELNET vx.xx
IP 172.16.1.2 mask 255.255.0.0
Press ESC or ctrl-C to exit


00:01:36
```

The time value is artificially low because my prototype system uses a clock without battery backup, and the means to set the clock (an HTML form) isn't included in this chapter.

Using the -t (TCP) and -v (verbose) Telnet options produces a more informative display.

```
TELNET vx.xx
IP 172.16.1.2 mask 255.255.0.0
Press ESC or ctrl-C to exit

    /ack 00000000 seq bef52a5b port 1025->13 <SYN> MSS 966 dlen 0h
Tx0 /len 44 ------SLIP------- IP 172.16.1.2 -> 172.16.1.1 TCP
Rx0 \len 40 ------SLIP------- IP 172.16.1.1 -> 172.16.1.2 TCP
    \seq 0003ffff ack bef52a5c port 1025<-13 <SYN><ACK> dlen 0h
    /ack 00040000 seq bef52a5c port 1025->13 <ACK> dlen 0h
Tx0 /len 40 ------SLIP------- IP 172.16.1.2 -> 172.16.1.1 TCP
Rx0 \len 50 ------SLIP------- IP 172.16.1.1 -> 172.16.1.2 TCP
    \seq 00040000 ack bef52a5c port 1025<-13 <FIN><ACK> dlen Ah
    /ack 0004000b seq bef52a5c port 1025->13 <FIN><ACK> dlen 0h
Tx0 /len 40 ------SLIP------- IP 172.16.1.2 -> 172.16.1.1 TCP
00:08:30
Rx0 \len 40 ------SLIP------- IP 172.16.1.1 -> 172.16.1.2 TCP
    \seq 0004000b ack bef52a5d port 1025<-13 <ACK> dlen 0h
```

Using the standard Windows Telnet utility instead produces a slightly different result. To obtain a diagnostic trace, I've had to eavesdrop on the serial communications using my NET-MON utility, which is described in Chapter 3.

```
netmon -c slip2 -t -v

NETMON vx.xx
Net 1 SLIP
Net 2 SLIP
Press ESC or ctrl-C to exit
Rx0 \len 64 ------SLIP------- IP 172.16.1.2 -> 172.16.1.1 TCP
    \seq 0287e09f ack 00000000 port 13<-1115 <SYN> MSS 1460 dlen 0h
Rx1 \len 40 ------SLIP------- IP 172.16.1.1 -> 172.16.1.2 TCP
    \seq 0006ffff ack 0287e0a0 port 1115<-13 <SYN><ACK> dlen 0h
Rx0 \len 40 ------SLIP------- IP 172.16.1.2 -> 172.16.1.1 TCP
    \seq 0287e0a0 ack 00070000 port 13<-1115 <ACK> dlen 0h
Rx1 \len 50 ------SLIP------- IP 172.16.1.1 -> 172.16.1.2 TCP
    \seq 00070000 ack 0287e0a0 port 1115<-13 <FIN><ACK> dlen Ah
00:12:48\r\n
Rx0 \len 40 ------SLIP------- IP 172.16.1.2 -> 172.16.1.1 TCP
    \seq 0287e0a0 ack 0007000b port 13<-1115 <ACK> dlen 0h
```

There is now a pause; Windows Telnet displays the time string in a dialogue box, indicating that the "connection was lost." From the last transmission, you can see that the connection is, in fact, half-Closed; Telnet has yet to send a FIN.

Clicking on the OK button causes the connection to be closed.

```
Rx0 \len 40 ------SLIP------- IP 172.16.1.2 -> 172.16.1.1 TCP
    \seq 0287e0a0 ack 0007000b port 13<-1115 <FIN><ACK> dlen 0h
Rx1 \len 40 ------SLIP------- IP 172.16.1.1 -> 172.16.1.2 TCP
    \seq 0007000b ack 0287e0a1 port 1115<-13 <ACK> dlen 0h
```

Summary

You now have a fully functional TCP/IP stack on the PICmicro, which will form the basis for the miniature Web server discussed in Chapter 11. I defined the hardware and included some real-world signals in the form of switches, lamps, and a temperature sensor. There is also a real-time clock and 32Kb external memory for Web pages. The I^{2}C bus, used for interconnections, simplified the hardware while adding little extra complexity to the software.

Using the techniques discussed in Chapter 9, it was possible to incorporate a TCP stack — which is oriented toward server operation — without overflowing the small amount of RAM or the modest amount of ROM available.

Source Files

There is no linker, so library files are included in the main source file.

pweb.c	Main program (includes Web server in Chapter 11)
pslip.h	Low-level serial functions
ptcp.h	TCP functions

Chapter 11

PWEB: Miniature Web Server for the PICmicro®

Overview

In the previous two chapters, I looked at the theoretical aspects of miniaturizing a TCP/IP stack so that it fits onto a microcontroller and at a practical TCP implementation on a PIC16C76.

I now turn your attention to the PICmicro[1] Web server itself and look at the ways in which you can provide a range of Web pages for I/O control and monitoring, and I introduce dynamic content within the severe constraints imposed by the microcontroller architecture.

Web Server

In Chapter 10, I got as far as the implementation of a Daytime server. This accepts an incoming TCP connection from a client, discards any incoming data, generates a response string, and closes the connection. The principal addition required by a Web server is that it must accept a request string, indicating the document that is to be provided. This request is in hypertext transfer protocol (HTTP) format, and the Web server must send an HTTP response header, followed by the required document.

---

1. PICmicro® is the registered trademark of Microchip Technology Inc.

I'm using the term "document" loosely to describe a resource (file) on the server that is sent to the client. For the most part, the server doesn't care what the content of the documents is; it just puts an HTTP header on them and sends them out. There are a few exceptions to this rule.

Content-type.   It is important that the HTTP header indicate the type of the data (HTML text, GIF graphic, etc.). Older browsers were able to deduce the content by looking at the file, but modern browsers expect to be told what the content is, even when it seems incredibly obvious.

Gateway interface.   Some files may contain placeholder variables, which must be filled in with actual values as the file is sent out. The server must recognize such files and perform the necessary actions on them. The standard method for doing this on a multi-user system is called server-side includes (SSI) in the common gateway interface (CGI). In Chapter 8, I looked at the problems associated with using SSI and CGI on an embedded system and developed an embedded gateway interface (EGI) as an alternative.

Generation on the fly.   It is possible for a server to generate a document completely from scratch, without accesses to the file system. I'll start off by exploring this technique for generating simple documents, though it is rather too resource hungry to generate meaningful content on a PICmicro.

Simple Text Server

To start, I'll generate a simple Web page on the fly, using the text string from the Daytime server as my model. In Chapter 10, you saw how the HTTP handler is called by the TCP sequencer when the HTTP request string is received.

```
...
else if (locport==HTTPORT && rpdlen)
{                                   // HTTP 'get' method?
    if (http_rx())                  // Send HTTP data & close
        tx = 0;
    else                            // ..or just close connection
        tflags = TFIN+TACK;
    ...
}
```

The `http_rx()` function must decode the request string, send a response, and return a non-zero value to indicate it has done so.

The HTTP request string consists of several text lines, each terminated by carriage return (CR) and line feed (LF) characters. The first line specifies the required filename and the HTTP version being employed. Subsequent lines give additional information that may be of use to the server, such as the identity of the requester and the types of files that can be accepted, for example.

```
GET /index.htm HTTP/1.0
User-Agent: Mozilla/4.5 [en] (Win95; I)
Pragma: no-cache
Host: 10.1.1.11
Accept: image/gif, image/x-xbitmap, image/jpeg, image/pjpeg, image/png,
*/*
Accept-Encoding: gzip
Accept-Language: en
Accept-Charset: iso-8859-1,*,utf-8
```

My TCP implementation will attempt to buffer as much of this as possible, but because of severe RAM constraints, that isn't a lot! So long as the filenames are kept reasonably short (e.g., 8.3 format), they should be received without truncation. Mercifully, you can discard all the additional information. If the client has requested a file it can't display, there's not a lot you can do about it. Likewise, you can ignore the HTTP version number because you only support the simplest of transactions anyway.

To start simply, I'll look for a leading $ character in the requested filename.

```
GET /$...
```

When you see this, return a time string and close the connection. To keep the browser happy, you'll have to prefix the string with an HTTP response header, with a content type of "plain text."

```
HTTP/1.0 200 OK
Content-type: text/plain

12:34:56
```

It is essential to include a single blank line after the header to distinguish it from the data that follows.

The code to generate this response is very similar to the Daytime server. It must take the following actions:

- Match the GET string.
- Skip the following space.
- Check for dummy filename $.
- Get the current time.
- Calculate the total data length.
- Send out the IP header using this length.
- Clear the checksum.
- Print the response string to the output buffer, calculating its checksum.
- Output the TCP header, including the data checksum.

    The only surprise in the code below might be the use of a function call in printf().

```
    printf(put_byte, "Hello");
```

The string is passed to put_byte() one character at a time so that it can be stored in the output buffer and added to the checksum. This feature is incredibly useful because you can specify an arbitrary function to receive the output of printf(). However, I would have preferred it if they had chosen a new name for this nonstandard print function.

In case you're wondering why I declare the string length separately from the string, instead of using `sizeof()` or `strlen()`, the answer is that these functions don't seem to work on string constants in the current release of the compiler, which is a bit of a nuisance.

```
#define TEXT_OK_LEN      45  // HTTP header for plain text
#define TEXT_OK          "HTTP/1.0 200 OK\r\nContent-type: text/plain\r\n\r\n"
#define DAYTIME_LEN      10  // Format string for daytime response
#define DAYTIME_STR      "%02x:%02x:%02x\r\n"


/* Receive an incoming HTTP request ('method'), return 0 if invalid */
BOOL http_rx(void)
{
    int len, i;
    BOOL ret=0;
    char c;

    tpdlen = 0;                             // Check for 'GET'
    if (match_byte('G') && match_byte('E') && match_byte('T'))
    {
        ret = 1;
        skip_space();
        match_byte('/');                    // Start of filename
        if (rxbuff[rxout] == '$')           // If dummy file starting wth '$'
        {
                                            // ..put out simple text string
            tpdlen = TEXT_OK_LEN + DAYTIME_LEN;
            get_rtc_time();                 // ..consisting of current time
            put_ip(TCPHDR_LEN+tpdlen);
            checkhi = checklo = 0;
            txin = IPHDR_LEN + TCPHDR_LEN;
            printf(put_byte, TEXT_OK);  // ..with HTTP header
            printf(put_byte, DAYTIME_STR, rtc.b[2], rtc.b[1], rtc.b[0]);
            txin = IPHDR_LEN;
            tflags = TFIN+TPUSH+TACK;
            put_tcp();
        }
    }
}
```

TCP Transaction Log

This first Web page is pretty unspectacular. I haven't included a screen-shot, so you'll have to imagine those eight plain-text characters in all their glory. Of more interest is to check out the TCP transaction, for which purpose you need to use NETMON to spy on the serial link (as discussed in Chapter 3).

```
netmon -c slip2 -t -v
NETMON vx.xx
Net 1 SLIP
Net 2 SLIP
Press ESC or ctrl-C to exit

--CLIENT--
Rx0 \len 64 ------SLIP------- IP 172.16.1.2 -> 172.16.1.1 TCP
    \seq 015aef62 ack 00000000 port 80<-1180 <SYN> MSS 1460 dlen 0h

--SERVER--
Rx1 \len 40 ------SLIP------- IP 172.16.1.1 -> 172.16.1.2 TCP
    \seq 0000ffff ack 015aef63 port 1180<-80 <SYN><ACK> dlen 0h

--CLIENT - IGNORED BY SERVER--
Rx0 \len 40 ------SLIP------- IP 172.16.1.2 -> 172.16.1.1 TCP
    \seq 015aef63 ack 00010000 port 80<-1180 <ACK> dlen 0h
--CLIENT--
Rx0 \len 368 ------SLIP------- IP 172.16.1.2 -> 172.16.1.1 TCP
    \seq 015aef63 ack 00010000 port 80<-1180 <PSH><ACK> dlen 148h
GET /$ HTTP/1.1\r\nAccept: image/gif, image/x-xbitmap, image/jpeg, image

--SERVER--
Rx1 \len 95 ------SLIP------- IP 172.16.1.1 -> 172.16.1.2 TCP
    \seq 00010000 ack 015af0ab port 1180<-80 <FIN><PSH><ACK> dlen 37h
HTTP/1.0 200 OK\r\nContent-type: text/plain\r\n\r\n14:59:37\r\n
--CLIENT - IGNORED BY SERVER--
Rx0 \len 40 ------SLIP------- IP 172.16.1.2 -> 172.16.1.1 TCP
    \seq 015af0ab ack 00010038 port 80<-1180 <ACK> dlen 0h  [ignored]
--CLIENT--
Rx0 \len 40 ------SLIP------- IP 172.16.1.2 -> 172.16.1.1 TCP
    \seq 015af0ab ack 00010038 port 80<-1180 <FIN><ACK> dlen 0h

--SERVER--
Rx1 \len 40 ------SLIP------- IP 172.16.1.1 -> 172.16.1.2 TCP
    \seq 00010038 ack 015af0ac port 1180<-80 <ACK> dlen 0h
```

I've inserted some labels for clarity. It is interesting to note that the client sends two ACK cycles that are unnecessary (though completely harmless) and that are ignored by the server.

The frame from the client to the server containing the Web page is 95 bytes long, and that from the serial output buffer is 96 bytes in size, which is the maximum contiguous value allowed by the PICmicro. Clearly, the one remaining byte gives you little scope for enhancing this Web page, so you must create a ROM file system if this Web server is to serve any useful purpose.

ROM File System

When specifying the hardware in the last chapter, I included a 32Kb E²ROM for Web pages. This is accessed serially using the I²C protocol, and I gave sample code that drives peripherals. ROM accesses aren't lightning fast, but they are adequate for the purposes here. About 8K bytes per second is a typical measured value when using a 7.3728MHz CPU clock.

E²ROM Access Cycle

The ROM has a built-in data pointer that auto-increments through the address range. To set the pointer, you need to send a write cycle, but, thereafter, a contiguous area of memory can be read by successive read cycles (Figure 11.1).

Figure 11.1   E²ROM accesses.

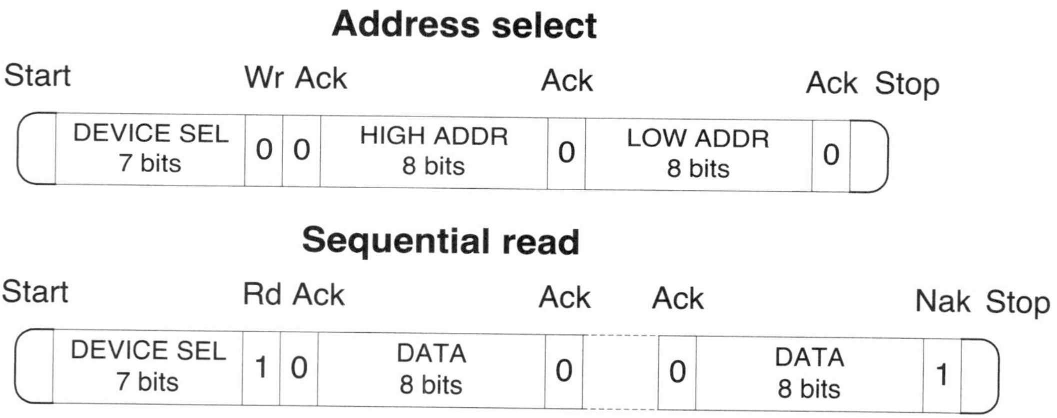

Address select

Start		Wr Ack			Ack		Ack Stop
DEVICE SEL 7 bits	0	0	HIGH ADDR 8 bits	0	LOW ADDR 8 bits	0	

Sequential read

Start		Rd Ack			Ack	Ack		Nak Stop
DEVICE SEL 7 bits	1	0	DATA 8 bits	0	0	DATA 8 bits	1	

The device selector (or I²C address) of a device is the value that distinguishes one device from all others on the I²C bus. It is a seven-bit value, followed by a single read/write bit, so it is convenient to combine both into a single byte.

```
#define EEROM_ADDR   0xa0   // i2c addr for eerom
```

This value is correct for write cycles, but the least-significant bit must be set for read cycles. The acknowledgment bit is asserted (pulled low) by the *recipient* of the data, i.e., the CPU for read cycles and the E²ROM for write cycles. It is most important that the CPU does *not* acknowledge the last byte in any read cycle before sending the stop bit.

The most efficient way to read a file from ROM is to use one long cycle, that is, do not stop and start the cycle for each byte read. There is no lower limit on the speed of I²C transfers, so you can fetch the next byte from the file whenever it suits you, with the proviso that you must terminate the ROM access cycle before you can start accessing any other I²C device.

File System Structure

You'll need to find a file in the system using the filename as an identifier. This implies a file directory of some sort, preferably in a block at the start of the ROM. This would allow you to search the directory using sequential read cycles until the desired file was found and would minimize the number of address select cycles you would have to emit.

The elements required in the directory are

- the length of the file in bytes,
- a pointer to the start of the file in ROM,
- a TCP checksum of the file,
- flags to enable EGI variable substitution, and
- a filename (lowercase, 8.3 format).

This results in the following directory structure:

```
#define ROM_FNAMELEN      12        /* Maximum filename size */
typedef struct                      /* Filename block structure */
{
    WORD len;                       /* Length of file in bytes */
    WORD start;                     /* Start address of file data in ROM */
    WORD check;                     /* TCP checksum of file */
    BYTE flags;                     /* Embedded Gateway Interface (EGI) flags */
    char name[ROM_FNAMELEN];        /* Lower-case filename with extension */
} ROM_FNAME;

/* Embedded Gateway Interface (EGI) flag values */
#define EGI_ATVARS        0x01      /* '@' variable substitution scheme */
#define EGI_HASHVARS      0x02      /* '#' and '|' boolean variables */
```

The end of the directory area in ROM is identified by an entry with a dummy length value of FFFFh. The total address space is limited to 64Kb by the use of 16-bit values. If this ever becomes a problem, the unused flag bits could be used to enable some form of extended format.

There are some additional PWEB-specific requirements for the file system.

Default page.   The default Web page, index.htm, must be easy to find, so it should be the first file in the ROM.

HTTP header.   String manipulation is rather difficult on the PICmicro because of the strict segregation of string variables and constants and the acute lack of spare RAM. The job of identifying the file type and prefixing the file with the correct HTTP response header (including the content type) is hard work for a PICmicro, but it is very easy for a DOS utility. Hence, the files should all be stored in ROM with their HTTP headers already attached.

EGI flags.   The EGI flags need to be set if EGI variable substitution is to be performed on the file as it is sent out. The precise nature of these substitutions and the meaning of the flags will be described in "Variable Substitution" on page 351.

Programming the E²ROM

I originally intended to include the ability to download files into ROM over the network, but I ran out of time to implement this, so the ROM image is prepared using a DOS utility and blown into ROM using a conventional PROM programmer.

The utility is called WEBROM, and it takes the files in a given directory and prepares a single ROM image containing them all.

```
webrom pweb.rom webdocs
```

This takes all files in the webdocs directory and creates a ROM image, pweb.rom, from them. The utility must take the following actions:

- Scan the directory, make a list of files, find index.htm, and work out the length of each file, including an appropriate HTTP header.
- Create the directory, including pointers to where the files will be.
- Copy out the files.

A file extension of .egi is used to determine that the file requires EGI variable substitution, so the EGI flags are set accordingly.

Definitions

```
#define INDEXFILE    "index.htm"
#define MAXFILES     500
#define MAXFILELEN   32000


char filenames[MAXFILES][ROM_FNAMELEN+1];
char *filehdrs[MAXFILES];

char srcepath[MAXPATH+8], srcedir[MAXPATH+8], srcefile[MAXPATH+8];
char destfile[MAXPATH+8];
int netdebug;

ROM_FNAME romfname;
char filedata[MAXFILELEN+100];

/* HTTP and HTML text */
#define HTTP_OK     "HTTP/1.0 200 OK\r\n"
#define HTTP_HTM    "Content-type: text/html\r\n"
```

```
#define HTTP_TXT     "Content-type: text/plain\r\n"
#define HTTP_GIF     "Content-type: image/gif\r\n"
#define HTTP_XBM     "Content-type: image/x-xbitmap\r\n"
#define HTTP_BLANK   "\r\n"
```

Initial Command-Line Parsing

```
void main(int argc, char *argv[])
{
    char *p, *fname;
    FILE *out;
    int i, nfiles=1, err=0;
    unsigned len;
    long filelen, romoff;
    WORD endw=0xffff;
    ROM_FNAME *rfp;

    printf("WEBROM v" VERSION "\n");              /* Sign on */
    printf("File sizes include HTTP headers\n");
    rfp = &romfname;
    if (argc < 3)
    {
        if (argc < 2)
            printf("No destination file specified\n");
        else
            printf("No source filepath specified\n");
        printf("e.g. WEBROM test.rom c:\\temp\\romdocs\n");
        exit(1);
    }
    strcpy(destfile, argv[1]);
    if ((p=strrchr(destfile, '.'))==0 || !isalpha(*(p+1)))
        strcat(destfile, ".ROM");
    strlwr(destfile);
    if (argv[2][0]!='\\' && argv[2][1]!=':' && argv[2][0]!='.')
        strcpy(srcepath, ".\\");
    strcat(srcepath, argv[2]);
    if (srcepath[strlen(srcepath)-1] != '\\')
        strcat(srcepath, "\\");
    strlwr(srcepath);
    strcat(strcpy(srcedir, srcepath), "*.*");
```

First pass. Scan the source directory and compile a list of files and headers. Put index.htm at the start of the list.

```
/* First pass: get files in source directory, check for index.htm */
if ((fname = find_first(srcedir)) != 0) do
{
    if (strlen(fname) > ROM_FNAMELEN)
        printf("ERROR: long filename '%s' not included\n", fname);
    else if (!stricmp(fname, INDEXFILE))
    {
        strcpy(filenames[0], fname);
        strlwr(filenames[0]);
        filehdrs[0] = HTTP_OK HTTP_HTM HTTP_BLANK;
    }
    else if (strlen(fname) > 2)
    {
        strcpy(filenames[nfiles], fname);
        strlwr(filenames[nfiles]);
        filehdrs[nfiles] =
            strstr(fname, ".htm") ? HTTP_OK HTTP_HTM HTTP_BLANK :
            strstr(fname, ".egi") ? HTTP_OK HTTP_HTM HTTP_BLANK :
            strstr(fname, ".txt") ? HTTP_OK HTTP_TXT HTTP_BLANK :
            strstr(fname, ".gif") ? HTTP_OK HTTP_GIF HTTP_BLANK :
            strstr(fname, ".xbm") ? HTTP_OK HTTP_XBM HTTP_BLANK :
            HTTP_OK HTTP_BLANK;
        nfiles++;
    }
} while ((fname = find_next()) != 0 && nfiles<MAXFILES);
if (!filenames[0][0])
    printf("ERROR: default file '%s' not found\n", INDEXFILE);
else if (nfiles > MAXFILES)
    printf("ERROR: only %u files allowed\n", MAXFILES);
else if ((out=fopen(destfile, "wb"))==0)
    printf("ERROR: can't open destination file\n");
else
{
```

Second pass. Write out the directory, including pointers, to where the files will be; then write a dummy length (FFFFh) to mark the end of the directory,

```
/* Second pass: create ROM directory */
printf("Creating %s using %u files from %s\n",
    destfile, nfiles, srcepath);
romoff = nfiles * sizeof(ROM_FNAME) + 2;
for (i=0; i<nfiles && !err; i++)
{
    printf("%12s ", fname=filenames[i]);
    if ((len = readfile(filedata, fname, filehdrs[i])) == 0)
    {
        printf("ERROR reading file\n");
        err++;
    }
    else
    {
        printf("%5u bytes\n", len);
        rfp->start = (WORD)romoff;
        rfp->len = len;
        rfp->check = csum(filedata, (WORD)len);
        rfp->flags = strstr(fname, ".egi") ? EGI_ATVARS+EGI_HASHVARS:0;
        memset(rfp->name, 0, ROM_FNAMELEN);
        strncpy(rfp->name, fname, strlen(fname));
        fwrite(rfp, 1, sizeof(ROM_FNAME), out);
        romoff += rfp->len;
    }
}
fwrite(&endw, 1, 2, out);
```

Third pass. Copy the file contents.

```
/* Third pass: write out file data */
for (i=0; i<nfiles && !err; i++)
{
    if ((len = readfile(filedata, filenames[i], filehdrs[i])) == 0)
    {
        printf("ERROR reading '%s'\n", filenames[i]);
        err++;
    }
```

```
                else
                    fwrite(filedata, 1, len, out);
            }
        if (!err)
            {
            filelen = ftell(out);
            if (filelen == romoff)
                printf("\nTotal ROM    %5lu bytes\n", filelen);
            else
                printf("ERROR: ROM size %lu, file size %lu\n", romoff,filelen);
            if (ferror(out))
                printf("ERROR writing output file\n");
            }
        fclose(out);
    }
}
```

Utility Functions

```
/* Read the HTTP header and file into buffer, return total len, 0 if error */
unsigned readfile(char *buff, char *fname, char *hdr)
{
    unsigned len;
    FILE *in;

    strcpy(buff, hdr);
    len = strlen(hdr);
    strcat(strcpy(srcefile, srcepath), fname);
    in = fopen(srcefile, "rb");
    len += fread(&buff[len], 1, MAXFILELEN, in);
    if (ferror(in))
        len = 0;
    fclose(in);
    return(len);
}


/* Return size of opened file in bytes */
long filesize(FILE *stream)
{
    long curpos, length;
```

```
    curpos = ftell(stream);
    fseek(stream, OL, SEEK_END);
    length = ftell(stream);
    fseek(stream, curpos, SEEK_SET);
    return(length);
}
```

This program also uses the standard TCP/IP checksum function csum() in netutil.c.

The resulting ROM file can be programmed into the E^{2}ROM using a PROM programmer equipped with a suitable adaptor. I use the EMP20 parallel port programmer from Needham's Electronics with personality module 18.

Finding a File in ROM

The filename arrives as a string in the HTTP GET method.

```
GET /index.htm HTTP/1.0
```

The filename isn't necessarily delimited by a space; later on, I'll look at HTML forms, where the form response values are separated from the filename by a question mark.

```
GET /resp.egi?val=1 HTTP/1.0
```

You need to create a storage area for your filename and the associated data you'll get from the directory. To make the code cleaner, create a union of the file directory structure with a simple byte array, because the latter will be required when fetching the directory data byte by byte from the E^{2}ROM.

```
typedef union                     // ROM file directory entry format
{
    ROM_FNAME f;                      // Union of filename..
    BYTE b[sizeof(ROM_FNAME)];  // ..with byte values for i2c transfer
} ROM_DIR;

ROM_DIR romdir;                   // Storage for one directory entry
int fileidx;                      // Index of current file (1=first, 0=error)
```

Now add a simple filename copying function to http_rx().

```
for (i=0; i<ROM_FNAMELEN; i++)
{
    c = rxbuff[rxout];
    if (c>' ' && c!='?')        // Name terminated by space or '?'
        rxout++;
    else
        c = 0;
```

```
        romdir.f.name[i] = c;
}                               // If file found in ROM
if (find_file())
    ...
```

The `find_file()` function must trawl through the ROM directory to match the filename and fill in the structure elements (length, file pointer, checksum, flags). As an added convenience, you can give the function a null-length filename, and it will return the first file in ROM.

The quickest way to fetch directory entries is by sequential-byte reads, so I have used this as an excuse to employ a rather stodgy linear search technique. In practice, the greatest time constraint is imposed by the speed of the file transfer itself, so I doubt whether a greater degree of sophistication in this area would bring any real benefit.

```
/* Find a filename in ROM file system. Return false if not found
** Sets fileidx to 0 if ROM error, 1 if file is first in ROM, 2 if 2nd..
** and leaves directory info in 'romdir'
** If the first byte of name is zero, match first directory entry */
BOOL find_file(void)
{
    BOOL mismatch=1, end=0;
    int i;
    BYTE b;
    char temps[ROM_FNAMELEN];

    fileidx = 0;                            // Set ROM address pointer to 0
    i2c_start();
    i2c_write(EEROM_ADDR);
    i2c_write(0);
    i2c_write(0);
    i2c_stop();
    do
    {
        i2c_start();                        // Read next directory entry
        i2c_write(EEROM_ADDR | 1);
        if ((romdir.b[0] = i2c_read(1)) == 0xff)
        {                                   // Abandon if no entry
            end = 1;
            i2c_read(0);
        }
        else
```

```
        {                           // Get file len, ptr, csum and flags
        for (i=1; i<7; i++)
            romdir.b[i] = i2c_read(1);
        mismatch = 0;               // Try matching name
        for (i=0; i<ROM_FNAMELEN; i++)
        {
            temps[i] = b = i2c_read(i<ROM_FNAMELEN-1);
            if (b != romdir.f.name[i])
                mismatch = 1;
        }
        if (!romdir.f.name[0])       // If null name, match anything
            mismatch = 0;
    }
    i2c_stop();                     // Loop until matched
} while (!end && fileidx++<MAXFILES && mismatch);
if (mismatch)
    romdir.f.len = 0;
return(!mismatch);
}
```

ROM File Not Found

If the file can't be found, I re-search using a null-length filename so that the index file will be found instead. This search may fail as well. A blank ROM may have been fitted to the board by mistake, for example. In this case, http_rx() returns a simple error message to the client.

```
#define HTTP_FAIL_LEN   34  // HTTP string for internal error
#define HTTP_FAIL       "HTTP/ 200 OK\r\n\r\nPWEB ROM error\r\n"
```

It's a pity you can't use sizeof() or strlen() on a string constant.

```
if (find_file())
{
    ...
}
else                    // File not found, get index.htm
{
    romdir.f.name[0] = 0;
    find_file();
}
if (!fileidx)                   // No files at all in ROM - disaster!
{
    tpdlen = HTTP_FAIL_LEN;      // Inform user of failure..
```

```
    put_ip(TCPHDR_LEN+tpdlen);    // Output IP header to buffer
    checkhi = checklo = 0;        // Reset checksum
    strcpy(&txbuff[IPHDR_LEN+TCPHDR_LEN], HTTP_FAIL);
    txin = IPHDR_LEN + TCPHDR_LEN;
    check_txbytes(tpdlen);        // Checksum data
    txin = IPHDR_LEN;             // Go back to end of IP header
    tflags = TFIN+TPUSH+TACK;
    put_tcp();                    // Output TCP header to buffer
}
```

Maybe 200 OK isn't the most appropriate HTTP return code I could have chosen.

ROM File Transfer

Having found the file in ROM, you need to transfer it over the SLIP link. The file must first be opened, that is, the ROM file pointer is positioned at the correct location.

```
/* Open the previously-found file for transmission */
BOOL open_file(void)
{
    i2c_start();
    i2c_write(EEROM_ADDR);                  // Write start pointer to eerom
    i2c_write(romdir.f.start >> 8);
    i2c_write(romdir.f.start);
    i2c_stop();
    i2c_start();
    i2c_write(EEROM_ADDR | 1);              // Restart ROM access as read cycle
}
```

For now, the task of fetching the next byte from the I2C link to the serial port is quite mundane. Soon I'll liven things up by inserting on-the-fly EGI variable substitution here.

```
/* Transmit a byte from the current i2c file to the SLIP link
** Return 0 when complete file is sent */
BOOL tx_file_byte(void)
{
    int ret=0, idx;
    BYTE b;

    if (romdir.f.len)                       // Check if any bytes left to send
    {
        b = i2c_read(1);                    // Get next byte from ROM
        if (...)                            // EGI processing will be inserted here
            ...
```

```
        else                            // Non-EGI byte; send out unmodified
            tx_byte(b);
        romdir.f.len--;
        ret = 1;
    }
    return(ret);
}
```

File closure involves issuing a dummy read without an ACK cycle and then a stop bit.

```
/* Close the previously-opened file */
void close_file(void)
{
    i2c_read(0);                        // Dummy read cycle without ACK
    i2c_stop();
}
```

Revised HTTP Handler

The HTTP handler needs to be updated to reflect all these new elements.

```
/* Receive an incoming HTTP request ('method'), return 0 if invalid */
BOOL http_rx(void)
{
    int len, i;
    BOOL ret=0;
    char c;

    tpdlen = 0;                         // Check for 'GET'
    if (match_byte('G') && match_byte('E') && match_byte('T'))
    {
        ret = 1;
        skip_space();
        match_byte('/');                // Start of filename
        if (rxbuff[rxout] == '$')       // If dummy file starting wth '$'
        {
                                        // ..put out simple text string
            tpdlen = TEXT_OK_LEN + DAYTIME_LEN;
            get_rtc_time();             // ..consisting of current time
            put_ip(TCPHDR_LEN+tpdlen);
            checkhi = checklo = 0;
            txin = IPHDR_LEN + TCPHDR_LEN;
```

```
            printf(put_byte, TEXT_OK);  // ..with HTTP header
            printf(put_byte, DAYTIME_STR, rtc.b[2], rtc.b[1], rtc.b[0]);
            txin = IPHDR_LEN;
            tflags = TFIN+TPUSH+TACK;
            put_tcp();
        }
        else
        {                                  // Get filename into directory buffer
            for (i=0; i<ROM_FNAMELEN; i++)
            {
                c = rxbuff[rxout];
                if (c>' ' && c!='?')       // Name terminated by space or '?'
                    rxout++;
                else
                    c = 0;
                romdir.f.name[i] = c;
            }                              // If file found in ROM
            if (find_file())
            {                              // ..check for form arguments
                check_formargs();
            }
            else                           // File not found, get index.htm
            {
                romdir.f.name[0] = 0;
                find_file();
            }
            if (!fileidx)                  // No files at all in ROM - disaster!
            {
                tpdlen = HTTP_FAIL_LEN;    // Inform user of failure..
                put_ip(TCPHDR_LEN+tpdlen); // Output IP header to buffer
                checkhi = checklo = 0;     // Reset checksum
                strcpy(&txbuff[IPHDR_LEN+TCPHDR_LEN], HTTP_FAIL);
                txin = IPHDR_LEN + TCPHDR_LEN;
                check_txbytes(tpdlen);     // Checksum data
                txin = IPHDR_LEN;          // Go back to end of IP header
                tflags = TFIN+TPUSH+TACK;
                put_tcp();                 // Output TCP header to buffer
            }
            else                           // File found OK
```

```
          {
                  tpdlen = romdir.f.len;       // Get TCP data length
                  put_ip(TCPHDR_LEN+tpdlen);   // Output IP header to buffer
                  checkhi = checklo = 0;       // Reset checksum
                  check_byte(romdir.f.check);  // Add on checksum on ROM file
                  check_byte(romdir.f.check >> 8);
                  tflags = TFIN+TPUSH+TACK;     // Close connection when sent
                  txi2c = 1;                    // Set flag to enable ROM file O/P
                  put_tcp();                    // Output TCP header to buffer
          }
      }
  }
  return(ret);
}
```

I'll look at forms processing and `check_formargs()` a little later.

Using the PWEB Server

You now should have a working Web server. Now you need to create some HTML pages for it. The pages are essentially the same as those for any other server, but with a few constraints.

Content-type.   The WEBROM utility currently only inserts the correct HTTP headers for HTML, text, GIF, and X-bitmap files, although it could be modified easily to accept additional types.

Length.   The one-frame technique constrains the page size to 952 bytes (1,006-byte SLIP frame minus two 20-byte IP and TCP headers and a 44-byte HTTP header). There is no reason why two or more frames couldn't be sent, but the current software does not support this.

Forms and dynamic content.   In the second half of this chapter, I'll look at how to support forms and embedded gateway interface (EGI) variable substitution.

Complexity.   If a Web page contains several graphics, the browser will attempt to fetch them all at the same time by opening multiple connections to the server. Your server is hampered in its ability to handle multiple requests by the lack of buffer space. If it is currently transmitting the response to one request when another arrives, it has no buffer space left, so it must discard the second request. Of course, the browser will retry the second request, so it will eventually succeed, but the Web page update time will be prolonged. It is better to forestall the problem by restricting the amount of graphics on each page. This is a miniature Web server, so you can hardly expect it to support elaborate Web pages.

Figure 11.2   PWEB home page.

To create your own Web page ROM, you need to gather all the HTML and graphic files in one directory on your hard disk. Take a quick look at the file sizes. If the individual files or the total directory size is too large, some trimming will be necessary.

Run the WEBROM utility to create a ROM file in a different directory.

```
webrom test.rom webdocs
```

This creates the file test.rom from all the files in the webdocs subdirectory. The utility reports the total ROM size, which will be slightly larger than the individual file sizes because of the addition of the HTTP headers. The resulting ROM file can be blown into the E²ROM using a conventional PROM programmer with a suitable adaptor.

To contact the PWEB server from a PC running Windows, the PC needs to be configured for dial-up network access via a standard modem, as described in Appendix A. On activating the dial-up access, it should connect to the PWEB server extremely quickly, and then a conventional Web browser can access the PWEB pages (Figure 11.2).

Dynamic Content

So far, the miniature Web server has been of largely academic interest. It may be impressive that such a small, low-powered microcontroller can actually serve up Web pages, but to be any real use, the pages have to contain dynamic content — real-time data that would not be available to a conventional disk-based system. Also, you ought be able to interact with the miniature server: set its digital output lines and adjust its clock.

In attempting to implement dynamic content, you face three major obstacles:

String handling.   I've already referred several times to the difficulties of string-handling on the PICmicro, most notably because of the small amount of RAM and the awkward storage of string constants in program space.

Datagram length.   Because you don't have enough RAM to store a datagram before transmission, you have to prepare and transmit the IP and TCP headers in advance of the data. As a result, the overall datagram length, and hence the data length, is fixed before the data is prepared, so you can't accommodate variable-length output data.

TCP Checksum.   As with the datagram length, the TCP checksum is transmitted before the data is prepared, so you must know in advance what the checksum value will be; therefore, you must know in advance what the data will be.

The last of these obstacles is particularly grim because it calls for the software equivalent of a fortune-teller's crystal ball. The solution comes with the realization that the only way to correctly predict the future is to control the future so that your predictions will come true — every deviation from the prescribed path must be balanced by an equal and opposite deviation. Enough philosophy — it's time to start work on some real examples.

Variable Substitution

My previous Web server had what I, in all modesty, considered a really nice programming interface, for which I coined the term embedded gateway interface (EGI), that took an HTML comment tag

```
<!--#$switch1-->
```

and replaced it with a graphic URL.

```
<img src='switchu.gif'>
```

Complex string-manipulation techniques are not very suitable for the PICmicro. There must be a much more memory-efficient way of doing this. You could start with simple numeric variable substitution, where a nonstandard tag is replaced by a dynamic value, such as the current temperature or time. To make the tag incredibly easy to parse, it should consist of a single character, with a numeric argument indicating which variable is to be selected.

@0	current time, seconds
@1	current time, minutes
@2	current time, hours
@3	current temperature, degrees

The HTML text could contain the following string:

```
The time is @2:@1:@0
```

This would be translated by the PWEB server.

```
The time is 01:23:45
```

As mentioned before, the output string must be the same length as the original; otherwise, the length in the IP header will be incorrect. If the temperature measurement is a four-digit string, sufficient space must be allocated for it so that

```
The temperature is @3   C
```

can become the following output string:

```
The temperature is 24.0C
```

Checksum Balancing

Initially, I could think of only two ways to solve the checksum problem.

- Scan the Web page once to generate the TCP checksum, emit the header with that checksum, and then scan the page a second time to send it down the serial link. If the Web page ROM were closely coupled to the CPU, this would be tempting, but scanning the E²ROM twice over an I²C interface would reduce throughput considerably.

- Emit some extra characters at the end of the Web page to restore the checksum to the correct value.

It may seem strange tacking extra characters onto the end of a page, but this could be done as an HTML comment that wouldn't show up on screen. The technique is easiest to explain by example. Suppose that you have a page consisting solely of the words "Web Test," and you have been so foolish as to predict that the checksum is FFFFh, when it really is 815Fh. How do you know the real value? Use this tiny DOS program, together with the checksum utility in netutil.h.

```
#define DLEN 8
BYTE data[DLEN+1] = {"Web Test"};

void main()
{
    int i;
    WORD sum;

    for (i=0; i<DLEN; i++)
        printf("%02X ", data[i]);
    sum = csum(data, DLEN);
    printf("checksum %02X%02Xh\n", sum&0xff, sum>>8);
}
```

This produces the following output.

```
57 65 62 20 54 65 73 74 checksum 815Fh
```

To balance out the checksum, add an HTML comment tag with five spaces in it (for reasons which I hope will become clear), and recompute the checksum.

```
Web Test<!--     -->
57 65 62 20 54 65 73 74 3C 21 2D 2D 20 20 20 20 20 2D 2D 3E checksum 7859h
```

Now you must replace the first four spaces in the comment field with characters that will restore the checksum to its predicted value. Don't forget that the checksum is computed by adding together 16-bit words, with an additional carry from the most significant byte to the least significant. What two characters will change 78h into FFh?

```
FFh - 78h => 87h
        87h => 43h + 44h
43h + ' ' => 'c'
44h + ' ' => 'd'
```

So, inserting c and d will fix the most-significant byte, s and s will fix the least-significant byte (work it out for yourself). Because the calculation adds 16-bit words, the new characters must be interleaved, so the following line shows the final string that produces a checksum of FFFFh.

```
Web Test<!--csds -->
```

I must confess that I've cheated somewhat in my original choice of string, so as to simplify the analysis. In practice, you have to insert eight characters (four for each byte) to ensure that they don't inadvertently include control characters (values less than 20h) or invalid network virtual terminal (NVT) characters (values greater than 7Fh). However, it is possible to make this technique work, but it isn't really simple enough for use on the PICmicro.

Counterbalanced Variables

Using the balancing technique as a starting point, it is possible to think of a third way to correct the checksum after variable values have been substituted, which is to substitute both the value and a counterbalancing value when the file is sent out so that the net effect is the checksum is unchanged.

To achieve this, two tags must be inserted in the file for every one variable.

```
@2 hours   <!--@N-->
```

The value @2 will be changed to the current hours count, and the value @N will be changed to the value needed to counterbalance the hours count. If the hours count is six, the above will be changed to

```
06 hours   <!--PJ-->
```

because of the following relationships.

```
'@' - '0' = 'P' - '@'
'2' - '6' = 'J' - 'N'
```

This is one of those algorithms that is easier for computers to understand than humans because we insist on ascribing meaning to characters. In another example using a four-digit temperature value, the entry in the unprocessed HTML file

```
Temperature @3  <!--@M``-->&deg;C
```

might be converted into the following:

```
Temperature 24.5<!--NLRK-->&deg;C
```

The PWEB code to do this substitution is ridiculously simple. The following code fragment handles the time variables:

```
b = i2c_read(1);                // Get next byte from ROM
if ((romdir.f.flags&EGI_ATVARS) && b=='@')
{                               // If EGI var substitution..
    b = i2c_read(1);            // ..get 2nd byte
    romdir.f.len--;
    inv_byte = b > '@';         // Get variable index
    idx = inv_byte ? 0x50-b : b-0x30;
    if (idx>=0 && idx<=2)       // Index 0-2 are secs, mins, hrs
        printf(tx_byte_inv, "%02X", rtc.b[idx]);
    ...
}
```

The simple function `tx_byte_inv()` emits the byte value or its inverse, depending on the state of the global flag.

```
/* Transmit a SLIP byte or its 'inverse' i.e. 80h minus the value
** Used by EGI functions to ensure data and its inverse add up to 80h */
void tx_byte_inv(BYTE b)
{
    tx_byte(inv_byte ? 0x80-b : b);
}
```

I have cunningly chosen the values in the variable tag and its inverse so that they add up to 80h (e.g., '@' + '@' = 80h, '2' + 'N' = 80h), which simplifies the arithmetic considerably.

With regard to the PICmicro code, that's essentially all there is to it. You have gained a significant increase in capability for very little extra work. A minor disadvantage is that you have to edit the HTML files to insert the double tags. At the moment, I am hand-editing the HTML to insert both sets of tags, but it would be relatively easy to write a Perl script to do the job or modify WEBROM to do it. In the mean time, here are some guidelines on doing the job manually.

Look at my examples.   Make sure you understand the process before creating your own.

Counterbalance. The tag characters you insert must add up to 80h: '@' + '@', '0' + 'P', '1' + '0', '2' + 'N', '3' + 'M', and so on. A space is counterbalanced by a back-tick, ASCII 60h.

Avoid odd numbers of characters between tags. For a tag and its inverse to counter-balance, they must be separated by an even number of characters.

Watch out for TCP checksum errors. If you get this technique wrong, the data will be rejected by the browser because it will contain a TCP checksum error.

I hope that by the time you read this, I will have created a utility to automate this rather tricky procedure. What makes this effort worthwhile is the display of live data in Web pages, as shown below.

Dynamic Web Pages

The following example contains the original file in ROM, followed by the dynamic Web page derived from it, and the associated display.

```
<html><meta http-equiv="refresh" content="3">
<head><title>PWEB - time and temperature now</title></head>
<body><h3>PWEB miniature Web server</h3>
<h4>Ambient temperature now</h4>
<table border=2>
<tr><td><font FACE='Arial,Helvetica' SIZE=4>Temperature
</font></td>
<td><font FACE='Arial,Helvetica' SIZE=5>@3  <!--@M``-->&deg;C
</font></td></tr>
<tr><td><font FACE='Arial,Helvetica' SIZE=4>Time
</font></td>
<td><font FACE='Arial,Helvetica' SIZE=5>@2<!--@N-->:@1<!--@O-->:@0<!--@P-->
</font></td></tr>
</table>
<br>
<br><a href='status.egi'>Status page</a>
<br><a href='index.htm'>Home page</a>
</body></html>
```

The `meta` statement causes the page to be refetched every three seconds (client pull). The server converts this page as follows.

```
<html><meta http-equiv="refresh" content="3">
<head><title>PWEB - time and temperature now</title></head>
<body><h3>PWEB miniature Web server</h3>
<h4>Ambient temperature now</h4>
<table border=2>
<tr><td><font FACE='Arial,Helvetica' SIZE=4>Temperature
</font></td>
<td><font FACE='Arial,Helvetica' SIZE=5>23.5<!--NMRK-->&deg;C
</font></td></tr>
<tr><td><font FACE='Arial,Helvetica' SIZE=4>Time
</font></td>
<td><font FACE='Arial,Helvetica' SIZE=5>20<!--NP-->:44<!--LL-->:19<!--OG-->
</font></td></tr>
</table>
<br>
<br><a href='status.egi'>Status page</a>
<br><a href='index.htm'>Home page</a>
</body></html>
```

This produces the display in Figure 11.3.

You can also use the dynamic variables to fill in the fields of a form. For example, you need an HTML form to set the real-time clock, and it would be handy if the form was already filled in with the current time (Figure 11.4).

```
<html>
<head>
<title>PWEB - set time</title>
</head>
<body>
<h3>PWEB miniature Web server</h3>
<h4>Set time (HH:MM:SS)</h4>
<form action='setime.egi'>
<input type=text name=hrs size=2 value='@2'><!--@N--> :
<input type=text name=min size=2 value='@1'><!--@0--> :
<input type=text name=sec size=2 value='@0'><!--@P-->
```

```
<br><br>
<input type=submit name=sub value='Set clock'>
</form><br>
<a href='index.htm'>Home page</a>
</body>
</html>
```

Figure 11.3   PWEB page with dynamic data.

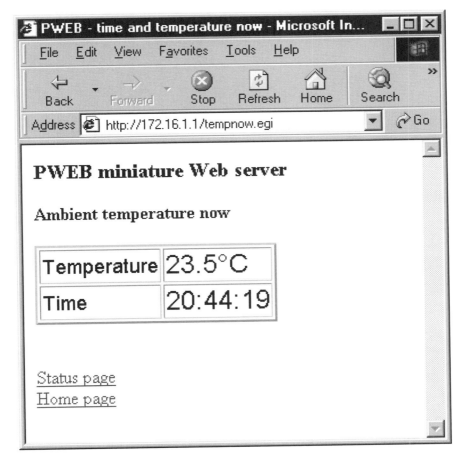

Figure 11.4 PWEB form to set clock.

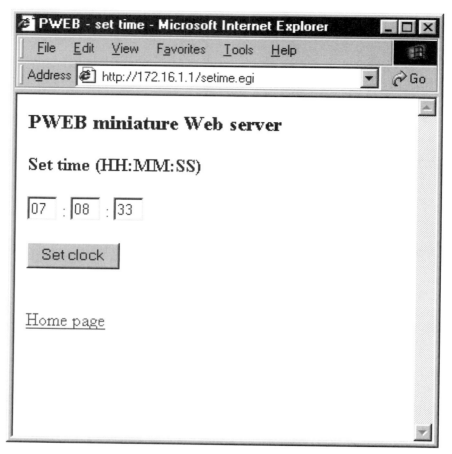

The user can edit the existing values, then press the Set Clock button, and the new time is entered, or rather, will be entered as soon as I describe the way the variables from an HTML form are processed.

HTML Form Variables

For the time-setting form to work, you need to decode the response and use it to change the real-time clock registers. If you alter the fields to 10:21:0 and hit the Set Clock button, you get something like the following HTTP request.

```
GET /setime.egi?hrs=10&min=21&sec=0&sub=Set+clock HTTP/1.0
```
... and all the extra stuff ...

The HTTP handler calls the following function to parse the string after the filename.

```
/* Check for form arguments in HTTP request string */
void check_formargs(void)
{
    BOOL update=0;
    char c, d, temps[5];

    portval = ~PORTB;                    // Read O/P port, just in case
    while (rxout < rxcount)
        {                                // Each arg starts with '?' or '&'
        c = rxbuff[rxout++];
        if (c=='?' || c=='&')
            {                            // Copy string const from ROM to RAM
            strcpy(temps, "hrs=");
            if (match_str(temps))        // ..before matching it
                {
                                         // ..and updating clock value
                update |= get_hexbyte(rtc.b[2]);
                continue;
                }
            strcpy(temps, "min=");  // RTC minutes?
            if (match_str(temps))
                {
                update |= get_hexbyte(rtc.b[1]);
                continue;
                }
            strcpy(temps, "sec=");  // RTC secs?
            if (match_str(temps))
                {
                update |= get_hexbyte(rtc.b[0]);
                continue;
                }                        // O/P port bit change?
            strcpy(temps, "out");
            if (match_str(temps) && get_byte(c) && isdigit(c) &&
                match_byte('=') && get_byte(d) && isdigit(d))
                {
                if (d == '0')            // If off, switch bit on
                    portval |= 1 << (c-'0');
                else                     // If on, switch bit off
```

```
                    portval &= ~(1 << (c-'0'));
            d = ~portval;        // Update hardware port
            PORTB = d;
        }
      }
  }
  if (update)                    // Update clock chip if time changed
      set_rtc_time();
}
```

It is worth bearing the following points in mind when reviewing the code:

String matching. The string constants are stored in program memory and must be copied into RAM before they can be matched. I've tried to make the code as clean as possible, but the compiler defeated all my attempts at creativity, which would otherwise have worked in standard C.

```
if match_str(strcpy(temps, "hrs="))     // Doesn't work!
    ...
```

Hexadecimal arithmetic. It seems strange using hexadecimal arithmetic to set the clock, which is showing time as a decimal value. The reason is that the clock registers use BCD (binary-coded decimal), which is a subset of hexadecimal values.

Digital outputs. In anticipation of doing something with the digital outputs, I include parsing of form strings.

```
out2=0
```

This sets bit 2 in the digital output port and can be tested by manually entering the form response into the browser's address window.

```
172.16.1.1/index.htm?out2=0
```

The third LED on the output port should light. In case you are wondering why a value of zero turns the LED on, the next section explains all.

Digital Outputs

Eight digital output lines are connected to LEDs, so it would be nice to make a point-and-click interface for them. In Chapter 8, I implemented such an interface using pretty toggle-switch graphics, but it behooves me to be a bit more restrained on a humble PICmicro. An array of buttons would be a reasonable compromise (Figure 11.5).

Figure 11.5   PWEB digital output display.

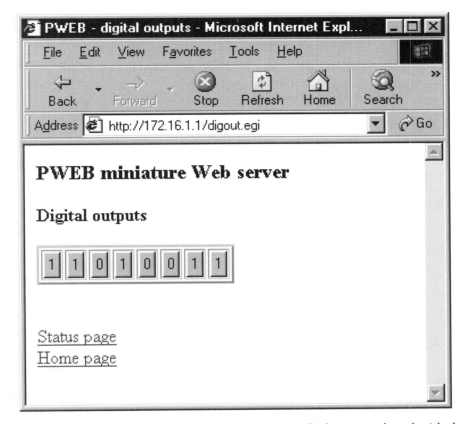

The HTML code uses an array to hold the buttons, which are numbered with the least significant bit on the right (out7 ... out0).

```
<html>
<head><title>PWEB - digital outputs</title></head>
<body><h3>PWEB miniature Web server</h3>
<h4>Digital outputs</h4>
<form action='digout.cgi'>
<table border=2><tr>
<td><input type=submit name=out7 value=1></td>
<td><input type=submit name=out6 value=1></td>
<td><input type=submit name=out5 value=0></td>
<td><input type=submit name=out4 value=1></td>
<td><input type=submit name=out3 value=0></td>
<td><input type=submit name=out2 value=0></td>
<td><input type=submit name=out1 value=1></td>
```

```
<td><input type=submit name=out0 value=1></td>
</tr></table></form>
<br><a href='status.egi'>Status page</a>
<br><a href='index.htm'>Home page</a>
</body></html>
```

When a button is clicked, it returns its name and current value, not the value it is changing to. If the first button (bit 7) is clicked, it would return the following value:

```
GET /digout.egi?out7=1
```

The form variable processing function `check_formargs()` already includes an inversion that asserts the bit 7 output.

It shouldn't come as a surprise that the next job is to insert special tags in place of the static values shown above. You could use a continuation of the @0, @1, @2 scheme described earlier, but there are two problems with this.

Number of characters. Each Boolean value is a single character: 1 or 0. The @ variable scheme needs a minimum of two characters. You could pad the Boolean value with spaces, but that is inconvenient.

Number of variables. It would be neater to treat the Boolean values as a block, rather than just pushing the variable count up.

I adopted a solution that creates two new, special, single-character tags for Boolean values.

Boolean Variable Substitution

Each digital output bit is replaced with a unique character — I chose #. Using the counterbalancing principle discussed earlier, a complementary tag is also inserted as a comment — I chose |. This produces the following new digital output page:

```
<html>
<head><title>PWEB - digital outputs</title></head>
<body><h3>PWEB miniature Web server</h3>
<h4>Digital outputs</h4>
<form action='digout.egi'>
<table border=2><tr>
<td><input type=submit name=out7 value=#><!--|--></td>
<td><input type=submit name=out6 value=#><!--|--></td>
<td><input type=submit name=out5 value=#><!--|--></td>
<td><input type=submit name=out4 value=#><!--|--></td>
<td><input type=submit name=out3 value=#><!--|--></td>
<td><input type=submit name=out2 value=#><!--|--></td>
<td><input type=submit name=out1 value=#><!--|--></td>
```

```
<td><input type=submit name=out0 value=#><!--|--></td>
</tr></table></form>
<br>
<a href='/index.htm'>Home page</a>
</body></html>
```

The PICmicro Web server has to translate each # into a character representing the corresponding bit value and each | into a value that will balance the checksum. If the # is converted to a 0, then the | must become an o; if the # becomes a 1, then the | becomes an n, because of the following relationships:

```
'0' - '#' = '|' - 'o'
'1' - '#' = '|' - 'n'
```

If bits 0 and 2 of the output port are set, then the button array becomes the following:

```
<td><input type=submit name=out7 value=0><!--o--></td>
<td><input type=submit name=out6 value=0><!--o--></td>
<td><input type=submit name=out5 value=0><!--o--></td>
<td><input type=submit name=out4 value=0><!--o--></td>
<td><input type=submit name=out3 value=0><!--o--></td>
<td><input type=submit name=out2 value=1><!--n--></td>
<td><input type=submit name=out1 value=0><!--o--></td>
<td><input type=submit name=out0 value=1><!--n--></td>
```

Hand-insertion of the counterbalancing tags is a nuisance and wouldn't be that difficult to automate. It would also be possible for such a utility to group all the counterbalancing tags together in a single comment, saving space in the Web page.

Digital Inputs

Port A on the PICmicro has provision for six digital inputs, which must also be accommodated in the code. It is tempting to create yet more special-purpose tag characters for the purpose, but if you continued in that vein, you'd soon run out of usable characters. Instead, I created two special-purpose @ variables that can be used to switch the Boolean variables between inputs (@?) and outputs (@!).

```
<!--@?-->
Input  bit 5 = # <!--|-->
Input  bit 4 = # <!--|-->
<!--@!-->
Output bit 7 = # <!--|-->
Output bit 6 = # <!--|-->
```

When the special-purpose variable is encountered, the PICmicro will reset its bit mask and load the value from the input or output port.

PWEB Variable Substitution Code

A relatively minor addition to the PWEB variable substitution code will support Boolean variables, as shown in the file transfer function, complete with all the features I have discussed.

```
/* Transmit a byte from the current i2c file to the SLIP link
** Return 0 when complete file is sent
** If file has EGI flag set, perform run-time variable substitution */
BOOL tx_file_byte(void)
{
    int ret=0, idx;
    BYTE b;

    if (romdir.f.len)                       // Check if any bytes left to send
    {
        b = i2c_read(1);                    // Get next byte from ROM
        if ((romdir.f.flags&EGI_ATVARS) && b=='@')
        {                                   // If EGI var substitution..
            b = i2c_read(1);                // ..get 2nd byte
            romdir.f.len--;
            inv_byte = b > '@';             // Get variable index
            idx = inv_byte ? 0x50-b : b-0x30;
            if (idx>=0 && idx<=2)           // Index 0-2 are secs, mins, hrs
                printf(tx_byte_inv, "%02X", rtc.b[idx]);
            else if (idx == 3)              // Index 3 is temperature
            {
                printf(tx_byte_inv, "%2u", temphi);
                if (templo & 0x80)
                    printf(tx_byte_inv, ".5");
                else
                    printf(tx_byte_inv, ".0");
                i2c_read(1);                // Discard padding in ROM
                i2c_read(1);
                romdir.f.len -= 2;
            }
            else if (b == '?')              // @? - read input port
            {
                tx_byte('@');
                tx_byte(b);
                hashmask = barmask = 0x20;
                portval = ~PORTA;
```

```
        }
        else if (b == '!')              // @! - read output port
        {
            tx_byte('@');
            tx_byte(b);
            hashmask = barmask = 0x80;
            portval = ~PORTB;
        }
        else                            // Unrecognized variable
            printf(tx_byte_inv, "??");
    }                                   // '#' and '|' are for boolean values
    else if (romdir.f.flags&EGI_HASHVARS && (b=='#' || b=='|'))
    {
        if (b=='#')                     // Replace '|' with '1' or '0'
        {
            tx_byte(portval&hashmask ? '1' : '0');
            hashmask >>= 1;
        }
        else                            // Replace '|' with inverse
        {
            tx_byte(portval&barmask ? '|'+'#'-'1' : '|'+'#'-'0');
            barmask >>= 1;
        }
    }
    else                                // Non-EGI byte; send out unmodified
        tx_byte(b);
    romdir.f.len--;
    ret = 1;
    }
    return(ret);
}
```

Status Page

The PWEB status page (Figure 11.6) demonstrates all the dynamic elements I have discussed. It uses client pull to update itself every three seconds.

Figure 11.6   PWEB dynamic status display.

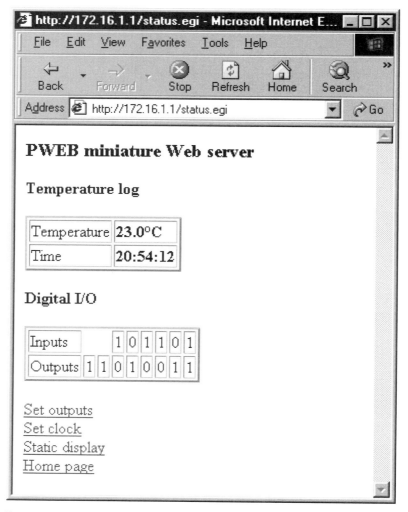

The following HTML code creates the page:

```
<html><meta http-equiv="refresh" content="3">
<head><title>PWEB</title></head>
<body><h3>PWEB miniature Web server</h3>
<h4>Temperature log</h4><table border=2>
<tr><td>Temperature</td><td><b>@3  <!--@M``-->&deg;C</b></td>
<tr><td>Time</td><td><b>@2<!--@N-->:@1<!--@O-->:@0<!--@P--></b></td></tr>
</tr></table>
<h4>Digital I/O</h4><table border=2>
```

```
<tr><td> Inputs </td><!--@?-->
<td> </td><td> </td><td>#</td><td>#</td>
<td>#</td><td>#</td><td>#</td><td>#</td>
<!--| | | | | |--></tr>
<tr><td> Outputs </td><!--@!-->
<td>#</td><td>#</td><td>#</td><td>#</td>
<td>#</td><td>#</td><td>#</td><td>#</td>
<!--| | | | | | | |--></tr></table>
<br><a href='digout.egi'>Set outputs</a>
<br><a href='setime.egi'>Set clock</a>
<br><a href='static.egi'>Static display</a>
<br><a href='index.htm'>Home page</a>
</body></html>
```

The `meta http-equiv="refresh"` tag makes the client refetch the page after the given delay. Three seconds is a reasonable time for the given page over a direct 38,400-baud serial link. This would need to be increased for a more complicated page or a slower network connection.

Summary

Despite the acute memory limitations (especially RAM), I've managed to create a surprisingly sensible Web server for the PICmicro that provides dynamic Web pages for controlling and monitoring the digital I/O and displaying the current temperature and time. I've also included some simple HTML forms handling for setting the real-time clock.

The compiler reports that 71% of the 8Kb on-chip ROM was used, between 73% and 84% of the RAM (depending on location) was used, and seven of the eight stack levels were used. There is plenty of room for expansion; for example, an Ethernet version of the PICmicro Web server is under development at the time of writing. See the Iosoft Ltd. Web site (www.iosoft.co.uk) for details.

PWEB

PWEB is a miniature Web server for the PICmicro, with interfaces to digital I/O, a real-time clock, and a temperature sensor. The compiler output file `pweb.hex` should be programmed into a PIC16C76 device. A diagnostic LED should be connected to port C bit 2, which will flash briefly on power-up and briefly thereafter every five seconds or when a serial message arrives. The serial link is set to 38,400 baud, eight data bits, no parity, and one stop bit, and it accepts SLIP frames with uncompressed headers and any destination IP address. Simple emulation of a standard 28,800-baud modem is also included.

The compiler doesn't use a linker, so library files are included in the main source file.

pweb.c	Main program
picslip.h	Low-level serial functions
pictcp.h	TCP functions
webrom.h	ROM file system definitions

WEBROM Source Files

webrom.c	ROM image utility
webrom.h	ROM file system definitions

WEBROM Utility

Utility	Create ROM image of all the files in a directory
Usage	webrom rom_file directory_name
Example	webrom test.rom c:\temp\romdocs
Notes	Adds HTTP header for the following extensions:
	.htm
	.egi
	.txt
	.gif
	.xbm

Chapter 12

ChipWeb — Miniature Ethernet Web Server

Overview

So far, the microcontroller Web server has been using a Serial Line Internet Protocol (SLIP) network interface. The next step is to adapt the server for use with Ethernet, and this chapter describes those adaptations.

Unfortunately, do-it-yourself hardware-construction techniques are no longer applicable. A few chips hand-soldered on a prototyping board may have been adequate so far, but the use of 100-pin flatpack components puts you in a different hardware league. To provide a standard hardware base, I'll use the Microchip PICDEM.net™[1] for all Ethernet experimentation, although the software could be adapted to other microcontroller networking platforms.

The Microchip PICDEM.net board also has a small amount of analog and digital I/O and an LCD display. These will be used for diagnostic purposes, although they are not essential to the operation of the networking software.

---

1. PICDEM.net™ is the trademark of Microchip Technology Inc.

Hardware

Demonstration Board

The demonstration board (Figure 12.1) has the following peripherals:

- two-line by 16-character alphanumeric LCD display
- two analog potentiometers
- user push button
- three light-emitting diodes (LEDs): one marked "system," two marked "user"
- 32Kb serial E2ROM
- Ethernet interface
- RS232 serial interface

Figure 12.1   Ethernet demonstration board.

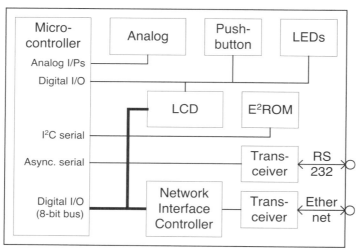

It would be difficult to drive all these peripherals using the same PIC16C76 as before. Instead, the 40-pin PIC16F877 is employed, which has

- 8K × 14-bit words Flash program memory
- 368 bytes data memory
- 256 bytes nonvolatile E2ROM
- eight-level hardware stack
- 10-bit analog-to-digital converter
- Synchronous and asynchronous communication ports
- three timers

It'd be nice if the device also had an on-chip Ethernet interface, but in its absence, an external Network Interface Controller (NIC) chip is used instead, with an eight-bit parallel interface to the microcontroller.

The device has the 31 I/O pins needed to drive all the peripherals and is compatible with the software tools used so far. It is unfortunate that the RAM and ROM sizes remain just as small as before, but there is the possibility of upgrading to the Microchip PIC18xxx devices, which are pin-compatible, so they can be plugged into the same prototyping board without any hardware changes. These devices have much more on-chip memory, but do require a new set of software tools because there are significant changes to the internal architecture.

Ethernet Hardware

The design of most NICs owes much to the constraints of the PC architecture. The original PC eight-bit bus, used by all add-on cards, was capable of handling just less than 1Mbyte per second, yet the Ethernet interface transfers 10Mbits per second. This means that an old-style (Industry Standard Architecture, or ISA) PC network card had to buffer all incoming and outgoing messages in a *packet buffer*, to avoid overloading the PC bus.

Modern PC cards need less buffering because they can interface directly with the PC main memory over a fast 32-bit bus. Such techniques are unsuitable for microcontrollers, which have slower I/O rates than the original PC, so an old-style ISA buffered interface is needed.

Figure 12.2   Ethernet hardware block diagram.

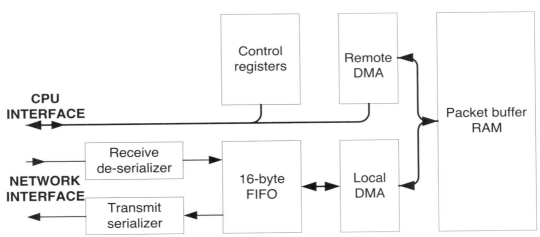

The Ethernet hardware I'll use is a derivative of the original National Semiconductor DP8390 NIC. Figure 12.2 shows the main functional blocks. Note the following points:

- The CPU can read and write the control registers. These registers are organized in banks (known as register *pages*), so that a large number of registers can be accessed using only four address bits.
- In this design, the CPU cannot access the packet buffer RAM directly; data accesses are done using a remote Direct Memory Access (DMA) controller in the NIC. The control registers are used to set up the desired packet buffer address and byte count; then the CPU repeatedly reads or writes a data latch to transfer a block of data to or from the packet buffer RAM. The latch is not included in the set of control registers, so an additional address line is needed, making a total of five (A0–A4, see Figure 12.3).

- Once set up, the NIC can automatically transfer data between the network interface and the packet buffer RAM using a small first in, first out (FIFO) buffer, and a second (local) DMA channel. The NIC is sufficiently intelligent that it can receive several network messages (Ethernet *frames*) in succession without any intervention from the host CPU. To achieve this, the NIC has a sophisticated ring buffer management capability based on 256-byte pages in RAM.

The terms *local* and *remote DMA* are used to keep in line with the NIC data sheets. It is important to note that this refers to the way the NIC handles internal transfers to and from its packet buffer RAM and has nothing to do with the way the CPU (in this case, the microcontroller) handles the data.

The Novell NE2000 was a very well-known PC card employing the DP8390 chip set. Such was its success that it spawned a wide variety of clone cards and the creation of highly integrated chips that incorporate the NIC, serializer, buffer RAM, and support for PC plug-and-play configuration. One such chip is the Realtek RTL8019AS, which achieves a remarkably high level of integration; very little external logic is needed to implement an ISA bus PC networking card.

There are a sufficiently large number of options on the Ethernet controller that it is usually equipped with an external nonvolatile serial E2ROM device. This device is programmed with option values that the controller loads into its configuration registers on reset. In the absence of an external ROM, the controller boots up with sensible default settings, so it is possible to omit the ROM.

Figure 12.3　Ethernet controller interface.

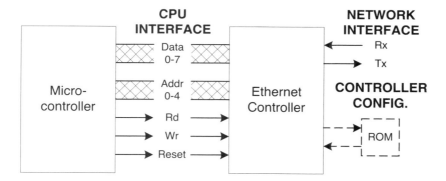

Microcontroller Interface

To drive the Ethernet Controller, the microcontroller has to simulate the read and write cycles of a PC ISA bus. This is not as difficult as it sounds, as shown by the steps for a single cycle (Figure 12.4).

1. **Set the address.** Although the Ethernet controller has 20 address lines (A0–A19), all but five of these can be fixed by being hard-wired to supply or ground. The remaining five lines are used to select the register of interest.

2. **Set the data lines.** In a read cycle, the microcontroller data lines must act as inputs; in a write cycle, they must be outputs and be set with the data byte value.

3. **Assert the Read or Write signals.** The signals are active-low, so either Read or Write must be set low.

4. **If read cycle, fetch the data.** If the Read line is asserted and the address is correct, the Ethernet controller will be driving the data bus, so the data can be read.

5. **Negate both Read and Write signals.** If a read cycle, the Ethernet controller will stop driving the data bus. If a write cycle, it will latch the data received from the micro-controller.

6. **Unset the data lines.** The microcontroller output drivers should be disabled to free up the data bus for other uses.

Figure 12.4   Ethernet controller access cycles.

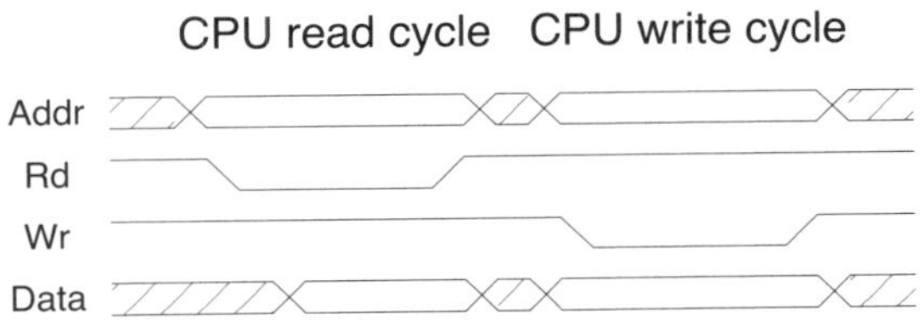

The following I/O lines are needed on the microcontroller:

- one read output and one write output.
- five address outputs. These are only significant while the Ethernet chip is being accessed, so they could be used for other duties as well.
- eight bidirectional data lines. Again, these are only significant during Ethernet chip accesses.

It is remarkable that only 15 I/O lines are needed to drive the 100-pin Ethernet controller, and only two of those lines are dedicated fulltime to driving this interface.

LCD Interface

An alphanumeric liquid crystal display (LCD) has also been included on the demonstration board (Figure 12.5). This uses a different bus structure with a single read/write line, a four-bit data bus (connected to D4–D7 on the LCD), and an Enable that is used to synchronize the accesses.

Figure 12.5   Liquid crystal display (LCD) interface.

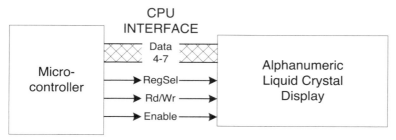

The Register Select line is essentially a single-bit address set low to select the instruction (control) register and high for the data (character) register. The read/write line is low for writes to the LCD and high for reads from the LCD. The cycles are synchronized by the Enable signal. On the rising edge, the Read/Write and Register Select signals are latched; on the falling edge, the data is transferred (Figure 12.6).

Figure 12.6   LCD access cycle.

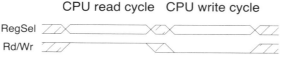

The LCD can be treated as a write-only device, since the main purpose of the read-back is to check whether the LCD is ready to accept the next byte, and this can be achieved using simple time delays instead. The steps for a write cycle are as follows:

1. set the data lines as outputs, and the data on them
2. set the Register Select and Read/Write lines
3. set the Enable line high, and ensure it remains high for at least 500ns
4. set the Enable line low
5. unset the data lines

Despite their differences, the Ethernet controller and LCD can share a common data bus, but it is important to establish a correct idle state for both devices before attempting any communications.

Many LCD interfaces are equipped with a potentiometer to adjust the display contrast and viewing angle. If this is set incorrectly, the display will always remain blank — a problem that can cause hours of fruitless debugging, trying to find a nonexistent hardware or software fault. The PICDEM.net board avoids this danger by using a fixed-resistor potential divider for the LCD contrast.

Other Peripherals

The prototype board has the following peripheral devices:

- two potentiometers connected to analog inputs. The voltage generated by these (approximately 0–5V) can be sampled by the on-chip analog-to-digital converter (ADC).
- one push button connected to a digital input
- two LEDs connected to digital outputs
- RS232 interface with handshake lines
- 32Kb I2C serial E2ROM memory device

There is also a convenient prototyping area for adding your own peripheral devices.

Ethernet Driver

NIC Initialization

The fundamental building blocks of the Ethernet device driver are a single read or write cycle to the NIC.

```
#BIT    NIC_IOW_    = PORTE.1
#BIT    NIC_IOR_    = PORTE.0
#BYTE   NIC_DATA    = PORTD
#define DATA_TO_NIC   set_tris_d(ALL_OUT);
#define DATA_FROM_NIC set_tris_d(ALL_IN);


/* Input a byte from a NIC register */
BYTE innic(int reg)
{
    BYTE b;

    DATA_FROM_NIC;
    NIC_ADDR = reg;
    NIC_IOR_ = 0;
    delay_cycles(1);
    b = NIC_DATA;
    NIC_IOR_ = 1;
    return(b);
}

/* Output a byte to a NIC register */
void outnic(int reg, int b)
{
    NIC_ADDR = reg;
```

```
    NIC_DATA = b;
    DATA_TO_NIC;
    NIC_IOW_ = 0;
    delay_cycles(1);
    NIC_IOW_ = 1;
    DATA_FROM_NIC;
}
```

The DATA_FROM and DATA_TO macros set the microcontroller data port direction (tris) register. To ensure the Read and Write signals remain low for a sufficient time, a delay of a single CPU cycle has been introduced into both the read and write functions.

Initializing the NIC consists of long strings of outnic() calls.

```
outnic(CMDR, 0x21);          /* Stop, DMA abort, page 0 */
delay_ms(2);                 /* ..wait to take effect */
outnic(DCR, DCRVAL);         /* Data configuration */
outnic(RBCR0, 0);            /* Clear remote byte count */
outnic(RBCR1, 0);
outnic(RCR, 0x20);           /* Rx monitor mode */
outnic(TCR, 0x02);           /* Tx internal loopback */
```

There is little point in analyzing these settings in great detail. As with any complex chip, it is largely a fill-in-the-blanks exercise, complicated by the overlapping register pages and the state machines for reception, transmission, and DMA (hence the stop, DMA abort command in the code above).

A few initialization parameters are worthy of note.

Ethernet media access control (MAC) address   This six-byte value must be supplied to the NIC so it can filter incoming packets. Normally, this is in a nonvolatile configuration ROM attached to the NIC, but this has been omitted from the board to save on component costs. It is stored in the microcontroller's on-chip E2ROM, from whence it is copied into RAM and then into the NIC.

Address filtering   Putting the NIC into *promiscuous* mode disables address filtering. This is useful during initial development and testing of the driver software; it is useful to receive some packets, even if they are not actually intended for this node. Promiscuous mode must be disabled when working on a real network because the large number of packets would tend to swamp the network controller.

RAM size   The size of the NIC packet buffer RAM is 16KB, but I have consistently experienced problems when attempting to use the whole RAM. A data sheet for another NE2000-compatible controller suggests that only 8KB should be used when in eight-bit mode. Of this, 1.5KB (six pages of 256 bytes) is needed for the transmit buffer, leaving 6.5KB (26 pages) for the receive buffer.

Accessing the Packet Buffer

Normally, the CPU would transfer the incoming Ethernet packets to its RAM before attempting to process them. Because one packet is much larger than the microcontroller RAM, it will be necessary to fetch and process the incoming packets in small chunks. The NIC remote DMA controller is quite useful for this because it can accept an arbitrary packet buffer address and byte count, even as small as one byte.

```c
/* Set the 'remote DMA' address in the NIC's RAM to be accessed */
void setnic_addr(WORD addr)
{
    outnic(ISR, 0x40);              /* Clear remote DMA interrupt flag */
    outnic(RSAR0, (BYTE)addr);      /* Data addr */
    outnic(RSAR1, addr>>8);
}


/* Get data from NIC's RAM into the given buffer (up to 255 bytes) */
void getnic_data(BYTE *data, int len)
{
    BYTE b;

    outnic(RBCR0, len);             /* Byte count */
    outnic(RBCR1, 0);
    outnic(CMDR, 0x0a);             /* Start, DMA remote read */
    while (len--)                   /* Get bytes */
    {
        b = innic(DATAPORT);
        *data++ = b;
    }
}
```

You may think that the setting of the byte count is largely redundant; why not set this to an arbitrarily large value, so that data bytes can be fetched as-and-when necessary? I have found the NIC to be highly intolerant of mismatched DMA settings, to the extent that it can lock up a PC if too many bytes are fetched. For this reason, I always set the length correctly before attempting any transfer.

Writing to the packet buffer is a very similar exercise.

```c
/* Put the given data into the NIC's RAM (up to 255 bytes) */
void putnic_data(BYTE *data, int len)
{
    outnic(ISR, 0x40);              /* Clear remote DMA interrupt flag */
    outnic(RBCR0, len);             /* Byte count */
```

```
      outnic(RBCR1, 0);
      outnic(CMDR, 0x12);                  /* Start, DMA remote write */
      while (len--)                        /* O/P bytes */
          outnic(DATAPORT, *data++);
      len = 255;                           /* Done: must ensure DMA complete */
      while (len && (innic(ISR)&0x40)==0)
          len--;
  }
```

The last few lines of code ensure that the DMA transaction is complete before any other NIC accesses are made. This has proved necessary on high-speed PCs, so it has been included for safety. It is probably unnecessary on a microcontroller because of the delays inherent in the bus interface.

Packet Reception

Packet reception involves:

- checking for overruns of the packet buffer,
- detecting that one or more packets have been received,
- checking the error status of the packet,
- establishing the start address of the packet in the buffer,
- establishing the length of the packet, and
- freeing up the buffer RAM used by the packet.

The process is made more complicated by the splitting of the RAM space into 256-byte pages and the wrapping of these pages to form a circular (*ring*) buffer. There is little point in exploring the algorithms here. For more information, refer to the source code and the Ethernet controller manufacturer's data sheets. Any modifications to the software must be tested very thoroughly under the following conditions:

- high packet rates
- excessive packet rates (recovery from a buffer overflow)
- mix of packet sizes
- fetching one packet from the buffer while others are being received

It might be thought that using the drivers in a Web server represents a good test, but this is not true for simple Web pages, because the transactions with the Web client can be simple lock-step exchanges with unvarying block sizes. The fetching of a Web page containing a variety of other objects (e.g., graphics) is more taxing because, the browser tends to issue several parallel requests to fetch the embedded content.

On the front of every Ethernet packet (*frame*) is a header containing the six-byte destination and source addresses (Figure 12.7). These are known as Media Access and Control (MAC) addresses.

Figure 12.7   Ethernet frame.

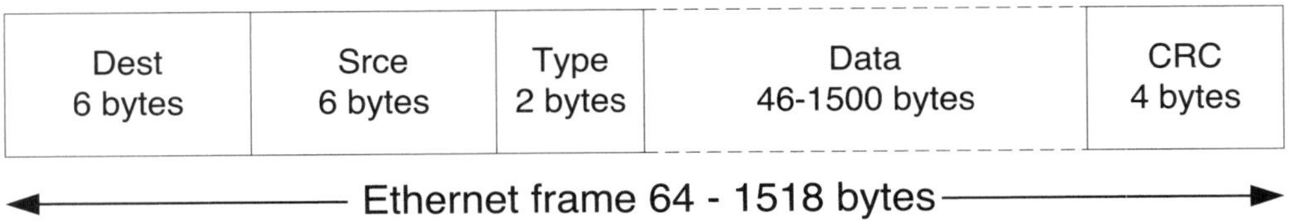

Dest 6 bytes	Srce 6 bytes	Type 2 bytes	Data 46-1500 bytes	CRC 4 bytes

← ——————— Ethernet frame 64 - 1518 bytes ——————— →

A structure is used to reference the Ethernet header.

```
#define MACLEN    6

typedef struct {              // Ethernet frame header
    BYTE dest[MACLEN];        //     Dest & srce MAC addresses
    BYTE srce[MACLEN];
    WORD pcol;                //     Protocol
} ETHERHEADER;
```

The total frame size, including the header and a four-byte CRC, is between 64 and 1,518 bytes, so the actual data length is between 46 and 1,500 bytes. If less than the minimum, the data is padded to fit.

When receiving a frame, the NIC adds its own hardware-specific header to the front so that the length and error status are known.

```
typedef struct {              // NIC hardware packet header
    BYTE stat;                //     Error status
    BYTE next;                //     Pointer to next block
    WORD len;                 //     Length of this frame incl. CRC
} NICHEADER;
```

As a result, the frame in the NIC buffer RAM has two headers in front of the data. To simplify access to the headers, the driver copies them into local RAM.

```
typedef struct {              // NIC and Ethernet headers combined
    NICHEADER nic;
    ETHERHEADER eth;
} NICETHERHEADER;
...
NICETHERHEADER nicin;         // Buffer for incoming NIC & Ether hdrs
...
WORD get_ether()
{
    WORD len=0, curr;
    ...
```

```
    getnic_data((BYTE *)&nicin, sizeof(nicin));
    len = nicin.nic.len;             /* Take length from stored header */
    ...
    len -= MACLEN+MACLEN+2+CRCLEN;
    ...
    return(len);                     /* Return length excl. CRC */
}
```

Packet Analysis

Because of a shortage of local on-chip RAM, the incoming packet must be analyzed in-place, in the NIC buffer RAM. The packet data is fetched a byte at a time, and a checksum is computed for Internet Protocol (IP) verification.

```
/* Get a byte from network buffer; if end, set flag */
BYTE getch_net(void)
{
    BYTE b=0;

    atend = rxout>=rxin;
    if (!atend)
    {
        b = getnic_byte();
        rxout++;
        check_byte(b);
    }
    return(b);
}


BYTE ungot_byte;
BOOL ungot;


/* Get an incoming byte value, return 0 if end of message */
BOOL get_byte(BYTE &b)
{
    if (ungot)
        b = ungot_byte;
    else
```

```
        b = getch_net();
    ungot = 0;
    return(!atend);
}
```

A simple 1-byte *pushback* buffer is included so that a byte may be returned to the buffer for rematching.

```
/* Unget (push back) an incoming byte value */
void unget_byte(BYTE &b)
{
    ungot_byte = b;
    ungot = 1;
}
```

The ampersand (&) in the argument of get_byte() and unget_byte() signifies a call-by-reference: any changes to the parameter are returned to the caller rather than discarded. This allows the CCS compiler to generate shorter code than the more conventional technique of passing a pointer to the value.

The packet analysis is performed using get_, match_, and skip_ calls as in the previous microcontroller implementation.

```
/* Get an incoming word value, return 0 if end of message */
BOOL get_word(WORD &w)
{
    BYTE hi, lo;

    hi = getch_net();
    lo = getch_net();
    w = ((WORD)hi<<8) | (WORD)lo;
    return(!atend);
}

/* Match an incoming word value, return 0 not matched, or end of message */
BOOL match_word(WORD w)
{
    WORD inw;

    return(get_word(inw) && inw==w);
}

/* Skip an incoming word value, return 0 if end of message */
BOOL skip_word(void)
```

```
{
    getch_net();
    getch_net();
    return(!atend);
}
```

Packet Transmission

Packet transmission is much simpler than reception. The driver has to

- write the Ethernet header (destination and source addresses and protocol word) into the packet buffer,
- write the packet data into the buffer,
- set the length of the packet in the NIC registers, rounding up if less than 64 bytes, and
- start the NIC state machine.

The NIC automatically retries the transmission if it fails because of a collision, but the transmission could still fail if the network is heavily loaded. The low-level drivers take no action in the event of failure, since it is usual for the higher layers (TCP) to initiate a retry.

The packet data comes from a series of put_ calls, and these could write directly into the NIC buffer area set aside for transmit, but this causes a problem when get_ and put_ calls are intermixed; the NIC DMA controller has to be reprogrammed continually to read or write the packet buffer. A compromise is to maintain a small local transmit data buffer and use this to store outgoing packet headers while they are being assembled.

```
#define TXBUFFLEN 64
BYTE txbuff[TXBUFFLEN];          // Tx buffer
int txin, txout;


/* Put a byte into the network buffer */
void putch_net(BYTE b)
{
    if (txin < TXBUFFLEN)
        txbuff[txin++] = b;
    check_byte(b);
}
```

The put_ether() function copies the NIC header and a data block into the packet buffer. Often, the data block will be the transmit buffer.

```
/* Send Ethernet packet given payload len */
void put_ether(void *data, WORD dlen)
{
    outnic(ISR, 0x0a);                 /* Clear interrupt flags */
    setnic_addr(TXSTART<<8);
```

```
    putnic_data(nicin.eth.srce, MACLEN);
    putnic_data(myeth, MACLEN);
    swapw(nicin.eth.pcol);
    putnic_data(&nicin.eth.pcol, 2);
    putnic_data(data, dlen);
}
```

If the volume of data to be transferred is larger than the transmit buffer, it is copied from the source (e.g., the Web page ROM) directly to the NIC packet buffer.

LCD Driver

LCD Data Transfers

The four LCD data bits and the two select lines (Read/Write, and Register Select) are all on port D (the Ethernet address bus), so it is convenient to create a bitfield structure and map it onto the port.

```
struct {
    BYTE data:4;
    BYTE regsel:1;
    BYTE read:1;
} LCD_PORT;
#BYTE LCD_PORT       = 8
#BIT    LCD_E        = PORTA.5
#define LCD_RD         LCD_PORT.read
#define LCD_RS         LCD_PORT.regsel
#define LCD_DATA       LCD_PORT.data
```

Bytes are transferred to the LCD in the form of two four-bit nybbles. After the transfer, the microcontroller data port is returned to being an input, so the bus is free for use by other devices. To simplify the interface, crude time delays have been used to space the byte transactions apart.

```
/* Send a command byte to the LCD as two nybbles */
void lcd_byte(BYTE &b)
{
    lcd_nybble(b >> 4);
    lcd_nybble(b);
    DATA_FROM_LCD;
    delay_us(40);
}
```

```
/* Send a byte to the LCD as two nybbles */
void lcd_nybble(BYTE b)
{
    DATA_TO_LCD;
    LCD_E = 1;
    LCD_DATA = b;
    LCD_E = 0;
}
```

To write a command or character to the LCD, it is only necessary to set or clear the Register Select line. A check is made to see if a display initialization command is being sent; if so, a delay is introduced to allow the display time to process the command.

```
/* Send a command byte to the LCD as two nybbles */
void lcd_char(BYTE b)
{
    DATA_TO_LCD;
    LCD_RD = 0;
    LCD_RS = 1;
    lcd_byte(b);
}

/* Send a command byte to the LCD as two nybbles */
void lcd_cmd(BYTE b)
{
    DATA_TO_LCD;
    LCD_RD = LCD_RS = 0;
    lcd_byte(b);
    if ((b & 0xfc) == 0)
        delay_ms(2);
}
```

LCD Initialization

After power-up, the display must be given time to stabilize before the transactions can begin. The initialization is rather long-winded because the LCD must be forced into eight-bit mode before being set to four bits and initialized.

```
/* Initialize the LCD */
void init_lcd(void)
{
    int i;
```

```
    LCD_E = 0;                      /* Clear LCD clock line */
    DATA_FROM_LCD;                  /* Ensure RS and RD lines are O/Ps */
    LCD_RD = LCD_RS = 0;
    delay_ms(15);                   /* Ensure LCD is stable after power-up */
    for (i=0; i<4; i++)             /* Force into 8-bit mode */
    {
        lcd_nybble(0x3);
        delay_ms(5);
    }
    lcd_cmd(0x28);                  /* Set 4-bit mode, 2 lines, 5x7 dots */
    lcd_cmd(0x06);                  /* Incrementing cursor, not horiz scroll */
    lcd_cmd(0x0e);                  /* Display on, cursor on, not blinking */
    lcd_cmd(0x01);                  /* Clear display, home cursor */
}
```

LCD Characters

Once initialized, it is easy to position the cursor and display a character.

```
#define LCD_SETPOS    0x80
#define LCD_LINE2     0x40

/* Go to an X-Y position on the display, top left is 1, 1 */
void lcd_gotoxy(BYTE x, BYTE y)
{
    if (y != 1)
        x += LCD_LINE2;
    lcd_cmd(LCD_SETPOS - 1 + x);
}

/* Send a character to the LCD */
void lcd_char(BYTE b)
{
    DATA_TO_LCD;
    LCD_RD = 0;
    LCD_RS = 1;
    lcd_byte(b);
}
```

Other Drivers

Analog Inputs

The PCM compiler provides library functions for driving the ADCs. Initialization consists of setting the appropriate pins as analog inputs, the clock rate for the conversion, and the channel to be sampled.

```
setup_adc_ports(RA0_ANALOG);
setup_adc(ADC_CLOCK_DIV_32);
SET_ADC_CHANNEL(0);
```

It would be better to set both RA0 and RA1 as analog inputs; however, this is not possible without sacrificing some other I/O pins as well — see the Microchip data sheet for more information. To access both channels, you have to briefly sacrifice digital output RA3, which is connected to an LED.

```
WORD adc1, adc2;                // Current ADC values

/* Read both ADC values */
void read_adcs(void)
{
    adc1 = READ_ADC();          // Read 1st channel
    setup_adc_ports(RA0_RA1_RA3_ANALOG);
    SET_ADC_CHANNEL(1);         // Change channel & settle
    delay_us(10);
    adc2 = READ_ADC();          // Read 2nd channel
    setup_adc_ports(RA0_ANALOG);
    SET_ADC_CHANNEL(0);         // Back to 1st channel
}
```

The LED connected to RA3 will flicker very briefly every time it becomes an analog input, but in practice, this isn't at all noticeable. If RA3 were connected to a switching device, such as a solid-state relay, this transient behavior could be a problem, and it would be necessary to rearrange the microcontroller I/O pin assignments.

Protocols

Address Resolution Protocol: ARP

All Ethernet messages use a six-byte MAC address, yet IP addresses are commonly expressed as four-byte numbers in dotted notation (e.g., 123.45.67.89). The client needs to be able to translate (*resolve*) the IP address into a MAC address, and it uses Address Resolution Protocol (ARP) for this purpose.

Chapter 3 discusses ARP in the context of a PC implementation. Until now, it hasn't been necessary to include it in the microcontroller implementation because the SLIP protocol uses IP addresses alone. With Ethernet, it is necessary to include an ARP server, but this is relatively simple. Each ARP transaction consists of a single resolution request, which generates a single reply (Figure 12.8).

Figure 12.8   ARP packet format.

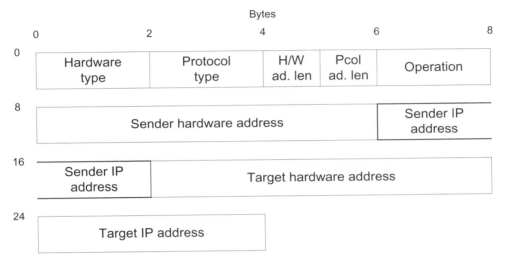

Analysis of the incoming message is implemented using a series of match_ and skip_ calls, as in the previous server; then the response is formulated using a series of put_ calls, culminating in a call to put_ether() to copy the transmit buffer into the NIC packet buffer, and a call to xmit_ether() to send the packet.

```
/* Handle an ARP message */
BOOL arp_recv(void)
{
    BOOL ret=0;

    if (match_byte(0x00) && match_byte(0x01) &&      // Hardware type
        match_byte(0x08) && match_byte(0x00) &&      // ARP protocol
        match_byte(6) &&  match_byte(4) &&           // Hardware & IP lengths
        match_word(ARPREQ) &&                        // ARP request
        skip_lword() && skip_word() &&               // Source MAC addr
        get_lword(remip.l) &&                        // Source IP addr
        skip_lword() && skip_word() &&               // Dest MAC addr
        match_lword(myip.l))
    {
```

```
            ret = 1;
            txin = 0;
            put_word(0x0001);
            put_word(0x0800);
            put_byte(6);
            put_byte(4);
            put_word(ARPRESP);
            put_data(myeth, MACLEN);
            put_lword(myip.l);
            put_data(nicin.eth.srce, MACLEN);
            put_lword(remip.l);
            put_ether(txbuff, txin);
            xmit_ether(txin);
        }
    return(ret);
}
```

Internet Protocol: IP

The IP decoder remains unchanged.

```
/* Get an IP datagram */
BOOL ip_recv(void)
{
    BYTE b, hi, lo;
    int n=0;
    BOOL ret=1;

    checkflag = 0;                                  // Clear checksum
    checkhi = checklo = 0;
    if (match_byte(0x45) && skip_byte() &&          // Version, service
        get_word(iplen) && skip_word() &&           // Len, ID
        skip_word() &&  skip_byte() &&              // Frags, TTL
        get_byte(ipcol) && skip_word() &&           // Protocol, checksum
        get_lword(remip.l) && match_lword(myip.l) && // Addresses
        checkhi==0xff && checklo==0xff)             // Checksum OK?
    {
        if (ipcol == PICMP)                         // ICMP?
            icmp_recv();                            // Call ping handler
        else if (ipcol == PTCP)                     // TCP segment?
            tcp_recv();                             // Call TCP handler
```

```
        else
            discard_data();                        // Unknown; discard it
    }
    else
        discard_data();
    return(ret);
}
```

Because of ROM space restrictions, only two higher-level protocols are supported: ICMP and TCP.

Internet Control Message Protocol: ICMP

The ICMP echo request (generally known as *ping*) is an important diagnostic tool, so it should be included in this discussion. On receipt of an echo request, the Web server has to send an echo reply with a duplicate of the incoming data. The data may be larger than the available RAM so it is copied directly from transmit to receive packet buffer. Because the ICMP request and reply headers are almost identical, the checksum need not be recomputed, but it can be adjusted to take account of the changes.

```
/* Respond to an ICMP message (e.g. ping) */
BOOL icmp_recv(void)
{
    BOOL ret=0;
    WORD csum;

    rpdlen = 0;
    if (match_byte(8) && match_byte(0) && get_word(csum))
    {
        while (skip_byte())                     // Check data
            rpdlen++;
        ret = (checkhi==0xff) && (checklo==0xff);
        if (ret && rpdlen<=MAXPING_LEN)
        {                                       // If OK and not bcast..
            DEBUG_PUTC('>');
            checkhi = checklo = 0;              // Clear checksum
            put_ip();                           // IP header
            put_word(0);                        // ICMP type and code
            csum += 0x0800;                     // Adjust checksum for resp
            if (csum < 0x0800)                  // ..including hi-lo carry
                csum++;
            put_word(csum);                     // ICMP checksum
```

```
            put_ether(txbuff, txin);          // Send ICMP response
            copy_rx_tx(txin, IPHDR_LEN+ICMPHDR_LEN, rpdlen);
            xmit_ether(IPHDR_LEN+ICMPHDR_LEN+rpdlen);
        }
    }
    return(ret);
}
```

The received ICMP data is copied a byte at a time into the transmit packet buffer. This crude method serves its purpose (after all, ICMP is only being used for diagnostics) but results in a slower-than-necessary ping response time.

```
/* Copy a block from NIC Rx to Tx buffers */
void copy_rx_tx(BYTE dest, BYTE srce, BYTE len)
{
    BYTE b;

    outnic(ISR, 0x40);
    dest += sizeof(ETHERHEADER);
    srce += sizeof(NICETHERHEADER);
    while (len--)
    {
        outnic(RSAR0, srce);
        outnic(RSAR1, curr_rx_page);
        b = getnic_byte();
        outnic(RSAR0, dest);
        outnic(RSAR1, TXSTART);
        putnic_byte(b);
        srce++;
        dest++;
    }
}
```

Transmission Control Protocol: TCP

The TCP decoder remains largely unchanged.

```
/* Respond to an TCP segment */
BOOL tcp_recv(void)
{
    int hlen, n;
    BOOL ret=0;
    WORD addr;
```

```
    checkhi = checklo = 0;
    if (get_word(remport) && get_word(locport) &&      // Source & dest ports
        get_lword(rseq.l) && get_lword(rack.l) &&       // Seq & ack numbers
        get_byte(hlen) && get_byte(rflags) &&           // Header len & flags
        skip_word() && skip_lword())                    // Window, csum, urgent ptr
    {
        iplen -= IPHDR_LEN;                             // Get TCP segment length
        check_word(iplen);                              // Check pseudoheader
        check_lword(myip.l);
        check_lword(remip.l);
        check_byte(0);
        check_byte(PTCP);
        rxout = (hlen>>2) + IPHDR_LEN;                  // Skip over options
        rpdlen = iplen - rxout + IPHDR_LEN;
        addr = getnic_addr();
        check_rxbytes(IPHDR_LEN+TCPHDR_LEN, iplen-TCPHDR_LEN);
        setnic_addr(addr);
        ret = (checkhi==0xff) && (checklo==0xff);
        if (ret)
            tcp_handler();
    }
    return(ret);
}
```

Note the calls to getnic_addr() and setnic_addr(), which are used to temporarily save and restore the NIC address pointer. This allows the code to scan ahead for the checksum computation and then go back for the detailed analysis of the TCP data.

```
/* Handle an incoming TCP segment */
void tcp_handler(void)
{
    BOOL tx=1;                          // Set transmission flag

    tpdlen = tpxdlen = 0;               // Assume no Tx data
    d_checkhi = d_checklo = 0;
    checkflag = 0;
    tflags = TACK;                      // ..and just sending an ack
    if (rflags & TRST)                  // RESET received?
        tx = 0;                         //..do nothing
    else if (rflags & TSYN)             // SYN received?
```

```
    {
        add_lword(rseq.l, 1);                // Adjust Tx ack for SYN
        if (locport==DAYPORT || locport==HTTPORT)
        {                                    // Recognized port?
            rack.w[0] = 0xffff;
            rack.w[1] = concount++;
            tflags = TSYN+TACK;              // Send SYN ACK
        }
        else                                 // Unrecognized port?
            tflags = TRST+TACK;              // Send reset
    }
    else if (rflags & TFIN)                  // Received FIN?
        add_lword(rseq.l, rpdlen+1);         // Ack all incoming data + FIN
    else if (rflags & TACK)                  // ACK received?
    {
        if (rpdlen)                          // Adjust Tx ack for Rx data
            add_lword(rseq.l, rpdlen);
        else                                 // If no data, don't send ack
            tx = 0;
        if (locport==HTTPORT && rpdlen)
        {                                    // HTTP 'get' method?
            http_recv();                     // Call handler..
            tx = 0;                          // ..which does its own Tx
        }
        else if (locport==DAYPORT && rack.w[0]==0)
        {                                    // Daytime request?
            daytime_handler();               // Prepare daytime data
            tx = 1;                          // ..and send it
        }
    }
    if (tx)                                  // If transmission required
        tcp_xmit();                          // ..do it
}
```

The TCP implementation supports two higher-level protocols:

- HTTP (for the Web server)
- Daytime (to demonstrate a simple TCP transaction)

HTTP is described in the next section.

The Daytime protocol returns a string with the current date and time.

```c
/* Respond to an Daytime request */
BOOL daytime_handler(void)
{
    checkhi = checklo = 0;
    txin = IPHDR_LEN + TCPHDR_LEN;      // O/P data to buffer, calc checksum
    printf(put_byte, DAYMSG);
    tpdlen = DAYMSG_LEN;                 // Data length of response
    d_checkhi = checkhi;                 // Save checksum
    d_checklo = checklo;
    tflags = TFIN+TACK;                  // Ack & close connection
}
```

The nonstandard CCS compiler call with a function argument `printf(put_byte, ...)` has been used so that the function `put_byte()` is called once for every character in the output string, which stores the string in the transmit buffer and computes its checksum. Unfortunately, there's no real-time clock chip, so a dummy message is put out instead.

```c
#define DAYMSG       "No daytime msg\r\n"
#define DAYMSG_LEN   16
```

The transmit buffer is 64 bytes in length, so is sufficient for short messages such as these (the total length of the above response is 20 + 20 + 16 = 56 bytes). Longer messages, such as the HTTP page data, are transferred direct from ROM to the NIC packet buffer; this will be described later.

Because of the absence of a state machine, the TCP implementation is known as *stateless*, that is, the server stores no state information about any of the clients. As described in Chapter 11, the server can send only one frame in response to each request from the client; otherwise, the lock-step nature of the transaction is broken, and the server would need to implement a retry strategy. Aside from simplicity, the huge advantage of the stateless approach is that the server can accommodate very large numbers of simultaneous connections, which makes it very responsive when handling multiple requests.

When responding to an incoming message, the server should perform an ARP cycle to resolve

- the return IP address, if in the current domain, or
- the router's IP address, if the return address is outside the current domain.

A much more compact approach is to reuse the incoming IP and MAC addresses for the return message. Providing the network routing has been kept simple, this approach will work fine.

```c
/* Put out a TCP segment. Data checksum must have already been computed */
void put_tcp(void)
{
    WORD len;
```

```
    checkflag = 0;                          // Ensure we're on an even byte
    checkhi = d_checkhi;                    // Retrieve data checksum
    checklo = d_checklo;
    put_word(locport);                      // Local and remote ports
    put_word(remport);
    put_lword(rack.l);                      // Seq & ack numbers
    put_lword(rseq.l);
    put_byte(tflags&TSYN ? TCPSYN_LEN*4 : TCPHDR_LEN*4);   // Header len
    put_byte(tflags);
    put_byte(0x0b);                         // Window size word
    put_byte(0xb8);
    len = tpdlen + tpxdlen + TCPHDR_LEN;
    if (tflags & TSYN)                      // If sending SYN, send MSS option
    {
        txin += 4;                          // Put MSS in buffer after TCP header
        len += TCPOPT_LEN;
        put_byte(2);
        put_byte(4);
        put_word(TCP_MSS);
        txin -= TCPOPT_LEN + 4;             // Go back to checksum in header
    }
    check_lword(myip.l);                    // Add pseudoheader to checksum
    check_lword(remip.l);
    check_byte(0);
    check_byte(PTCP);
    check_word(len);
    put_byte(~checkhi);                     // Send checksum
    put_byte(~checklo);
    put_word(0);                            // Urgent ptr
    if (tflags & TSYN)                      // Adjust Tx ptr if sending MSS option
        txin += TCPOPT_LEN;
}
```

Hypertext Transfer Protocol: HTTP

The TCP/IP Lean microcontroller HTTP implementation was dominated by the problem of the TCP checksum. It must be computed before the Web page goes out, yet there is insufficient RAM to buffer the page prior to transmission. As a result, a complex scheme of counterbalancing HTTP tags was employed, so that the checksum was known in advance, even when the page was being changed on the fly.

The good news is that the Ethernet controller stores a complete page in its packet buffer prior to transmission, so the need for a precomputed checksum is removed, using the following method.

1. Copy data (i.e., Web page) to the packet buffer, computing checksum as it is copied.
2. Prepare the TCP header in local RAM using the data checksum value.
3. Copy the TCP header from local RAM to the packet buffer.
4. Transmit the whole packet in the buffer.

```c
/* Receive an incoming HTTP request ('method'), return 0 if invalid */
BOOL http_recv(void)
{
    int len, i;
    BOOL ret=0;
    char c;
    WORD blen;

    tpxdlen = 0;                        // Check for 'GET'
    DEBUG_PUTC('h');
    if (match_byte('G') && match_byte('E') && match_byte('T'))
    {
        ret = 1;
        match_byte(' ');
        match_byte('/');                // Start of filename
        memset(romdir.f.name, 0, ROM_FNAMELEN);
        for (i=0; i<ROM_FNAMELEN && get_byte(c) && c>' ' && c!='?'; i++)
        {                               // Name terminated by space or '?'
            romdir.f.name[i] = c;
        }                               // If file found in ROM
        if (find_file())
        {                               // ..check for form arguments
            check_formargs();
        }
        else                            // File not found, get index.htm
        {
            romdir.f.name[0] = 0;
            find_file();
        }
        checkhi = checklo = 0;
        checkflag = 0;
        txin = IPHDR_LEN + TCPHDR_LEN;
```

```
        if (!fileidx)                  // No files at all in ROM - disaster!
        {
            setnic_addr((TXSTART<<8)+sizeof(ETHERHEADER)+IPHDR_LEN+TCPHDR_LEN);
            printf(putnic_checkbyte, HTTP_FAIL);
            tflags = TFIN+TACK;
            d_checkhi = checkhi;
            d_checklo = checklo;
            tcp_xmit();
        }
        else                           // File found OK
        {
            open_file();                   // Start i2c transfer
            setnic_addr((TXSTART<<8)+sizeof(ETHERHEADER)+IPHDR_LEN+TCPHDR_LEN);
            while (tx_file_byte())         // Copy bytes from ROM to NIC
                ;
            close_file();
            tflags = TFIN+TPUSH+TACK;   // Close connection when sent
            d_checkhi = checkhi;        // Save checksum
            d_checklo = checklo;
            tcp_xmit();                 // Do header, transmit segment
        }
    }
    return(ret);
}
```

The `tx_file_byte()` function copies the data from ROM to packet buffer, performing Embedded Gateway Interface (EGI) variable substitution as it goes.

```
/* Transmit a byte from the current i2c file to the NIC
** Return 0 when complete file is sent
** If file has EGI flag set, perform run-time variable substitution */
BOOL tx_file_byte(void)
{
    int ret=0, idx;
    BYTE b;

    if (romdir.f.len)                      // Check if any bytes left to send
    {
        b = i2c_read(1);                   // Get next byte from ROM
        if ((romdir.f.flags&EGI_ATVARS) && b=='@')
        {                                  // If '@' and EGI var substitution..
```

```
            b = i2c_read(1);              // ..get 2nd byte
            romdir.f.len--;
            idx = b - 0x30;
            if (idx == 1)                 // Scaled ADC value for slider 1
                printf(putnic_checkbyte, "%u", (BYTE)(adc1/11)+6);
            else if (idx == 2)            // Scaled ADC value for slider 2
                printf(putnic_checkbyte, "%u", (BYTE)(adc2/11)+6);
            else if (idx == 3)            // Voltage value for ADC 1
                putnic_volts(adc1);
            else if (idx == 4)            // Voltage value for ADC 2
                putnic_volts(adc2);
            else if (idx == 5)            // User O/P LED 1 state
                putnic_checkbyte(USERLED1 ? '0' : '1');
            else if (idx == 6)            // User O/P LED 2 state
                putnic_checkbyte(USERLED2 ? '0' : '1');
            else if (idx == 7)            // I/P button state
                putnic_checkbyte(USER_BUTTON ? '0' : '1');
            else                          // Unknown variable
                printf(putnic_checkbyte, "??");
        }
        else                          // Non-EGI byte; send out unmodified
            putnic_checkbyte(b);
        romdir.f.len--;               // Decrement length
        ret = 1;
    }
    return(ret);
}

/* Send the voltage string for the given ADC to the NIC */
void putnic_volts(WORD val)
{
    BYTE v;

    v = (BYTE)(val / 21);
    putnic_checkbyte(v/10 + '0');
    putnic_checkbyte('.');
    putnic_checkbyte(v%10 + '0');
}
```

The EGI variables are prefixed with an @ character and are used as follows:

- @1, @2—scaled ADC values for slider animation (see next section)
- @3, @4—voltage values for ADCs
- @5, @6—O/P LED states
- @7, @8—I/P button states

When one of these variables is encountered in the file, it is changed into the corresponding real-time data value with the use of the nonstandard CCS implementation of `printf(putnic_checkbyte, ...)` to copy the characters into the packet buffer RAM while keeping track of the checksum.

Protocol Debugging

To help in debugging the TCP/IP stack, a simple diagnostic output can be enabled on the serial port using the DEBUG compile-time definition.

```
#define DEBUG        1           // Set non-zero to enable diagnostic printout
```

The output owes more to brevity than clarity; the display for a simple HTTP request might look as follows.

```
Rx46>a>A Rx46>i>t>T Rx46>i>t> Rx56>i>t>h index.htm>T Rx46>i>t>T Rx72>i
```

Each received packet is indicated by Rx, the length (in decimal), and a right arrow (>). Subsequent protocol decodes are indicated by lowercase letters.

a	ARP
i	IP
c	ICMP
t	TCP
h	HTTP

Transmissions are indicated by the corresponding uppercase letters, so

```
Rx46>a>A
```

indicates that a 46-byte ARP request was received, and a response was sent. For HTTP requests, the filename is displayed, so

```
Rx56>i>t>h index.htm>T
```

shows that `index.htm` was requested, and a TCP (strictly speaking, HTTP–TCP–IP) response was sent.

Although it is a poor substitute for a Protocol Analyzer, this diagnostic capability can be very useful for resolving simple protocol problems.

User Interface

Dynamic Web Pages: HTML

Because of the SLIP link, the TCP/IP Lean Web server couldn't handle complicated Web pages (those with a lot of embedded graphics). The browser will issue multiple parallel get requests, one for each graphic, which causes an overflow of the small serial buffer. Although the browser's retry mechanism ensures that all the graphics arrive, it does make for frustratingly slow Web page updates.

The Ethernet controller removes the bottleneck caused by the serial buffer, so it is possible to create quite elaborate displays, with the constraint that each individual file fit within one Ethernet frame (file plus HTTP header must fit in 1,460 bytes). The home page for the PIC-DEM.net board is shown in Figure 12.9.

Figure 12.9   Home page for PICDEM.net board.

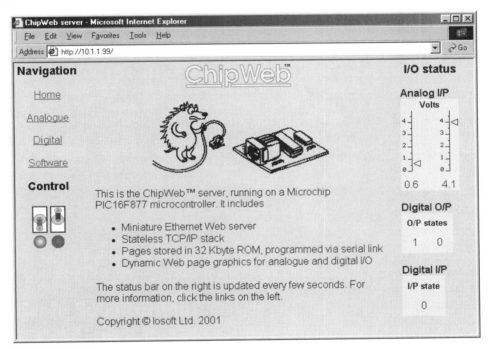

Aside from using this opportunity to include cute graphics, you can use HTTP *frames* to segregate the user's screen into three areas: the main center section, which changes according to the user's selections, and the two sidebars, one for navigation and control and the other to show the I/O status. This achieves several objectives:

- It provides a clearer layout for user navigation.
- It splits the screen area into several smaller files rather than one large file.
- It allows the use of simple techniques for the toggle-switch controls.
- It limits the size of the dynamic update area, since this can flicker during updates.

The home page `index.htm` is just a shell that contains the three frames.

```
<HTML>
<HEAD>
<TITLE>ChipWeb server</TITLE>
</HEAD>
<FRAMESET cols="120,*,120" border=0>
   <FRAME name="left"   src="digout00.htm" marginheight=2 marginwidth=2>
   <FRAME name="middle" src="main1.htm" marginheight=5 marginwidth=5>
   <FRAME name="right"  src="stat1.egi" marginheight=5 marginwidth=5>
</FRAMESET>
</HTML>
```

Status Display Frame

The right-hand status frame uses a *client pull* instruction so that the browser refetches it from the server every five seconds.

```
<html><meta http-equiv="refresh" content="5">
<head><title>ChipWeb status</title></head>
<body bgcolor=#d0d0ff><font face=helvetica>
<h3><center>I/O status</center></h3>
...
```

This technique is acceptable for Local Area Network (LAN) use, but it is unusable over the Internet; random delays in the path between client and server can cause a backlog of requests at the client and annoyingly inconsistent update times.

The analog display uses variable GIF block sizes to vertically position the indicators, as discussed in Chapter 7. Some browsers flicker when updating dynamic graphics, so it is best to keep the area small and offer the user a more accurate (and flicker-free) text indication of the value as well.

```
...
<b>Analog I/P</b>
<table border=0 cellpadding=1 cellspacing=0 bgcolor=#e0e0ff>
<tr><th colspan=5 align=center><small>Volts</small></th></tr>
<tr align=center valign=bottom>
<td><img src="v5_100n.gif"><br><img src="clr.gif" height=10 width=10></td>
<td><img src="lptr.gif"><br><img src="clr.gif" height=@1 width=10></td>
<td><img src="clr.gif" height=1 width=25></td>
<td><img src="v5_100n.gif"><br><img src="clr.gif" height=10 width=10></td>
<td><img src="lptr.gif"><br><img src="clr.gif" height=@2 width=10></td>
</tr><tr>
<td colspan=2 align=center>@3</td>
```

```
<td> </td>
<td colspan=2 align=center>@4</td>
</tr>
</table><br>
```

The display of digital inputs and outputs is text only.

```
<b>Digital O/P</b><br>
<table cellpadding=6 bgcolor=#e0e0ff vspace=0>
<tr><th align=center colspan=2><small>O/P states</small></th></tr>
<tr><td align=center>@5</td> <td align=center>@6</td> </tr>
</table><br>

<b>Digital I/P</b><br>
<table cellpadding=6 bgcolor=#e0e0ff vspace=0>
<tr><th align=center colspan=1><small>I/P state</small></th></tr>
<tr><td align=center>@7</td></tr>
</table>
</body></html>
```

Navigation Frame

This frame contains links that select what is to be displayed in the central area. This is achieved using `href` entries that specify a target location. The possible targets are `left`, `middle`, and `right`, as specified within the frameset definition of `index.htm`.

```
<HTML><HEAD></HEAD>
<BODY bgcolor=#d0d0ff><font face=helvetica>
<h3><center>Navigation</center></h3>
<CENTER>
<A href="main1.htm" target=middle>Home</A><P>
<A href="maina.htm" target=middle>Analogue</A><P>
<A href="maind.htm" target=middle>Digital</A><P>
<A href="mains.htm" target=middle>Software</A><P>
</CENTER>
...
```

The navigation frame also has the toggle switches and LEDs for controlling the outputs. This grouping not only helps the user to navigate using a minimum of mouse movement, but also helps by permitting the use of a very simple technique for animating the toggle switches. At the start of Chapter 8, I describe how a toggle switch and lamp can be animated using two Web pages, one showing "off" and the other "on." Clicking the switch on one page causes the other page to load. Because the PICDEM.net board has two outputs, this principle has been expanded to four pages: one for each state. The switch URLs are cross-linked, as shown below.

```
File digout00.htm:
...
<h3><center>Control</center></h3>
<center><table>
<tr valign=middle>
<td><a href="digout10.htm"><img src="sw0.gif"></a></td>
<td><a href="digout01.htm"><img src="sw0.gif"></a></td>
</tr><tr valign=middle>
<td align=center><img src="led0.gif"></td>
<td align=center><img src="led0.gif"></td>
</tr></table>
...


File digout01.htm:
...
<table>
<tr valign=middle>
<td><a href="digout11.htm"><img src="sw0.gif"></a></td>
<td><a href="digout00.htm"><img src="sw1.gif"></a></td>
</tr><tr valign=middle>
<td align=center><img src="led0.gif"></td>
<td align=center><img src="led1.gif"></td>
</tr></table>
...


Same logic for digout10.htm and digout11.htm...
```

Rudimentary parsing of the input filename is used to propagate the on/off states to the hardware, bearing in mind the hardware inversion: the LED is on when the output line is low.

```
#bit USERLED1 = PORTA.2              // User LEDs
#bit USERLED2 = PORTA.3
...
/* Check for arguments in HTTP request string
** Simple version: just check last 2 digits of filename, copy to 2 LEDs */
void check_formargs(void)
{
    if (romdir.f.name[6]=='0' || romdir.f.name[6]=='1')
```

```
        USERLED1 = (romdir.f.name[6] == '0');
    if (romdir.f.name[7]=='0' || romdir.f.name[7]=='1')
        USERLED2 = (romdir.f.name[7] == '0');
}
```

LCD Display

The two-line LCD is used to display a sign-on message (including the software version number) and the IP address of the server. In configuration mode, the second display line is changed to reflect the off-line state of the server.

To save space, a single "print" character handler is used, with flags to determine whether the output will be directed to the serial port, the LCD, or both.

```
BOOL disp_lcd, disp_serial;      // Flags to enable display O/Ps
...
disp_lcd = disp_serial = TRUE;      // Set display flags
printf(displays, SIGNON);           // ..and sign on

...
/* Display handler; redirects to LCD and/or serial */
void displays(BYTE b)
{
    if (disp_lcd)
    {
        if (b == '\r')
            lcd_cmd(LCD_SETPOS);
        else if (b == '\n')
            lcd_cmd(LCD_SETPOS + LCD_LINE2);
        else
            lcd_char(b);
    }
    if (disp_serial)
    {
        if (b == '\n')
            putchar('\r');
        putchar(b);
    }
}
```

Configuration

The following configuration items need to be stored in nonvolatile memory.

- four-byte IP address
- six-byte MAC address
- Web pages

The first two are stored in the PIC16F877 on-chip E^2ROM, whereas the last is stored in the separate serial E^2ROM. To simplify production and to allow the user to alter the values on-site, they need to be programmable over the serial link.

To configure the system, the user connects a serial terminal (e.g., Windows HyperTerminal) to the board. A simple checksum is used to see if the CPU is uninitialized, and if so, it automatically enters configuration mode. If not, the user has to manually initiate the mode by holding down the user I/P push button while the CPU is coming out of reset. As with the LED outputs, there is an I/O inversion in the hardware, so the input reads zero when the button is pressed.

```c
#bit USER_BUTTON=PORTB.5          // User pushbutton
...
while (!read_nonvol() || !USER_BUTTON)  // If csum error, or button
{
    printf(displays, "Config ");
    user_config();                          // ..call user config
}
...
/* Read in the nonvolatile parameters, return 0 if error */
BOOL read_nonvol(void)
{
    int i;

    myeth[4] = read_eeprom(0);
    myeth[5] = read_eeprom(1);
    myip.b[3] = read_eeprom(2);
    myip.b[2] = read_eeprom(3);
    myip.b[1] = read_eeprom(4);
    myip.b[0] = read_eeprom(5);
    return (csum_nonvol() == read_eeprom(6));
}


/* Do a 1's complement checksum of the non-volatile data */
BYTE csum_nonvol(void)
{
```

```
    int i;
    BYTE sum=0;

    for (i=0; i<6; i++)
        sum += read_eeprom(i);
    return(~sum);
}
```

Address Configuration

The first three bytes of the six-byte Media Access and Control (MAC) address identify the board vendor. For Microchip, these are 00 04 A3. It is important that the remaining three bytes are unique among all Microchip products, so each board is allocated a serial number, and the user is prompted to enter this number on the serial interface. This value is used directly as the last three MAC bytes. So long as all the serial numbers are unique and the user enters the number properly, there is no danger of two boards having the same MAC address. Although acceptable on a prototype board, this method could not be used on production equipment. The MAC address should be machine-programmed on manufacture and should not be changeable by the user.

The IP address is the other user-configured item. It must be set to reflect the network (*domain*) the Web server is connected to and must be a unique value within that domain.

To configure both these settings, the user enters configuration mode and is then prompted for the required values. The escape key can be used to bypass a setting if it is already correct.

Originally, the user interface was coded using a conventional get-line-then-parse approach, but a parse-on-the-fly approach was much smaller. The absence of a loop might also be surprising, but it must be understood that the compiler can resolve an array reference with a constant index (such as myip.b[1]) to a fixed memory location, so the "unrolled" loop can be more economical in terms of code and data space.

```
BOOL escaped;
...
/* User initialization code; get serial number and IP address
** Skip if user hits ESC */
void user_config(void)
{
    WORD w;
    BYTE t1, t2;

    escaped = 0;
    printf("\r\nSerial num? ");
    w = getnum();
    if (!escaped)
    {
```

```
        myeth[4] = w >> 8;
        myeth[5] = w;
    }
    escaped = 0;
    printf("\r\nIP addr? ");
    USERLED1 = USERLED2 = 1;
    myip.b[3] = getnum();
    putchar('.');
    if (!escaped)
        myip.b[2] = getnum();
    putchar('.');
    if (!escaped)
        myip.b[1] = getnum();
    putchar('.');
    if (!escaped)
        myip.b[0] = getnum();
    if (!escaped)
        write_nonvol();
    printf("\r\nXmodem? ");
    xmodem_recv();
}
```

Web Page Download

The WEBROM utility takes all the files out of a given directory, adds the HTTP headers, and merges them into a single ROM file, ready for transfer to the Web page E2ROM.

```
webrom webpage.rom c:\chipweb\pcm\romdocs
```

To transfer the file down the serial link, the simplest file-transfer protocol is employed: XMODEM. This sends the file in 128-byte blocks with a simple header and a checksum trailer (which is ignored in this implementation). A minor complication is that the E2ROM isn't able to accept the whole 128 bytes at once. It must be broken into chunks, with a pause between them to allow the device to program itself. Fortunately, the programming time (5ms) allows the use of a relatively small serial buffer to cover this delay. A detailed discussion of the XMODEM protocol is outside the scope of this document. There are copious references on the Web that analyze the byte structure and timing in detail. This implementation employs a few Boolean flags and a timer to construct a simple state machine. The timer handles any interruptions in the transfer, but other errors (such as single-character corruption) are not handled. It is assumed that there is a short cable link to the PC, not a noisy modem connection.

```c
#define SOH 0x01
#define EOT 0x04
#define ACK 0x06
#define NAK 0x15
#define CAN 0x18

#define ROMPAGE_LEN 32
#define XBLOCK_LEN  128

/* Handle incoming XMODEM data block */
void xmodem_recv(void)
{
    BYTE b, len=0, idx, blk, i, oset;
    BOOL rxing=FALSE, b1=FALSE, b2=FALSE, b3=FALSE;

    timeout(ledticks, 0);
    while (1)
    {
        while (!kbhit())
        {
            restart_wdt();              // Kick watchdog
            geticks();                  // Check for timeout
            if (timeout(ledticks, LEDTIME))
            {
                SYSLED = !SYSLED;
                USERLED1 = 1;
                if (!rxing)             // Send NAK if idle
                {
                    len = 0;
                    b1 = b2 = b3 = FALSE;
                    putchar(NAK);
                }
                rxing = FALSE;
            }
        }
        b = getchar();                  // Get character
        rxing = TRUE;
        if (!b1)                        // Check if 1st char
        {
```

```
        if (b == SOH)                // ..if SOH, move on
            b1 = TRUE;
        else if (b == EOT)           // ..if EOT, we're done
        {
            putchar(ACK);
            break;
        }
    }
    else if (!b2)                    // Check if 2nd char
    {
        blk = b;                     // ..block num
        b2 = TRUE;
    }
    else if (!b3)                    // Check if 3rd char
    {
        if (blk == ~b)               // ..inverse block num
        {
            b3 = TRUE;
            blk--;
        }
    }
    else if (len < XBLOCK_LEN)       // Rest of chars up to block len
    {                                // Buffer into ROM page
        idx = len & (ROMPAGE_LEN - 1);
        len++;
        txbuff[idx] = b;             // If end of ROM page..
        if (idx == ROMPAGE_LEN-1)    // ..write to ROM
        {
            i2c_start();
            i2c_write(EEROM_ADDR);
            i2c_write(blk >> 1);
            oset = len - ROMPAGE_LEN;
            if (blk & 1)
                oset += 0x80;
            i2c_write(oset);
            for (i=0; i<ROMPAGE_LEN; i++)
                i2c_write(txbuff[i]);
            i2c_stop();
        }
```

```
        }
        else                         // End of block, send ACK
        {
            putchar(ACK);
            timeout(ledticks, 0);
            SYSLED = !SYSLED;
            len = 0;
            b1 = b2 = b3 = FALSE;
        }
    }
}
```

Conclusion

A Web server is an ideal candidate as an initial application to run on the Microchip PIC-DEM.net board because it allows the board to be exercised thoroughly. The software in this chapter is even used to test the hardware for production.

The Ethernet interface is an easy-to-use method of transporting data at relatively high speed. Although more complex than a serial interface, the Ethernet controller offers many useful features, especially the buffering of incoming data. This allows the processor to poll the network interface when convenient, rather than use an interrupt handler to handle every incoming byte.

In the following chapters, I explore other TCP/IP protocols that can be used with a microcontroller Ethernet interface.

Source Files

Development of the ChipWeb TCP/IP package is ongoing, so the CD-ROM accompanying this book has various software versions on it, each with specific attributes. See the `readme.txt` file in the appropriate directory for more information. The most up-to-date files are available on the Iosoft Ltd. Web site `www.iosoft.co.uk`.

The original source files used in this chapter were compiled with Custom Computer Services "PCM" C compiler v2.693. Later ChipWeb versions offer compatibility with a wider range of compilers.

The compiler doesn't use a linker, so library files are included in the main source file.

p16web.c	Main program
p16_drv.h	Low-level driver
p16_eth.h	Ethernet interface
p16_ip.h	IP and TCP
p16_lcd.h	LCD interface
p16_usr.h	User interface and configuration
p16_http.h	HTTP and Web page processing
webrom.h	ROM file system definitions

The source files should reside in the directory \chipweb\pcm.

To maintain compatibility with a variety of software tools, p16web.c has absolute file paths for the include files. These have to be changed if a different directory is used.

The compiler output file, chipweb.hex, should be programmed into a PIC16F77 device.

The Web documents are in \chipweb\pcm\romdocs. To reload them into the Web page ROM, they must first be merged using the utility from TCP/IP Lean.

```
cd \chipweb\pcm
\tcplean\webrom webpage.rom romdocs
```

They must then be downloaded using XMODEM protocol (not XMODEM-1k) using a PC terminal emulator program running at 9600 baud.

Chapter 13

Point-to-Point Protocol: PPP

Overview

So far, all TCP/IP serial communications have used a very simple serial encapsulation protocol, SLIP. Many applications require a more versatile serial interface, so they employ PPP (Point-to-Point Protocol) instead. It is particularly useful in establishing a dial-up link to an ISP (Internet Service Provider) because it includes a framework for the validation of the user's identity and negotiates the client's IP address.

PPP provides much more than just a user name and addressing capability; it is a container for various other protocols, each of which is negotiated independently. To create a PPP connection, LCP (Link Control Protocol), IPCP (Internet Protocol Control Protocol), and probably a user-authentication protocol such as PAP (Password Authentication Protocol) have to be negotiated. There are no shortcuts around this. If any one of the three negotiations fails, the PPP link will fail, and no data can be transferred.

Whole books have been written about PPP. In this chapter, I'll look at the basic protocol elements and how they can be implemented. It would have been nice to include a complete set of source code for the PIC16F877 on the Microchip PICDEM.net ™ board, but unfortunately, the small RAM limits the applications that can be accommodated. Instead, I'll write software that takes advantage of the increased ROM and RAM sizes of the Microchip PIC18 series microcontrollers. At the time of writing, these devices are relatively new, so this chapter represents work in progress, rather than a completed application.

Design of PPP

As its name implies, PPP is intended for use in one-to-one links between two network nodes. Although I concentrate on RS232 dial-up connections in this chapter, PPP can also be employed in other situations where a dedicated connection between two nodes is needed; for example, PPP-over-Ethernet provides a PPP "pipe" between two specific nodes on an Ethernet network. PPP provides the following:

Framing   Subdivision of a serial data stream into blocks, with start and end markers.

Error Checking   Cyclic redundancy code (CRC) verifies the integrity of a message.

Escape Sequences   Common control characters can be eliminated from the data stream with the inclusion of specified escape sequences.

Negotiation   The two nodes can negotiate to achieve a mutually acceptable set of protocols and options. These often include authentication (verification of identity) and addressing (allocation of a dynamic IP address from a pool of addresses).

It is important to note that parameters and protocols are *negotiated*, not fixed, although a negotiator might have a fixed idea as to what is acceptable and reject all other possibilities. To misquote George Orwell's *Animal Farm*: "All options are negotiable — but some are more negotiable than others."

Framing

As with SLIP, start/end marker bytes subdivide a data stream into blocks. Unlike SLIP, the framing scheme is borrowed from an ISO standard communications protocol called HDLC (High-level Data Link Control), which in turn is based on the pioneering work by IBM in their SNA (Systems Network Architecture) SDLC (Synchronous Data Link Control) protocol layer. Because an asynchronous serial link is being used, most of the capabilities of HDLC are redundant, so a subset called AHDLC (asynchronous HDLC) is employed (Figure 13.1).

Figure 13.1   HDLC frame.

The HDLC *flag* character marks the start and end of each *frame*. The *link header* also has *address* and *control* fields, whereas the *link trailer* also has an FCS (Frame Check Sequence).

The HDLC address field is unused, so it is fixed at FFh (broadcast). The control field is fixed as unnumbered information with a poll/final bit set to zero (i.e., a value of 03h). This means you can expect a PPP frame to start with the byte sequence 7E FF 03h, which is very convenient when attempting to analyze the communications between two PPP nodes. However, the nodes are allowed to negotiate the removal of the address and control bytes under some circumstances, so they might not appear.

The *information* field contains the protocol data; by default, this can be up to 1,502 bytes.

Error Checking

The FCS allows the recipient to check the integrity of the incoming frame; it defaults to a 16-bit CRC, but an alternative may be negotiated.

The 16-bit CRC generator uses a starting value of FFFFh, which is updated for each byte in the address, control, and information fields. A byte-wide table-lookup algorithm is usually employed for speed. Here is an equivalent function that loops across each data bit.

```
/* Return new PPP CRC value, given previous CRC value and new byte */
WORD ppp_crc16(WORD crc, BYTE b)
{
    BYTE i;

    for (i=0; i<8; i++)
    {
        if ((crc ^ b) & 1)
            crc = (crc >> 1) ^ 0x8408;
        else
            crc >>= 1;
        b >>= 1;
    }
    return(crc);
}
```

The 16-bit FCS is appended to the information field with the high-order byte first.

On receipt, the whole frame, excluding flag bytes, is checked. The resulting CRC value should be F0B8h.

Escape Sequences

The HDLC information or FCS fields may contain a byte value of 7Eh, which could be mistaken for a start/end flag, so an escape sequence is sent in its place. This consists of the escape byte 7Dh, followed by the original value exclusive OR 20h (i.e., 7D 5Eh). The escape character may also occur among the data, in which case it is sent as 7D 5Dh.

This escape scheme also protects control characters, which can have special meanings on serial links. For example, software handshaking may be enabled, in which case XOFF and XON characters (control-S and control-Q) are used by the low-level drivers to suspend and resume serial transmissions. To avoid any confusion between XOFF and a data byte value of 14h, the latter may be sent as the escape sequence 7D 34h. By default, all values between 00 and 1Fh are escaped, but (as you've probably already guessed) the nodes can negotiate to remove the escape sequences from some or all of these values.

It is important to remember that the escape sequence doesn't just apply to the data field, but also the control and FCS fields, so the control field would become the two-byte value 7D 23h.

Negotiation

Negotiation is a critical component of PPP. A set of default options are used when a specific point-to-point link is started; then the two nodes attempt to agree on the parameters to be used, such as data compression, addressing, and authentication, with the following characteristics:

Symmetry   Two complete sets of negotiations take place, one for each direction.

Options   The things to be negotiated are known as *options*; an option can be

- a value (e.g., maximum packet size),
- a Boolean flag (e.g., enable/disable compression), or
- a protocol (e.g., which authentication protocol to use).

Request/response   A series of *configure-request* and associated response negotiations.

ACK/NAK/REJ   An option can be accepted (configure-acknowledge, *config-ACK*) or the recipient can

- not agree to the requested value (configure-negative-acknowledge, *config-NAK*),
- not recognize the option (configure-reject, *config-REJ*),
- not recognize the message (code reject, *code-REJ*), or
- not recognize the protocol (protocol reject, *proto-REJ*).

Transmission of a NAK or REJ does not imply that the negotiation has failed; the sender can choose another option or value and try again. To help in this process, the NAK can contain a *hint* as to what option value would be acceptable.

Initiation/termination   The transmission of a configure-request implies that the sender wants to initiate a PPP connection. Acknowledgment of all requested options for all required protocols in both directions implies that a PPP connection has been established. This persists until explicitly terminated by sending a terminate-request (*term-REQ*) and receiving a terminate-acknowledge (*term-ACK*).

Protocols   PPP is not one protocol, but a framework containing several *component protocols* that share a (nearly) common set of negotiation methods (command/response codes and state machines); for example,

- Link Control Protocol (LCP) establishes base communications parameters,
- IP Control Protocol (IPCP) negotiates IP-specific options, and
- Password Authentication Protocol (PAP) provides security.

The two nodes must individually negotiate each component protocol as to
- whether it should be used or not (only LCP is compulsory) and
- what options are to be used.

Network layer   Some *network layer* component protocols, such as IPCP, are further subdivided into two parts: a *control protocol* for negotiation, and a *network protocol* for data transmission.

State machine   The negotiation process is controlled by a standardized state machine. One instance is needed for each component protocol in use.

Convergence   The negotiation should *converge* as the two nodes reach agreement on mutually acceptable options. If a particular option is essential to one node but unacceptable to the other, convergence will not be achieved.

Reliability   The underlying data transport is assumed unreliable, so all transmissions must be repeated if no response is received.

Connection   If the nodes agree on a mutually acceptable set of protocols, and options within those protocols, then a PPP connection will be established.

Data   As part of the negotiation, the nodes agree on the data transmission (network layer) protocols to be used, such as IPCP for carrying IP datagrams or NBFCP for carrying NetBIOS frames. The network-layer protocols provide a wrapper around the network data, which allows one PPP connection to handle several concurrent network protocols.

Protocol Components

Component Header

Figure 13.2 shows the format of the PPP component protocols within the information field of a PPP (AHDLC) frame. The sizes are based on the default maximum size of 1,502 bytes; this can be negotiated higher or lower. All protocol components start with a protocol word to identify which component is in use.

c021	Link control protocol (LCP)
8021	IP Control Protocol (IPCP)
80fd	Compression Control Protocol (CCP)
c023	Password Authentication Protocol (PAP)

The basic format of most components is the same.

Code	config-REQ, config-ACK, config-NAK, etc.
ID	a sequence number that allows responses to be matched to specific requests
Length	the total size in bytes of the code, ID, length, and options fields
Options	a variable-length field containing zero or more options

Figure 13.2　　Protocol component format.

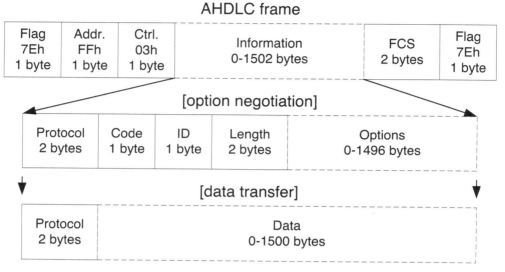

The network layer component protocols (such as IPCP) have two formats: one for negotiation and the other for data transfer. The latter is identified by a different protocol word value (0021 for IPCP data) followed by the IP data itself.

Option Format

Each option has a one-byte *type* that identifies which option is required, its total length as a one-byte value, and a data field up to 253 bytes (Figure 13.3).

Figure 13.3　　Option format.

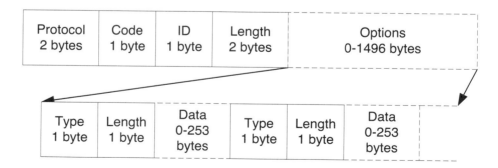

The following are frequently used option types:

LCP

01	Maximum Receive Unit — maximum size of PPP information field
02	Asynchronous Control Character Map — control characters that must be escaped
03	Authentication Protocol — choice of protocol for authentication
05	Magic Number — random number for loop-back detection
07	Protocol Field Compression — reduction of protocol field to one byte
08	Address and Control Field Compression — elimination of HDLC header

IPCP

02	IP Compression Protocol — compression of IP headers
03	IP Address — address of local system

State Machine

A standard state machine is used for tracking LCP and IPCP negotiations. In the standard, it is expressed as a two-dimensional table of actions, the rows representing events and the columns representing states. It easily translates into a two-dimensional C array, which allows instant lookup of the correct action, given the current state and an event.

Initial	Closed state on startup
Starting	Opened state, before lower levels are up
Closed	after closure, awaiting the interface going down
Stopped	awaiting closure
Closing	active close: attempting to close connection
Stopping	passive close: peer has started closure
Request-sent	configuration request has been transmitted
ACK-received	request has been ACKed, but remote request not yet accepted
ACK-sent	remote request ACKed, but local request not yet accepted
Opened	connection established

Events

Up	the lower layer is ready
Down	the lower layer is no longer ready
Open	a request for the connection process to start
Close	a request for the disconnection process to start
TO+, TO–	restart timer expired (restart counter >0 or =0)
RCR+, RCR–	received config-request (accepted or not accepted)
RCA	received config-ACK
RCN	received config-NAK
RTR	received terminate-request
RTA	received terminate-ACK
RUC	received unknown code
RXJ+, RXJ–	received code/protocol reject (acceptable or catastrophic)
RXR	received echo-request, echo-reply, or discard-request

Actions

XXX	illegal event; should not occur
TLU	signal to upper layer that this layer is up
TLD	signal to upper layer that this layer is down
TLS	signal to lower layer that this layer has started
TLF	signal to lower layer that this layer is finished
IRC	initialize the restart counter to the maximum value
ZRC	zero the restart counter
SCR	send a config-request
SCA	send a config-ACK
SCN	send a config-NAK or config-reject
STR	send a terminate-request
STA	send a terminate-ACK
SCJ	send code-reject
SER	send echo-reply

Each location in the state table has one or more actions, plus an optional state change indicated by a numeric value. For example, the entry IRC+SCR+6 means "initialize the restart count, send a config-request, and go to state 6 (request-sent)."

The C array is too wide to print here, so it has been chopped into two halves and the braces removed for clarity.

0 Initial	1 Starting	2 Closed	3 Stopped	4 Closing	
2,	IRC+SCR+6,	XXX,	XXX,	XXX,	// Up
XXX,	XXX,	0,	TLS+1,	0,	// Down
TLS+1,	XXX,	IRC+SCR+6,	3,	5,	// Open
0,	TLF+0,	2,	2,	4,	// Close
XXX,	XXX,	XXX,	XXX,	STR+4,	// TO+
XXX,	XXX,	XXX,	XXX,	TLF+2,	// TO-
XXX,	XXX,	STA+2,	IRC+SCR+SCA+8,	4,	// RCR+
XXX,	XXX,	STA+2,	IRC+SCR+SCN+6,	4,	// RCR-
XXX,	XXX,	STA+2,	STA+3,	4,	// RCA
XXX,	XXX,	STA+2,	STA+3,	4,	// RCN
XXX,	XXX,	STA+2,	STA+3,	STA+4,	// RTR
XXX,	XXX,	2,	3,	TLF+2,	// RTA
XXX,	XXX,	SCJ+2,	SCJ+3,	SCJ+4,	// RUC
XXX,	XXX,	2,	3,	4,	// RXJ+
XXX,	XXX,	TLF+2,	TLF+3,	TLF+2,	// RXJ-

5 Stopping	6 ReqSent	7 AckRcvd	8 AckSent	9 Opened	
XXX,	XXX,	XXX,	XXX,	XXX	// Up
1,	1,	1,	1,	TLD+1	// Down
5,	6,	7,	8,	9	// Open
4,	IRC+STR+4,	IRC+STR+4,	IRC+STR+4,	TLD+IRC+STR+4	// Close
STR+5,	SCR+6,	SCR+6,	SCR+8,	XXX	// TO+
TLF+3,	TLF+3,	TLF+3,	TLF+3,	XXX	// TO-
5,	SCA+8,	SCA+TLU+9,	SCA+8,	TLD+SCR+SCA+8	// RCR+
5,	SCN+6,	SCN+7,	SCN+6,	TLD+SCR+SCN+8	// RCR-
5,	IRC+7,	SCR+6,	IRC+TLU+9,	TLD+SCR+6	// RCA
5,	IRC+SCR+6,	SCR+6,	IRC+SCR+8,	TLD+SCR+6	// RCN
STA+5,	STA+6,	STA+6,	STA+6,	TLD+ZRC+STA+5	// RTR
TLF+3,	6,	6,	8,	TLD+SCR+6	// RTA
SCJ+5,	SCJ+6,	SCJ+7,	SCJ+8,	SCJ+9	// RUC
5,	6,	6,	8,	9	// RXJ+
TLF+3,	TLF+3,	TLF+3,	TLF+3,	TLD+IRC+STR+5	// RXJ-

The RXR event (received echo-request/reply) has been omitted from the table because it is conveniently handled within the code. If the state is opened, respond to the echo-request; otherwise, ignore it.

Sample PPP Negotiation

The following negotiation between a Windows 95 system and an ISP, using modem dial-up, shows serial traffic between the PC (*Data Terminal Equipment*, or DTE) and modem (*Data Communications Equipment*, or DCE) that was intercepted and decoded.

First, look at the modem commands to dial up the ISP. The PC (DTE) commands are in bold type.

DTE 1	**AT<CR>**	Establish contact with modem
DCE 1	<CRLF>OK<CRLF>	Modem response
DTE 2	**ADTD 0845 ... <CR>**	Dial
DCE 2	<CRLF>CONNECT 37333 ... <CRLF>	Modem response when connected

LCP

As soon as the modem connects to the remote system, the PC starts Link Control Protocol negotiation.

DTE 3	**7e ff 7d 23**	Flag and HDLC header
	c0 21 7d 21 7d 21 7d 20 7d 37	LCP **Config-REQ** ID 01 length 23 options:
	7d 22 7d 26 7d 20 7d 2a 7d 20 7d 20	Async-Control-Character-Map 0a00h
	7d 25 7d 26 7d 20 2c 6f 99	Magic-Number Len 6: 002c6f99h
	7d 27 7d 22	Protocol-Field-Compression
	7d 28 7d 22	Address-and-Control-Field-Compression
	7d 2d 7d 23 7d 26	Callback
	ad 50 7e	CRC and flag

Note the copious use of the 7dh escape byte for all values below 20h. The requested options are

- only escape XON/XOFF characters,
- use a magic number for loop-back detection,
- compress the PPP protocol field,
- eliminate the HDLC address and control field, and
- call back the remote system for security purposes.

The response from the ISP shows it isn't yet ready to accept PPP frames because it is still in the process of initializing the protocol stack. It emits some text messages and an HDLC flag character followed by a backspace-space-backspace sequence, as if seeking to take it back again. This fragment is unlikely to confuse a real PPP implementation but does show the kind of garbage it must discard.

DCE		
	`0d 0a 0d 0a 6c 6f 67 69 6e 3a 20`	`<CRLF><CRLF>"login: "`
	`7e 08 20 08 7d 23`	HDLC flag?
	`0d 0a 50 50 50 20 73 65 73 73 69`	`<CRLF>"PPP session from (x.x.x.15)"`
	`6f 6e 20 66 72 6f 6d 20 28 32 31`	
	`32 2e 38 37 2e 38 38 2e 31 35 29`	
	`20 74 6f 20 4e 65 67 6f 74 69 61`	
	`74 65 64 20 62 65 67 69 6e 6e 69`	`"to Negotiated beginning …"`
	`6e 67 2e 2e 2e 2e`	

Having started its PPP stack, the ISP does not acknowledge the config-request but sends one of its own instead.

DCE 3		
	`7e ff 7d 23`	Flag and HDLC header
	`c0 21 7d 21 7d 21 7d 20 7d 38`	LCP **config-REQ** ID 01 length 24 options:
	`7d 22 7d 26 7d 20 7d 20 7d 20 7d`	Async-Control-Character-Map: 0a00h
	`20 7d 25 7d 26 2f 4a e8 9d`	Magic-number: 2f4ae89dh
	`7d 27 7d 22`	Protocol-Field-Compression
	`7d 28 7d 22`	Address-and-Control-Field-Compression
	`7d 23 7d 24 c0 23`	Authentication-Protocol c023 (PAP)
	`aa 96 7e`	CRC and flag

The requested options are

- only escape XON/XOFF characters,
- use a magic number for loop-back detection,
- compress the PPP protocol field,
- eliminate the HDLC address and control field, and
- employ Password Authentication Protocol.

The PC responds by accepting all the ISP options, so agreement has been reached on its incoming LCP connection; however, three seconds have elapsed since it sent an initial config-request without any response, so it is resent with a new sequence number.

DTE 4
```
7e ff 7d 23
c0 21 7d 22 7d 21 7d 20 7d 38
7d 22 7d 26 7d 20 7d 20 7d 20 7d 20
7d 25 7d 26 2f 4a e8 9d
7d 27 7d 22
7d 28 7d 22
7d 23 7d 24 c0 23
66 7b 7e
```
Flag and HDLC header
LCP **config-ACK** ID 01 length 24 options:
  Async-Control-Character-Map: 0a00h
  Magic-number: 2f4ae89dh
  Protocol-Field-Compression
  Address-and-Control-Field-Compression
  Authentication-Protocol c023 (PAP)
CRC and flag

DTE 5
```
7e ff 7d 23
c0 21 7d 21 7d 21 7d 20 7d 37
7d 22 7d 26 7d 20 7d 2a 7d 20 7d 20
7d 25 7d 26 7d 20 2c 6f 99
7d 27 7d 22
7d 28 7d 22
7d 2d 7d 23 7d 26
5b a3 7e
```
Flag and HDLC header
LCP **config-REQ** ID 02 length 23 options:
  Async-Control-Character-Map 0a00h
  Magic-Number Len 6: 002c6f99h
  Protocol-Field-Compression
  Address-and-Control-Field-Compression
  Callback
CRC and flag

Unsurprisingly, the ISP doesn't understand the callback option and rejects it. Implicitly, it does agree with the other options, but PPP has no mechanism for saying "I reject this but accept all the rest," so the PC has to resend all the nonrejected options, and then the ISP can agree (ACK) to the revised list.

DCE 5
```
7e ff 7d 23
c0 21 7d 24 7d 22 7d 20 7d 27
7d 2d 7d 23 7d 26
fb 3e 7e
```
Flag and HDLC header
LCP **config-reject** ID 02 length 7 options:
  Callback
CRC and flag

DTE 6
```
7e ff 7d 23
c0 21 7d 21 7d 23 7d 20 7d 34 7d 22
7d 26 7d 20 7d 2a 7d 20 7d 20
7d 25 7d 26 7d 20 2c 6f 99
7d 27 7d 22
7d 28 7d 22
ce 7d 3e 7e
```
Flag and HDLC header
LCP **config-REQ** ID 03 length 20 options:
  Async-Control-Character-Map 0a00h
  Magic-Number: 002c6f99h
  Protocol-Field-Compression
  Address-and-Control-Field-Compression
CRC and flag

DCE 6
```
7e ff 7d 23
c0 21 7d 22 7d 23 7d 20 7d 34 7d 22
7d 26 7d 20 7d 2a 7d 20 7d 20
7d 25 7d 26 7d 20 2c 6f 99
7d 27 7d 22
7d 28 7d 22
25 77 7e
```
Flag and HDLC header
LCP **config-ACK** ID 03 length 20 options:
  Async-Control-Character-Map 0a00h
  Magic-Number: 002c6f99h
  Protocol-Field-Compression
  Address-and-Control-Field-Compression
CRC and flag

PAP

LCP has been established in both directions, but the ISP did include Password Authentication Protocol (PAP) in its options list, so the PC must now supply its username (*peer ID*) and password before proceeding. The PC could also have included an LCP option for the ISP to authenticate itself (i.e., two-way authentication), but this would probably have been rejected because the ISP isn't configured for this.

DTE 7	7e	Flag
	c0 23 01 01 00 17	PAP **Authenticate-Request** ID 1 length 23
	0b xx xx xx xx xx xx xx xx	Peer-ID length 11
	xx xx 06 xx xx xx xx xx xx	Password length 6
	25 a0 7e	CRC and flag
DCE 7	7e ff 7d 23	Flag and HDLC header
	c0 23 7d 22 7d 21 7d 20 7d 34	PAP **Authenticate-ACK** ID 1 length 20
	7d 2f 4c 6f 67 69 6e 20 53 75	Message Length 15 "Login Succeeded"
	63 63 65 65 64 65 64	
	a1 7c 7e	CRC and flag

PAP uses a simplified version of the LCP negotiation. The peer-ID (i.e., username) and password are sent in an authenticate-request message as plain text, prefixed with length bytes. The ACK or NAK response does not echo the peer-ID or password but does include some explanatory text.

PAP is one of several authentication options. Other protocols can provide better security and protection against eavesdroppers. The alternatives, and their relative merits, are outside the scope of this book.

IPCP

Now that the PC is authenticated, IPCP (IP Control Protocol) negotiation can begin. As it happens, both sides send their initial config-requests at roughly the same time. The PC takes advantage of the revised control-character map to eliminate unnecessary escape characters and compresses (eliminates) the HDLC header. The ISP still uses the escape sequences for the negotiation.

DCE 8	7e ff 7d 23	Flag and HDLC header
	80 21 7d 21 7d 21 7d 20 7d 30	IPCP **Config-REQ** ID 1 length 16 options:
	7d 22 7d 26 7d 20 2d 7d 2f 7d 20	IP Compression: Van Jacobson
	7d 23 7d 26 xx xx xx xx	IP Address length 6
	2f 4a 66	CRC and flag
DTE 8	7e	Flag
	80 21 01 01 00 22	IPCP **Config-REQ** ID 1 length 34 options:
	03 06 00 00 00 00	IP Address: 0.0.0.0
	81 06 00 00 00 00	Primary DNS Server Address: 0.0.0.0
	82 06 00 00 00 00	Primary NBNS Server Address: 0.0.0.0
	83 06 00 00 00 00	Secondary DNS Server Address: 0.0.0.0
	84 06 00 00 00 00	Secondary NBNS Server Address: 0.0.0.0
	bc b0 7e	CRC and flag

The ISP requests the use of VJ compression and gives its IP address value. Although the address is given in a config-request, it is expecting the value to be agreed on — don't forget, "... some are more negotiable than others."

The PC gives its IP address, DNS (Domain Name Server) address, and so on as being zero. This suggests that it doesn't know these values, and expects to get a NAK with a hint as to their true value. The ISP could be awkward and just ACK these values, in which case the PC is in trouble. An IP address of zero may be a problem, and a DNS address of zero would probably disable all name lookups.

In fact, the four DNS and NBNS (NetBIOS Name Server) options are Microsoft specific and not recommended for general use. The standard TCP/IP method of obtaining the DNS address is using DHCP (Dynamic Host Configuration Protocol), as described in Chapter 15. There is no necessity for the DNS address to be known at this early stage of PPP negotiation. A more flexible solution is to discover it using DHCP after the link is established.

As it happens, I have also disabled VJ TCP/IP header compression on the PC, so the next round of negotiation involves rejections from both sides and then resubmission of the revised option lists. The ISP is determined to use VJ compression, so it has another try to get it accepted, but it is rejected a second time.

DTE 9	7e	Flag
	80 21 04 01 00 0a	IPCP **config-reject** ID 1 length 10 options:
	02 06 00 2d 0f 00	IP Compression: Van Jacobson
	71 21 7e	CRC and flag
DCE 9	7e ff 7d 23	Flag
	80 21 7d 24 7d 21 7d 20 7d 36	IPCP **config-reject** ID 1 length 22 options:
	82 7d 26 7d 20 7d 20 7d 20 7d 20	Primary NBNS Server Address: 0.0.0.0
	83 7d 26 7d 20 7d 20 7d 20 7d 20	Secondary DNS Server Address: 0.0.0.0
	84 7d 26 7d 20 7d 20 7d 20 7d 20	Secondary NBNS Server Address: 0.0.0.0
	e7 ad 7e	CRC and flag
DCE 10	7e ff 7d 23	Flag and HDLC header
	80 21 7d 21 7d 22 7d 20 7d 2e	IPCP **config-REQ** ID 2 length 14 options:
	7d 22 7d 24 7d 20 2d	IP Compression: Van Jacobson
	7d 23 7d 26 xx xx xx 7d xx	IP Address length 6
	af 22 7e	CRC and flag
DTE 10	7e	Flag
	80 21 01 02 00 10	IPCP **config-REQ** ID 2 length 16 options:
	03 06 00 00 00 00	IP Address: 0.0.0.0
	81 06 00 00 00 00	Primary DNS Server Address: 0.0.0.0
	83 ab 7e	CRC and flag
DTE 11	7e	Flag
	80 21 04 02 00 08	IPCP **config-reject** ID 2 length 8 options:
	02 04 00 2d	IP Compression: Van Jacobson
	ff a0	CRC and flag

Now that the IPCP rejections are out of the way, the IP addresses can be agreed on. The ISP uses a NAK with a hint to persuade the PC to change its address (and DNS address) from 0.0.0.0 to a more sensible value.

DCE 11	`7e ff 7d 23`	Flag and HDLC header
	`80 21 7d 23 7d 22 7d 20 7d 30`	IPCP **Config-NAK** ID 2 length 16 options:
	`7d 23 7d 26 xx xx xx xx`	IP Address: $x.x.x.x$
	`81 7d 26 xx xx xx 7d xx`	Primary DNS Server Address: $x.x.x.x$
	`34 7d 29`	CRC and flag
DCE 12	`7e ff 7d 23`	Flag
	`80 21 7d 21 7d 23 7d 20 7d 2a`	IPCP **Config-REQ** ID 3 length 10 options:
	`7d 23 7d 26 xx xx xx 7d xx`	IP Address: $x.x.x.x$
	`5b d3 7e`	CRC and flag
DTE 12	`7e`	Flag
	`80 21 01 03 00 10`	IPCP **config-REQ** ID 3 length 16 options:
	`03 06 xx xx xx xx`	IP Address: $x.x.x.x$
	`81 06 xx xx xx xx`	Primary DNS Server Address: $x.x.x.x$
	`d2 e3 7e`	CRC and flag
DTE 13	`7e`	Flag
	`80 21 02 03 00 0a`	IPCP **config-ACK** ID 3 length 10 options:
	`03 06 xx xx xx xx`	IP Address: $x.x.x.x$
	`4c 49 7e`	CRC and flag
DCE 13	`7e ff 7d 23`	Flag
	`80 21 7d 22 7d 23 7d 20 7d 30`	IPCP **config-ACK** ID 3 length 16 options:
	`7d 23 7d 26 xx xx xx xx`	IP Address: $x.x.x.x$
	`81 7d 26 xx xx xx 7d xx`	Primary DNS Server Address: $x.x.x.x$
	`7d 2e cf 7e`	CRC and flag

IP Data

IPCP negotiation is complete, and there is a working IP data link. The PC uses this opportunity to send out a large number of (unnecessary) Microsoft-specific UDP requests. I eliminated these for clarity. Using `ping` on the PC results in the following transmissions. Note that the ISP is now taking advantage of the negotiated compression options.

DTE 14	`7e`	Flag
	`21`	PPP Internet Protocol version 4
	`45 00 00 3c 58 00 00 00 20`	IP V4, Hdr 5, Len 60 ID 5800h TTL 32
	`01 ef 7d 31`	Pcol ICMP, csum ef11
	`xx xx xx xx yy yy yy yy`	Source $x.x.x.x$ Dest $y.y.y.y$
	`08 00 47 5c 01 00 05 00`	ICMP **echo-REQ**, csum 475c ID 0100 Seq 0500
	`61 62 63 64 65 66 67 68 69 6a`	ICMP Data
	`6b 6c 6d 6e 6f 70 71 72 73`	
	`74 75 76 77 61 62 63 64 65`	
	`66 67 68 69`	
	`12 0f 7e`	CRC and flag

DCE 14	7e	Flag
	21	PPP Internet Protocol version 4
	45 00 00 3c 7b 1e 00 00 fe	IP V4, Hdr 5, Len 60 ID 7b1eh TTL 254
	01 ed f2	Pcol ICMP, csum edf2
	yy yy yy yy xx xx xx xx	Source y.y.y.y Dest x.x.x.x
	00 00 4f 5c 01 00 05 00	ICMP echo-reply, csum 4f5c ID 0100 Seq 0500
	61 62 63 64 65 66 67 68 69 6a	ICMP Data
	6b 6c 6d 6e 6f 70 71 72 73	
	74 75 76 77 61 62 63 64 65	
	66 67 68 69	
	fb a9 7e	CRC and flag

Closure

To close the PPP link, it is not necessary to close the component protocols individually. The PC just terminates the LCP connection and then does a modem hang-up.

DTE 15	7e	Flag
	c0 21 7d 25 7d 24 7d 20 7d 24	LCP **terminate-REQ** ID 04 length 4
	80 fe 7e	CRC and flag
DCE 15	7e ff 7d 23	Flag and HDLC header
	c0 21 7d 26 7d 22 7d 20 7d 24	LCP **terminate-ACK** ID 02 length 4
	94 7d 2d 7e	CRC and flag
DCE 16	<CRLF>OK<CRLF>	
DTE 16	ATH<CR>	
DCE 17	<CRLF>OK<CRLF>	

PPP Implementation

Implementing PPP on a microcontroller is quite a formidable task, particularly if one wishes to conform to the standards. Theoretically, the PPP frame can be up to 1,500 bytes, so a sizeable amount of RAM is needed to accommodate the transmit and receive buffers, not to mention the complexity of running four simultaneous state machines (modem, LCP, PAP, and IPCP). There is some scope for exploiting the commonality of the LCP and IPCP state machines, but it will be necessary to negotiate some options, and these are sufficiently different for the two protocols that the benefit from any shared code is reduced.

The main components of the PPP implementation are

- reception and transmission,
- message decoding and encoding,
- state machines, and
- event handlers.

Reception and Transmission

An interrupt-driven serial communications handler that stores incoming characters in a buffer is assumed. The characters are fetched from the buffer, the escape sequences are decoded, and the checksum is validated.

```c
#define PPP_END 0x7e
#define PPP_ESC 0x7d
#define PPP_START 0xff

/* Poll the serial interface for Rx and Tx characters */
void poll_net(void)
{
    BYTE b, saver=1;
    static BYTE lastb=0;
    static WORD crc;

    if (serial_kbhit())                     // Incoming serial byte?
    {
        b = serial_getch();
        if (lastb == PPP_ESC)               // Last char was Escape?
        {
            lastb = 0;
            b ^= 0x20;
        }
        else if (b == PPP_ESC)              // This char is Escape?
        {
            lastb = PPP_ESC;
            saver = 0;
        }
        else if (b == PPP_END)              // No escape; maybe End?
        {
            if (inframe)                    // If currently in a frame..
            {                               // Check CRC & save length
                if (crc==0xf0b8  && rxbuffin>2)
                    net_rxin = rxbuffin - 2;
                inframe = saver = 0;
            }
            else
                saver = 0;
            rxbuffin = 0;
```

```
        }
        if (!inframe)                       // If not in PPP frame..
        {
            if (b == PPP_START)             // ..check for start of frame
            {
                crc = 0xffff;
                rxbuffin = 0;
                inframe = 1;
            }
        }
        if (saver)                          // If saving the new byte
        {
            if (inframe)
                crc = ppp_crc16(crc, b);    // ..do CRC as well
            if (rxbuffin < RXBUFFLEN)
            {
                rxbuff[rxbuffin] = b;
                rxbuffin++;
            }
        }
    }
}
```

This puts complete, correct PPP frames in the receive buffer; incorrect frames are discarded.

Transmission of a PPP frame assumes that the frame data is stored in the transmit buffer. The HDLC header, protocol word, checksum, and escape sequences are added on immediately before transmission to minimize storage requirements.

```
/* Put out a byte using PPP escape codes, update the CRC */
void send_ppp_byte(BYTE b)
{
    poll_net();
    if (b==PPP_END || b==PPP_ESC || b<0x20)
    {
        serial_putch(PPP_ESC);
        poll_net();
        serial_putch(b ^ 0x20);
    }
    else
        serial_putch(b);
```

```
    poll_net();
    txcrc = ppp_crc16(txcrc, b);
}


/* Transmit the PPP frame */
void transmit(void)
{
    WORD n;
    BYTE hi, lo;

    txcrc = 0xffff;
    serial_putch(PPP_END);                              // Start flag
    send_ppp_byte(0xff);                                // HDLC address
    send_ppp_byte(3);                                   // ..and control byte
    send_ppp_byte((BYTE)(ppp_pcol >> 8));   // PPP protocol word
    send_ppp_byte((BYTE)ppp_pcol);
    for (n=0; n<net_txlen; n++)                   // Transmit PPP data
        send_ppp_byte(txbuff[n]);
    hi = ~ (BYTE)(txcrc >> 8);                      // Append CRC
    lo = ~ (BYTE)txcrc;
    send_ppp_byte(lo);
    send_ppp_byte(hi);
    serial_putch(PPP_END);                              // End flag
}
```

Message Decoding

An incoming frame needs to be dispatched to the appropriate component protocol handler or rejected if it is an unknown protocol.

```
#define PPP_LCP         0xc021  // Link control
#define PPP_IPCP        0x8021  // IP control
#define PPP_IP_DATA     0x0021  // IP data
#define PPP_CCP         0x80fd  // Compression control
#define PPP_PAP         0xc023  // Password authentication


/* Check for incoming PPP frame */
void get_net(void)
{
    WORD mlen;
```

```
...
    poll_net();                                 // If incoming frame..
    if (net_rxin)
    {
        init_rxbuff();                          // Prepare Rx buffer
        rx_checkoff = 1;
        if (skip_byte() && skip_byte() &&       // Skip HDLC addr & ctrl bytes
            get_word(&ppp_pcol))                 // Get PPP protocol
        {
            init_txbuff(0);
            if (ppp_pcol == PPP_IP_DATA)         // If IP data, just return
            {                                    // (data left for IP handler)
                return(1);
            }
            setpos_txin(PPP_HEADLEN);            // If LCP, IPCP or PAP..
            if (ppp_pcol==PPP_LCP || ppp_pcol==PPP_IPCP || ppp_pcol==PPP_PAP)
            {                                              // ..get header
                if (get_byte(&ppp_code) && get_byte(&ppp_rxid) &&
                    get_word(&mlen) && mlen>=4 && net_rxin>=mlen+2)
                {                                // ..and call handler
                    if (ppp_pcol == PPP_LCP)
                        lcp_rx_handler();
                    else if (ppp_pcol == PPP_PAP)
                        pap_rx_handler();
                    else
                        ipcp_rx_handler();
                }
            }
            else
            {
                put_word(ppp_pcol);              // Reject unknown protocol
                ppp_pcol = PPP_LCP;
                send_ppp(PPP_PCOL_REJ, ppp_rxid, 1);
            }
        }
        net_rxin = 0;
    }
}
```

The LCP handler illustrates the key points.

Config-request   If received, scan the associated options. In this implementation, only the authentication option is accepted; all the others are rejected.

Echo-request   This is an LCP-level equivalent of ICMP echo-request. The data is echoed back to the sender.

Others   All other requests and responses are processed by the state machine.

```c
/* Rx handler for Link Control Protocol */
void lcp_rx_handler(void)
{
    BYTE opt, optlen, code;
    BYTE rejects=0;
    WORD auth=0;
    LWORD lw;

    if (ppp_code == PPP_CFG_REQ)                    // LCP Config request?
    {                                               // Check option list
        while (get_byte(&opt) && get_byte(&optlen) && optlen>=2)
        {
            if (opt==LCP_OPT_AUTH && optlen>=4) // Authentication option?
            {                                       // (may be accepted)
                get_word(&auth);
                skip_data(optlen - 4);
            }
            else
            {
                put_byte(opt);                      // Skip other options
                put_byte(optlen);                   // (will be rejected)
                copy_rx_tx(optlen - 2);
                rejects++;
            }
        }
        if (rejects)                                // If any rejected options..
        {
            lcp_event_handler(EVENT_RCR_ERR);   // ..inform state machine
            if (lcp_action & SCN)                   // If OK to respond..
                send_ppp(PPP_CFG_REJ, ppp_rxid, 1); // ..send rejection
        }
        else if (auth)                              // If authentication request..
        {
```

```
                    client_auth = auth==PPP_PAP;              // ..only PAP is acceptable
                    lcp_event_handler(client_auth ? EVENT_RCR_OK : EVENT_RCR_ERR);
                    if (lcp_action & (SCA|SCN))               // If OK to respond..
                    {
                        put_byte(LCP_OPT_AUTH);               // ..send my response
                        put_byte(4);                          // (either ACK PAP..
                        put_word(PPP_PAP);                    // ..or NAK with PAP hint)
                        code = (lcp_action & SCA) ? PPP_CFG_ACK : PPP_CFG_NAK;
                        send_ppp(code, ppp_rxid, 1);
                    }
                }
            else
            {
                lcp_event_handler(EVENT_RCR_OK);       // Request is all OK
                if (lcp_action & SCA)                  // If OK to respond, do so
                    send_ppp(PPP_CFG_ACK, ppp_rxid, 0);
            }
        }
    else if (ppp_code == PPP_ECHO_REQ)                  // LCP echo request?
    {
        if ((lcp_state&0xf) == PPP_OPENED && get_lword(&lw))
        {                                              // Get magic num
            lw.l++;                                     // Return magic num + 1
            put_lword(&lw);
            copy_rx_tx(net_rxin-rxout);                 // Echo the data
            send_ppp(PPP_ECHO_REP, ppp_rxid, 1);
        }                                              // Others to state machine..
    }
    else if (ppp_code == PPP_TERM_REQ)                  // ..terminate request
        lcp_event_handler(EVENT_RTR);
    else if (ppp_code == PPP_CFG_ACK)                  // ..config ACK
        lcp_event_handler(EVENT_RCA);
    else if (ppp_code == PPP_CFG_NAK)                  // ..config NAK
        lcp_event_handler(EVENT_RCN);
    else if (ppp_code == PPP_TERM_ACK)                  // ..terminate ACK
        lcp_event_handler(EVENT_RTA);
    else if (ppp_code == PPP_CODE_REJ)                  // ..code reject
        lcp_event_handler(EVENT_RXJ_ERR);
    do_lcp_actions();
}
```

Summary

PPP is a surprisingly complex protocol, given its relatively humble role in TCP/IP communication. Having explored the essentials of PPP communication, I have provided some software that can be used as a basis for further experiment and development. Possibly by the time you read this, a full implementation will be available on the Iosoft Ltd. Web site at www.iosoft.co.uk.

Source Files

The PPP driver code is used in place of the Ethernet driver in creating the ChipWeb server.

  p16_ppp.c   PPP driver

Because of RAM constraints, a PIC16 microcontroller cannot be used, so the software is written for use with the PIC18xxx family of microcontrollers. See Appendix D for PICmicro-specific information.

Chapter 14

UDP Clients, Servers, and Fast Data Transfer

Overview

I have been accused of having an unhealthy preoccupation with Web servers, to the detriment of other more useful protocols. It has even been suggested that the title of this book was dictated by purely commercial considerations and "UDP/IP Lean: Data Transfers for Real Applications" or some such title, although unlikely to make the best-seller lists, would have been a much more worthwhile project

In a belated but enthusiastic attempt to address this imbalance, the next three chapters expand your microcontroller horizons to include UDP for simple data transfer, DCHP for auto-configuration, and TCP clients and email protocols.

The terms *client* and *server* are used a lot in the coming chapters, so it is necessary to clarify what is implied by the use of these labels. While on the subject, it is worth looking at an alternative scheme: *peer-to-peer* networking.

Client–Server Networking

Way back in Chapter 2, I suggested a client is the *requester* of a service, and the server is the *provider* of that service. These definitions can now be expanded.

- The client *initiates* contact with the server; the server *responds* to those contacts.
- There is generally a *one-to-many* relationship between a server and its clients.
- A server normally needs greater *resources* than a client.
- The labels *client* and *server* apply to *logical functions* within a node.
- A node can *simultaneously host* one or more clients and one or more servers.

The last points are probably the least well understood. It's easy to point at a box and say, "That is an email server," but the truth is probably more complex. As a server, it may offer email transfer and storage facilities to a group of email clients, but what happens if it needs to forward those emails to another server? One server must become a client to the other, so many servers have client functionality as well (Figure 14.1).

Figure 14.1   Client–server roles.

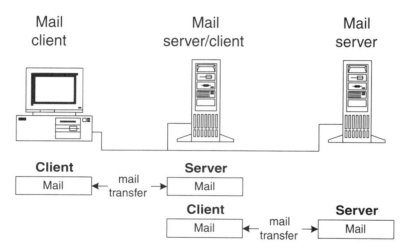

The following steps put the generic client and server attributes into a TCP/IP context:

- A background process in the server (a *daemon*), performs a *passive open* of a *well-known* port and waits for incoming traffic. The well-known port number reflects the *service* the daemon is providing.
- The client opens a socket and uses an *ephemeral* port to contact the server, choosing the port number on the server that corresponds to the required service.
- If the client attempts to use an unsupported service, an error indication is returned. For TCP, a *reset* response is used, whereas for UDP, an ICMP *destination unreachable* response is sent.

Note that the client–server terminology refers only to the way initial contact is established; after that, the data transfers proceed in either or both directions, as determined by the protocol. A Web server is a unidirectional information provider, but other servers, such as a file server, operate bidirectionally. When the transfer is complete, either party may close the connection. TCP closure is generally triggered by a command in the higher-level protocol, although complete connection failure (receipt of a TCP reset) must be handled as well.

Peer-to-Peer Networking

Is there an alternative to this client–server hierarchy? In some *peer-to-peer* networking protocols, all nodes have truly equal status.

You may recall that Ethernet is a peer-to-peer network. The higher TCP/IP layers are generally master–slave, although they could also be peer-to-peer, which might have advantages for some applications. To find a simple illustration of genuine peer-to-peer networking, I've had to stray into the realms of building management networks. In real life, the following example doesn't use TCP/IP protocols or Ethernet networking, but it certainly could.

Imagine an office where each light and switch is a network node. There is no electrical linkage between them, except that they are nodes on a common network. Each switch has a single variable that reflects whether it is on or off, and each light has a similar on/off variable. A particular switch is linked to a specific light by a process called *binding*, so that every time the switch changes state, the light changes to match (Figure 14.2). As well as one-to-one binding, there could be a one-to-many binding. A master switch can be linked to all the lamps, so that when the master is switched off, all the lamps are turned off.

Figure 14.2   Binding variables in peer-to-peer networking.

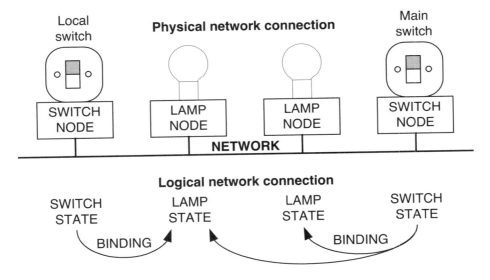

This is a *distributed processing* model, which treats the network links as transparent pipes for information between nodes. An object (such as a variable) on one node can be *bound* to one or more objects on other nodes. Any change to one object within this group will automatically be propagated to the others. In an exceedingly crude analogy of the human brain, the intelligence of the network is not only in the nodes themselves, but in the interconnections between them. The ability to unbind and rebind the variables (either locally or using a global management tool) gives a tremendous degree of flexibility.

Beyond the Web Server

So far, the microcontroller TCP/IP stack has been targeted at server applications, but now a more general-purpose network interface is needed that can be used for clients, servers, and peer-to-peer networking. The network stack must be able to

- initiate communications to a specific port on a remote system (client);
- respond to incoming communications on a specific port (server);
- detect whether the intended destination is on the local subnet or remote (subnet masking);
- send outgoing communications directly to the destination, if local, or via a router, if remote (routing);
- resolve the destination or router IP address to obtain the MAC address (ARP request).

To support this increased flexibility, I first need to improve on the rather clumsy buffering scheme used so far.

Buffer Enhancements

Using a microcontroller with 368 bytes of data memory coupled to a Network Interface Controller (NIC) with 8KB of packet buffer RAM makes for an interesting programming challenge. The maximum Ethernet frame size (1.5 KB) is far greater than the whole microcontroller RAM, so only a fragment of a transmitted or received frame can be in microcontroller RAM at any one time.

This causes problems when trying to create a general-purpose UDP programming interface. The usual technique of calling a function with a pointer to the user data is just unworkable. My initial work-around was to use direct-read and direct-write functions for the NIC RAM, but this approach is rather too hardware specific for comfort.

Shadow Buffers

A good compromise is to create two *shadow buffers* — one for transmit and the other for receive — which are used for all data transfers to and from the NIC. These buffers can accommodate only a small part of an Ethernet frame. As soon as the receive buffer is empty or the transmit buffer is full, a transfer to or from the NIC is initiated automatically. This is similar to the disk-buffering scheme employed by an operating system (OS). The user read and writes a buffer in RAM, while the OS manages the disk transfers, which keep the right amount of data in RAM.

When creating a datagram, it can be useful to skip forward or backward an arbitrary distance; for example, write the user data first and then go back to the UDP header to fill in the length word and checksum. Under an OS, this would be achieved by using a "seek" call to an appropriate point in a file, followed by a read or write operation. The code here can have an equivalent function to allow arbitrary positioning within an Ethernet frame prior to reading or writing the shadow buffers.

```
void init_rxbuff(void);         // Initialize the receive buffer
BOOL setpos_rxout(WORD newpos); // Move Rx O/P pointer to the given location
BOOL get_byte(BYTE *bp);        // Get the next input byte
```

```
BOOL match_byte(BYTE b);        // Match the next input byte
BOOL skip_byte(void);           // Skip the next input byte
BOOL get_word(WORD *wp);        // Get the next input word
... and so on ...
```

At any time during read/write operations, setpos_rxout() or its partner setpos_txin() can be used to skip forward or backward within the Ethernet frame.

Buffer Margin

One weakness of this buffering scheme is the difficulty in matching strings. Imagine that the incoming frame contains the letters of the alphabet, and you are attempting to match the string "GHI." If the buffer size happens to be eight, then the desired string will straddle two buffers and will not be recognized, as shown in Figure 14.3.

Figure 14.3   Shadow buffers.

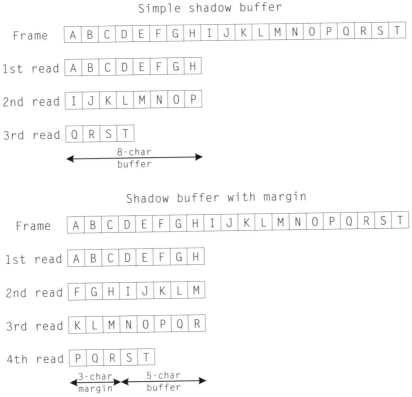

A simple way around this is to maintain a *margin* of data within the buffer every time it is loaded. If the maximum length of string to be matched is three characters, then each time the eight-byte buffer is reloaded, the oldest three characters are shifted to the front of the buffer, and five new characters are added. In this way, it can be guaranteed that any three-character string is always available for matching, regardless of its position within the input data stream.

Receive Buffer Implementation

The new buffering scheme adds a significant amount of code to the get_byte() function because the next byte might not be in local RAM, but in NIC RAM instead, so a NIC data transfer function has to be called.

```
#define RXBUFFLEN 42            // Size of Rx buffer
#define RXMARGIN  10            // Minimum number of pushback bytes

WORD net_rxin;                  // Length of incoming packet in NIC
WORD rxout;                     // Length of packet processed so far
BYTE rxbuffin, rxbuffout;       // I/O pointers for Rx process buffer
BYTE rxbuff[RXBUFFLEN];         // Rx buffer
BYTE nic_tx_transfer;           // Flag set if Tx data is being sent to NIC

/* Get an incoming byte value, return 0 if end of message */
BOOL get_byte(BYTE *bp)
{
    BYTE b;

    if (rxbuffout >= rxbuffin)
        load_rxbuff(RXMARGIN, RXBUFFLEN);
    if (rxbuffout >= rxbuffin)
        return(0);
    b = rxbuff[rxbuffout++];
    rxout++;
    if (!rx_checkoff)
        check_byte(b);
    *bp = b;
    return(1);
}
```

A further complication is that the NIC receive read cycles might be interleaved with the transmit write cycles, in which case the NIC DMA controller must be switched between the two.

```
/* Load a given number of bytes from the NIC into the Rx buffer
** Remove any processed data from the buffer, but allow a margin
** of data to be retained. Return the actual number of bytes loaded */
BYTE load_rxbuff(BYTE margin, BYTE count)
```

```
{
    BYTE in, n;
    WORD rxleft;

    if (rxbuffout > margin)
    {
        in = margin;
        while (rxbuffout<rxbuffin)
            rxbuff[in++] = rxbuff[rxbuffout++];
        rxbuffin = in;
        rxbuffout = margin;
    }
    rxleft = net_rxin - rxout;
    rxleft = MIN((WORD)count, rxleft);
    n = (BYTE)MIN((BYTE)rxleft, RXBUFFLEN-rxbuffin);
    if (n)
    {
        if (nic_tx_transfer)
            setnic_rx(rxout);
        getnic_rxbuff(n);
    }
    return(n);
}
```

The function which repositions the output pointer has to clear the receive buffer and reload it with the new data.

```
/* Move the Rx O/P pointer to the given location, return 0 if beyond data */
BOOL setpos_rxout(WORD newpos)
{
    if (newpos > net_rxin)
        return(0);
    setnic_rx(newpos);
    rxout = newpos;
    rxbuffin = rxbuffout = 0;
    return(1);
}
/* Set the 'remote DMA' address in the current NIC Rx packet buffer */
void setnic_rx(WORD addr)
{
    addr += curr_rx_addr;
```

```
    if (addr >= RXSTOP*256)
        addr += (RXSTART - RXSTOP)*256;
    outnic(ISR, 0x40);                    /* Clear remote DMA interrupt flag */
    outnic(RSAR0, (BYTE)addr);            /* Data addr */
    outnic(RSAR1, (BYTE)(addr>>8));
    nic_tx_transfer = 0;
}
```

Additional helper functions are also included to

- determine the number of bytes left in the receive buffer,
- push a fetched byte back into the receive buffer, and
- truncate the received data to a given length.

The last of these functions is useful when receiving a short datagram, which will have been padded out to the minimum Ethernet size. As soon as the real size is known (from the IP length field), this function discards the unwanted pad bytes.

```
/* Return the number of Rx data bytes left in buffer */
WORD rxleft(void)
{
    return(net_rxin - rxout);
}
/* Push back a byte value, return 0 if no room */
BOOL unget_byte(BYTE b)
{
    if (rxbuffout && rxout)
    {
        rxbuff[--rxbuffout] = b;
        rxout--;
        return(1);
    }
    return(0);
}
/* Truncate the remaining Rx data to the given length */
void truncate_rxout(WORD len)
{
    WORD end;

    if ((end = rxout+len) < net_rxin)
        net_rxin = end;
    if ((end = rxbuffout+len) < rxbuffin)
        rxbuffin = end;
}
```

Transmit Buffer Implementation

The transmit buffer code is similar in concept to the receive code, without the complication of buffer margins.

```c
#define TXBUFFLEN 42         // Size of Tx buffer
#define TXMARGIN  10         // Buffer to be emptied when this space remains
BANK3 WORD net_txlen;        // Max length of Tx data sent to NIC
BANK3 WORD txin;             // Current I/P pointer for Tx data
BANK3 BYTE txbuffin;         // I/P pointer for Tx process buffer
BANK3 BYTE txbuff[TXBUFFLEN];  // Tx buffer


/* Send a byte to the network buffer, then add to checksum
** Return 0 if no more data can be accepted */
BOOL put_byte(BYTE b)
{
    if (txin >= MAXFRAME)
        return(0);
    if (txbuffin >= TXBUFFLEN-TXMARGIN)
        save_txbuff();
    txbuff[txbuffin++] = b;
    txin++;
    check_byte(b);
    return(1);
}
/* Save the contents of the Tx buffer into the NIC */
void save_txbuff(void)
{
    if (txbuffin)
    {
        if (!nic_tx_transfer)
            setnic_tx(txin - txbuffin);
        putnic_txbuff();
        txbuffin = 0;
    }
    if (txin > net_txlen)
        net_txlen = txin;
}
/* Move the Rx O/P pointer to the given location, return 0 if beyond data */
BOOL setpos_txin(WORD newpos)
{
```

```
    if (newpos > MAXFRAME-ETHERHDR_LEN)
        return(0);
    save_txbuff();
    setnic_tx(txin = newpos);
    return(1);
}
/* Set the 'remote DMA' address in the NIC Tx packet buffer */
void setnic_tx(WORD addr)
{                                          /* Add on Tx buffer offset */
    addr += (TXSTART << 8) + ETHERHDR_LEN;
    outnic(ISR, 0x40);                     /* Clear remote DMA interrupt flag */
    outnic(RSAR0, (BYTE)addr);             /* Remote DMA addr */
    outnic(RSAR1, (BYTE)(addr>>8));
    nic_tx_transfer = 1;
}
```

As with the receive buffer, the transmit buffer size (42 bytes) might seem unrealistically small, but, in practice, it works remarkably well.

There is a specific function to copy data from the receive to the transmit buffer. Care has to be taken to avoid a double checksum computation because the get_ and put_ functions both compute checksums by default.

```
/* Copy data from Rx to Tx buffers, return actual byte count */
WORD copy_rx_tx(WORD maxlen)
{
    BYTE n, done=0;
    WORD count=0;

    save_txbuff();
    while (load_rxbuff(0, RXBUFFLEN) || rxbuffin>0)
    {
        n = rxbuffin;
        if (count+n >= maxlen)
        {
            done = 1;
            n = (BYTE)(maxlen - count);
        }
        for (rxbuffout=0; rxbuffout<n; rxbuffout++)
            put_byte(rxbuff[rxbuffout]);
        rxout += n;
        count += n;
```

```
        if (done)
            break;
    }
    net_txlen = MAX(txin, net_txlen);
    return(count);
}
```

IP and ICMP Processing

IP Handler

The new buffering scheme can simplify the IP datagram handler.

```
BYTE ipcol;                    // IP protocol byte
LWORD locip, remip;            // Local & remote IP addresses
WORD locport, remport;         // ..and TCP/UDP port numbers


/* Check an IP header, return the protocol in 'ipcol' */
SEPARATED BOOL ip_recv(void)
{
    rx_checkoff = checkflag = 0;            // Clear checksum
    checkhi = checklo = 0;
    if (match_byte(0x45) && skip_byte() &&  // Version, service
        get_word(&iplen) && skip_word() &&  // Len, ID
        skip_word() &&  skip_byte() &&      // Frags, TTL
        get_byte(&ipcol) && skip_word() &&  // Protocol, checksum
        get_lword(&remip) && get_lword(&locip) &&  // Addresses
        (locip.l==myip.l || locip.l==0xffffffffL) &&
        checkhi==0xff && checklo==0xff &&   // Checksum OK?
        iplen>IPHDR_LEN)                    // IP length OK
    {
        truncate_rxout(iplen-IPHDR_LEN);
        return(1);
    }
    return(0);
}
```

Note the use of `truncate_rxout()` to eliminate any padding on short datagrams and the acceptance of broadcast (all 1s) IP addresses. This capability will be needed for DHCP, as described in the next chapter.

Imagine the `get_()` `skip_()` and `match_()` calls moving a file pointer through the incoming Ethernet frame. At the end of the `ip_recv()` function, that pointer will be at the end of the UDP header, so processing of the next header (ICMP, UDP, TCP, or whatever) can start.

```
if (ip_recv())
{
    if (ipcol == PICMP)               // ICMP?
        icmp_recv();                  // ..call ping handler
    else if (ipcol == PUDP)           // UDP datagram?
        udp_recv();                   // ..call UDP handler
    else if (ipcol == PTCP)           // TCP segment?
    {
        tcp_recv();                   // ..call TCP handler
    }
}
```

Conventional wisdom dictates that the various protocol handlers should be called from within the IP handler. I have broken with this tradition to keep stack usage low, bearing in mind the PIC16xxx processors have only an eight-level call/return stack.

ICMP Handler

The ICMP handler resumes datagram processing where the IP handler left off. By use of `setpos_rxout()`, it can scan the ICMP payload data once to verify the checksum and then go back to copy it into the transmit buffer.

```
/* Respond to an ICMP message (e.g. ping) */
BOOL icmp_recv(void)
{
    WORD csum, addr, len;

    init_txbuff();                              // Initialize Tx, clear checksum
    len = iplen - IPHDR_LEN - 4;
    if (locip.l==myip.l && match_byte(8) && match_byte(0) && get_word(&csum))
    {
        addr = rxout;                           // Save current position
        skip_data(len);                         // Skip over the data
        setpos_rxout(addr);                     // Go back again
        if (checkhi==0xff && checklo==0xff)
        {                                       // If OK and not bcast..
            put_ip();                           // IP header
            put_word(0);                        // ICMP type and code
            csum += 0x0800;                     // Adjust checksum for resp
```

```
                if (csum < 0x0800)          // ..including hi-lo carry
                    csum++;
                put_word(csum);             // ICMP checksum
                copy_rx_tx(len);            // Copy data to Tx buffer
                transmit();                 // ..and send it
                return(1);
            }
        }
    return(0);
}
```

The sophisticated transmit and receive buffering scheme allows get_() and put_() calls to be intermixed freely; the receive pointer is unaffected by transmit and vice versa.

IP transmission assumes that, by default, a message is being returned to the sender.

```
/* Put an IP datagram header in the Tx buffer */
void put_ip(void)
{
    static BYTE id=0;

    checkhi = checklo = 0;          // Clear checksum
    checkflag = 0;
    put_byte(0x45);                 // Version & hdr len */
    put_byte(0);                    // Service
    put_word(iplen);
    put_byte(0);                    // Ident word
    put_byte(++id);
    put_word(0);                    // Flags & fragment offset
    put_byte(100);                  // Time To Live
    put_byte(ipcol);                // Protocol
    check_lword(&myip);             // Include addresses in checksum
    check_lword(&remip);
    put_byte(~checkhi);             // Checksum
    put_byte(~checklo);
    put_lword(&myip);               // Source & destination IP addrs
    put_lword(&remip);
}
```

The transmit function takes the total amount of data in the output buffer (as indicated by the net_txlen global variable), and sends it.

```
/* Transmit the Ethernet frame */
void transmit(void)
{
    WORD dlen;

    dlen = net_txlen;
    save_txbuff();
    setnic_tx(-ETHERHDR_LEN);               /* Go to start of Ether header */
    txbuffin = 0;
    put_data(host.eth.srce, MACLEN);        /* Destination addr */
    put_data(myeth, MACLEN);                /* Source addr */
    put_byte((BYTE)(host.eth.pcol>>8));     /* Protocol */
    put_byte((BYTE)host.eth.pcol);
    dlen += ETHERHDR_LEN;                    /* Bump up length for MAC header */
    putnic_txbuff();
    if (dlen < MINFRAME)
        dlen = MINFRAME;                     /* Constrain length */
    outnic(TBCR0, (BYTE)dlen);              /* Set Tx length regs */
    outnic(TBCR1, (BYTE)(dlen >> 8));
    outnic(CMDR, 0x24);                      /* Transmit the packet */
}
```

UDP Servers

Daytime and Echo Servers

The standard UDP servers, which I usually include for testing purposes, are Echo (port 7) and Daytime (port 13). Unfortunately, the prototype board doesn't possess a real-time clock chip, so the Daytime server just returns a static string, but this is adequate for test purposes.

```
#define UDPHDR_LEN     8           // UDP header length
#define UDPIPHDR_LEN  28           // UDP+IP header length
#define MAXUDP_DLEN (1500-20-8)    // Max length of UDP data

/* Receive an incoming UDP datagram, return 0 if invalid */
SEPARATED BOOL udp_recv(void)
{
    WORD addr;
```

```
checkhi = checklo = 0;
checkflag = udp_checkoff = 0;
DEBUG_PUTC('u');
DEBUG_PUTC('>');
if (get_word(&remport) && get_word(&locport) && // Source & dest ports
    get_word(&udplen) && get_word(&ucsum) &&     // Dgram length, checksum
    udplen>=UDPHDR_LEN && udplen<=MAXUDP_DLEN)
{
    DEBUG_PUTC('>');
    if (ucsum)
    {
        check_word(udplen);                       // Check pseudoheader
        check_lword(&locip);
        check_lword(&remip);
        check_byte(0);
        check_byte(PUDP);
        addr = rxout;
        skip_data(rxleft());
        setpos_rxout(addr);
    }                                             // If checksum OK
    if (ucsum==0 || (checkhi==0xff && checklo==0xff))
    {                                             // ..call UDP handler
        init_txbuff(0);
        checkhi = checklo = 0;                     // Clear checksum
        checkflag = 0;
        DEBUG_PUTC('*');
        if (locport == ECHOPORT)                   // Echo: return copy of data
        {
            setpos_txin(UDPIPHDR_LEN);
            copy_rx_tx(udplen - UDPHDR_LEN);
            udp_xmit();
            DEBUG_PUTC('U');
        }
        else if (locport == DAYPORT)               // Daytime: return string
        {
            print_lcd = print_serial = FALSE;
            print_net = TRUE;
            setpos_txin(UDPIPHDR_LEN);
            putstr(DAYMSG);
```

```
                    udp_xmit();
            }
        return(1);
        }
    }
    DEBUG_PUTC('!');
    return(0);
}
```

The Echo and Daytime servers have become remarkably simple. Blink and you'll miss them! The important steps are

- initialize the transmit buffer,
- position the transmit buffer pointer at the end of the UDP/IP header,
- write an arbitrary amount of data to the transmit buffer (checksum is updated every time a byte is written), and
- call udp_xmit() to encapsulate and send the UDP data.

UDP Transmission

The udp_xmit() function checks how much UDP data (if any) is in the transmit buffer, prefixes it with IP and UDP headers, and transmits the result. It assumes that the UDP data checksum has already been computed. This is normally true because put_byte(), put_word(), and others update the checksum for every byte written.

```
/* Transmit the UDP data that is in the Tx buffer */
void udp_xmit(void)
{
    WORD tpdlen;

    save_txbuff();
    d_checkhi = checkhi;                  // Save checksum
    d_checklo = checklo;
    tpdlen = 0;                           // Get data length
    if (txin > UDPIPHDR_LEN)
        tpdlen = txin - UDPIPHDR_LEN;
    udplen = tpdlen + UDPHDR_LEN;         // ..and UDP length
    iplen = udplen + IPHDR_LEN;           // ..and IP length
    setpos_txin(0);                       // Go to start of Tx buffer
    host.eth.pcol = PCOL_IP;              // Ether protocol is IP
    ipcol = PUDP;                         // ..and IP protocol is UDP
    put_ip();                             // Add IP header
    put_udp();                            // Add UDP header
```

```
    transmit();                            // Send it all!
    DEBUG_PUTC('U');
}


/* Put out a UDP datagram. Data checksum must have already been computed */
void put_udp(void)
{
    checkflag = 0;                 // Ensure we're on an even byte
    checkhi = d_checkhi;           // Retrieve data checksum
    checklo = d_checklo;
    put_word(locport);            // Local and remote ports
    put_word(remport);
    put_word(udplen);             // UDP length
    check_lword(&myip);           // Add pseudoheader to checksum
    check_lword(&remip);
    check_byte(0);
    check_byte(PUDP);
    check_word(udplen);
    if (udp_checkoff)             // If no checksum computed..
    {                             // ..force to null value
        checkhi = checklo = 0xff;
    }                             // (don't remove the braces!)
    put_byte(~checkhi);           // Send checksum
    put_byte(~checklo);
}
```

UDP Time Client

So far, the UDP software has responded only to incoming requests. It has not initiated any transactions. This has allowed some significant simplifications to the code — most notably, the "return to sender" approach to message transmission.

To demonstrate a UDP client in action, I will add the capability to fetch the current time from a UDP time server at regular intervals and display it on the LCD in the hours, minutes, and seconds format. The time server could be on the local network, or it could be remote (somewhere on the Internet), in which case communications have to be routed through a gateway. This is actually a very sensible way of adding a real-time clock capability to a simple embedded system. Not only does it eliminate the need for a clock chip and battery, it also ensures that all the systems work to a common time reference, rather than having to be set using individual user interfaces.

UDP Client Method

To obtain the time from a server, your client needs to

- check whether the server is on a local or remote network —
  - if local, send the ARP request to the server and
  - if remote, send the ARP request to the gateway;
- obtain the ARP response and save the six-byte MAC address;
- send a UDP request with null data to port 37 at the time server IP address using the MAC address obtained above; and
- receive the four-byte response and use it to update the real-time clock.

Of course, UDP and ARP are inherently unreliable protocols, so a retry strategy is needed in case of failure. In this case, the best approach is to resend the transmission after a suitably leisurely interval. If the server isn't responding, it could be because of an unreasonably large number of requests, and you don't want to make the problem worse.

To coordinate the ARP and UDP cycles, a single state variable is needed to show whether the server (or gateway) has been ARPed or not.

```c
#define TIME_ARP_INTERVAL       (SECTICKS * 2)
#define TIME_REFRESH_INTERVAL   (SECTICKS * 5)


WORD timeticks;                 // Tick count
LWORD tserver_ip;               // Time server IP and Ethernet addresses
BYTE tserver_eth[MACLEN];
BOOL tserver_arp_ok=0;          // Flag to show if ARPed

/* Poll the time client */
void check_time(void)
{
    if (!net_ok)                // Do nothing if network not available
        timeout(&timeticks, 0);
    else if (!tserver_arp_ok)
    {                           // If not ARPed..
        if (host_arped(&tserver_ip))
        {                       // ..get ARP response if arrived
            memcpy(tserver_eth, remeth, MACLEN);
            tserver_arp_ok = TRUE;
            request_time();     // ..and request time
        }                       // ..or resend ARP
        else if (timeout(&timeticks, TIME_ARP_INTERVAL))
            arp_host(&tserver_ip);
```

```
    }
    else if (timeout(&timeticks, TIME_REFRESH_INTERVAL))
        request_time();          // Re-request time
}
```

Client Address Resolution

First, it is necessary to initialize some global variables with the router (gateway) IP address and netmask. This would normally be done using standard C variable initialization, but the current compilers don't seem able to handle this because LWORD is a union of bytes, words, and a long word.

```
#define ROUTER_ADDR  10,1,1,100 // IP address for router
#define NETMASK_ADDR 255,0,0,0  // Subnet mask value

LWORD router_ip;                // Router IP address
BYTE router_eth[6];             // Router Ethernet address
BOOL router_arp_ok;             // Flag to show router has been ARPed
. . .
init_ip(&router_ip, ROUTER_ADDR);   // Initialize router addr
init_ip(&netmask, NETMASK_ADDR);    // ..and subnet mask
. . .
/* Initialize an IP address using the given constants */
void init_ip(LWORD *lwp, BYTE a, BYTE b, BYTE c, BYTE d)
{
    lwp->b[0] = d;
    lwp->b[1] = c;
    lwp->b[2] = b;
    lwp->b[3] = a;
}
```

Instead of sending the ARP request directly to the host, it might be necessary to redirect it to a router, which will forward UDP communications to the host. Furthermore, the router already might have been ARPed by another application, in which case there is no need to send an ARP request at all.

```
#define ARPREQ      0x0001    // ARP request & response IDs
#define ARPRESP     0x0002

/* Send an ARP request for the given host */
void arp_host(LWORD *hip)
{
    if (in_subnet(hip))
```

```
            remip.l = hip->l;
        else if (!router_arp_ok)
        {
            remip.l = router_ip.l;
            arp_xmit(ARPREQ);
        }
    }
}


/* Check if the given host is within my subnet (i.e. router not required) */
BOOL in_subnet(LWORD *hip)
{
    return(((hip->l ^ myip.l) & netmask.l) == 0);
}


/* Send an ARP request or response */
void arp_xmit(BYTE op)
{
    init_txbuff(0);
    if (op == ARPREQ)                          // If request, broadcast
        memset(host.eth.srce, 0xff, MACLEN);
    put_word(0x0001);                          // Hardware type
    put_word(0x0800);                          // ARP protocol
    put_byte(6);                               // Hardware & IP lengths
    put_byte(4);
    put_word((WORD)op);                        // ARP req/resp
    put_data(myeth, MACLEN);                   // My MAC addr
    put_lword(&myip);                          // My IP addr
    put_data(host.eth.srce, MACLEN);           // Remote MAC addr
    put_lword(&remip);                         // Remote IP addr
    host.eth.pcol = PCOL_ARP;
    transmit();
    DEBUG_PUTC('A');
}
```

Any incoming ARP messages (requests or responses) are handled by a single function.

```
/* Handle an ARP message */
BOOL arp_recv(void)
{
    WORD op;
```

```
    if (match_byte(0x00) && match_byte(0x01) &&       // Hardware type
        match_byte(0x08) && match_byte(0x00) &&       // ARP protocol
        match_byte(6) &&  match_byte(4) &&            // Hardware & IP lengths
        get_word(&op) &&                              // Operation
        get_data(remeth, MACLEN) &&                   // Sender's MAC addr
        get_lword(&remip) &&                          // Sender's IP addr
        skip_data(6) &&                               // Null MAC addr
        match_lword(&myip))                           // Target IP addr (me?)
    {
        if (op == ARPREQ)                             // Received ARP request?
        {
            arp_xmit(ARPRESP);
            return(1);
        }
        else if (op == ARPRESP)                       // Received ARP response?
        {
            if (remip.l == router_ip.l)
                memcpy(router_eth, remeth, MACLEN);
            arp_resp = 1;
        }
    }
    return(0);
}
```

Every now and then, a check is made to see if a matching ARP response has been returned.

```
/* Check if an ARP response has been received for the given host */
BOOL host_arped(LWORD *hip)
{
    if (in_subnet(hip))
        return(arp_resp && remip.l==hip->l);
    else if (router_arp_ok)
        return(1);
    else
        return(arp_resp && remip.l==router_ip.l);
}
```

Client UDP Messaging

Once the server (or gateway) has been ARPed, UDP transmissions can begin. The UDP request has no data, so it is extremely simple.

```
#define TIMECLIENT_PORT 1501      // Arbitrary source port for time client
#define TIMESERVER_PORT 37

/* Send a UDP request for the time */
void request_time(void)
{
    init_txbuff(0);
    remip.l = tserver_ip.l;
    memcpy(host.eth.srce, tserver_eth, MACLEN);
    locport = TIMECLIENT_PORT;
    remport = TIMESERVER_PORT;
    checkhi = checklo = 0;
    udp_xmit();
}
```

The generic UDP message handler needs an addition to ensure it doesn't discard the incoming response.

```
if (ucsum==0 || (checkhi==0xff && checklo==0xff))
{                                       // ..call UDP handler
    . . .
    if (locport == ECHOPORT)            // Echo: return copy of data
    {
        . . .
    }
    else if (locport == DAYPORT)        // Daytime: return string
    {
        . . .
    }
    else if (locport == TIMECLIENT_PORT)    // Time client
        time_handler();
}
```

The time message consists of a 32-bit long word, giving the number of seconds since January 1, 1970. A small amount of division and modulo arithmetic gives an LCD display of hours, minutes, and seconds.

```
/* Handle an incoming time response */
void time_handler(void)
{
    LWORD time;
    BYTE h, m, s;

    if (get_lword(&time))
    {
        print_lcd = TRUE;
        print_net = print_serial = FALSE;
        lcd_gotoxy(9, 1);
        s = (BYTE)(time.l % 60);
        time.l /= 60;
        m = (BYTE)(time.l % 60);
        time.l /= 60;
        h = (BYTE)(time.l % 24);
        printf("%02u:%02u:%02u", h, m, s);
    }
}
```

The display could be improved by using an onboard clock to update it every second, with the network providing periodic adjustments to keep it in sync with a master clock.

The advantage of the time protocol is that it provides a 32-bit value that is easy to use in time computations, such as measurement of elapsed time, or in time-zone conversions. An alternative is to use the Daytime port 13 on the server to provide a complete date and time string.

```
Tue Jan  1 16:37:20 2002
```

This can be useful for applications that display the current time and date, so long as they don't need to perform time calculations.

High-Speed Data Transfer

Having added a UDP capability to the microcontroller TCP/IP stack, I was asked what data throughput it could achieve. In these cases, it is traditional to mock up some spurious application that shows the TCP/IP stack in the best possible light. Instead, I wanted to create a real demo application that could be used as a test bench for various high-speed data strategies.

The candidate that sprang to mind was video transfer. By implementing a simple video digitizer (*frame grabber*), I could transfer video images in real time to a PC and display them using simple PC client software. The images are transmitted across the network in raw, uncompressed form, so this demonstration is not directly applicable to real-world applications, where image compression is required. Rather, the demonstration highlights the following points:

- **High data throughput** The video signal is transmitted at a *net* data rate of 2Mbits per second (when using a PIC18 microcontroller), demonstrating that high data throughput can be obtained from low-cost components.

- **Standard protocols** Although the raw video format is nonstandard, it is transferred using the UDP over Ethernet, so it is compatible with all the standard network tools and utilities.

- **Data pipe** The microcontroller is providing a high-speed, low-cost bidirectional data pipe to a PC running Windows. This concept can be applied in various industrial applications, transferring data at higher speed and lower cost than existing solutions.

Hardware

Figure 14.4 gives an overview of the demonstration hardware. The control, networking, and user-interface functionality is provided by the Microchip PICDEM.net™ board; the video-capture interface is in the form of a daughter board, and the dotted lines show how it interfaces to the microcontroller on the main board.

The original prototype used a PIC16F877 processor running at 16MHz, but the current design uses a PIC18C452 (or its Flash equivalent) running at 32MHz. This not only speeds up the data transfer rate (from 1 to 2Mbits per second) but also simplifies the timing logic.

Video Standards

The capture board is designed for compatibility with the two standard composite video formats, as generated by low-cost single-board cameras. These are *analog* video signals, as used in broadcast TV, and should not be confused with the parallel data produced by digital cameras. The U.K. and U.S. monochrome variants are known as 625 composite and 525 composite, whereas the color versions are PAL (Phase Alternation Line) and NTSC (National Television Standards Commission). Table 14.1 summarizes the timing data of the standards.

Table 14.1   Video standards.

	625 Composite	**525 Composite**
Total lines per frame	625	525
Line frequency (Hz)	15,625	15,734
Field frequency (Hz)	50	59.94
Color subcarrier (MHz)	4.4336	3.5795

Each picture, or *frame*, is composed of two *fields*, which are *interlaced*, so one field has all the odd and the other all the even picture lines. If a color camera is used, the color information is encoded onto a *subcarrier*: a high-frequency carrier that is superimposed onto the monochrome signal. This is filtered out by the capture board, and only the monochrome (luminance) signal is processed; therefore, the capture board should be compatible with worldwide video standards.

Figure 14.4   Video hardware.

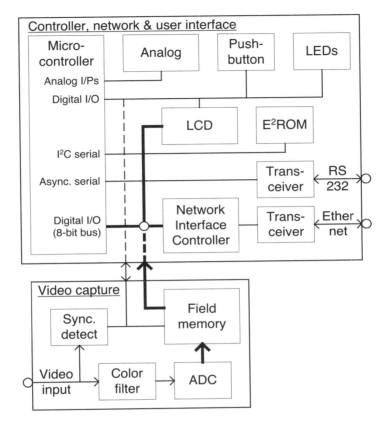

Video Capture

This daughter board takes in a standard composite video signal and digitizes it for transmission over the network.

The analog-to-digital converter (ADC) is equipped with a filter to remove the color information (color subcarrier) from the video signal, leaving only the monochrome signal to be digitized. The ADC clock frequency determines the aspect ratio (height-to-width ratio) of the captured image; 8MHz is an acceptable compromise for both U.S. and European standard video inputs.

The video signal includes horizontal and vertical synchronization pulses to indicate the start of a video line and video field. A sync detector chip extracts this timing information, which is then used by the ADC (to correct the amplitude and DC level of the video signal) and the controller (to determine when a complete picture has been captured).

Because of the disparity between the incoming and outgoing data rates, it is necessary for the board to buffer the video picture. A video picture (frame) consists of two interlaced fields, each containing half the video lines. The board captures only one of the two fields, so only half of the horizontal lines are stored. Table 14.2 gives the relationship between size of the raw data, effective display resolution (once the blank picture margins have been removed), and final picture data.

Table 14.2   Capture parameters.

	625 Composite	**525 Composite**
Stored lines	312	262
Raw data (bytes)	160,000	133,000
Picture size (H x V)	412 × 285	412 × 234
Picture data (bytes)	117,400	96,400

Although the video timing is standardized, there does seem to be some variation in the active picture area. Some low-cost cameras produce a slightly narrower picture than the standard. To avoid having a black bar on the right-hand side of the picture, the image is cropped slightly to produce the picture sizes in Table 14.2.

A 256KB memory device is needed to store the complete field. A specific type of RAM, field memory, is employed, which is essentially a large dual-port FIFO (first in, first out) device. The data is clocked in sequentially from the ADC on one eight-bit port and clocked out sequentially on another eight-bit port. Instead of address lines, there are two synchronous address counters, each with control signals for resetting and skipping unwanted bytes. The data acquisition is performed without any processor intervention, so the capture cycles can free-run until a picture is required; then the processor stops the acquisition at the end of the next field.

Microcontroller Interface

Having captured one video field, the processor finds the start of the first picture line by clocking the data out of the RAM and looking for the markers that indicate the start of each video line. After a specific number of markers and dummy bytes, as determined by the video timing, the picture data begins.

The microcontroller has insufficient RAM to store the image, so it is transferred straight from the field memory into the NIC. This direct-transfer method also gives a major improvement in data throughput.

The data transfers are complicated by the synchronous nature of the field-memory device. All transfer cycles are locked to a clock signal. Figure 14.5 shows some read cycles, which are very similar.

With the Output Enable line low, the data-output pins are high impedance. On the first rising clock edge after Output Enable is asserted, the current sample (n) is put out on the data bus. When the reset line is asserted by the microcontroller, the output data changes on the next rising clock edge to reflect the sample value at address zero (0). To move to the next sample, the Read Enable line is held high for one clock cycle.

Figure 14.5   Field memory timing.

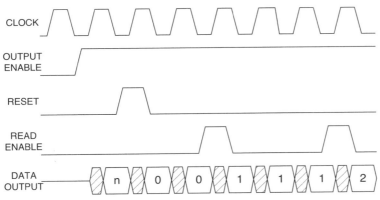

Although two clocks can be used (one for the input and another for the output cycles), avoid this if possible because of the risk of *wallpapering*, where clock interference causes undesirable patterns on the captured image. Even at low levels, these patterns are very noticeable, and the easiest way to avoid them is to lock all signals to the 8MHz data-acquisition clock. Because the PICmicro® has to pulse the Read Enable line high for a single 8MHz clock cycle, it must be clocked at 32MHz. This is impossible using the 16F877 supplied with the PICDEM.net kit, so the first prototype video capture board had to use a clock-switching scheme. Once the PIC18C452 became available with a multiply-by-four DPLL (digital phase-locked loop) clock and a maximum speed of 40MHz, the circuitry became much easier; the PICmicro could be clocked at 8MHz like all the other devices.

Circuit diagrams and PCB layouts are available from Iosoft Ltd. See www.iosoft.co.uk for more information.

Software

Two pieces of software are needed: one running on the PICmicro to take the video data and send it down the network and another running under Windows to take the data off the network and display the video image. Both applications are designed for speed so that a display rate of two images per second can be achieved.

Before creating these utilities, the format of the network data must be decided.

Data Format

Because the data is sent uncompressed, the video stream will be incompatible with all the standard Windows multimedia utilities, so there is little point in using a standard format for the video data itself. However, to simplify addressing and transport of the data around the network, it is best to encapsulate the data within a standard TCP/IP transport format. The simplest of these is UDP. On a conventional Ethernet network, the maximum UDP data size is 1,472 bytes, so three 412-pixel display lines will fit into a single UDP block. The display-line format has

- a 16-bit line number, most significant byte first;
- a 16-bit line width in pixels, most significant byte first; and

- uncompressed ADC data: black value 20, white value 127.

```
Typedef struct              // Data format for 1 video line (my Rx)
{
    WORD lnum;                  // Current line number
    WORD width;                 // Number of pixels per line
    BYTE data[IMAGE_WIDTH];     // Video data
} VIDLINE;
```

It is unrealistic to send uncompressed data around a sizeable network, let alone the Internet, so you can take some shortcuts for a small-scale demonstration. Ethernet is already protected by a 32-bit CRC code, so the checksum can be disabled, saving the PICmicro a large amount of computation. If an error occurs, refetching and resending the faulty data block would take a significant amount of time and is of limited value when another full image is being sent immediately afterwards. For the purposes of the PC demonstration, no error handling (apart from the continual resending of images) is included.

The remaining question is how the PICmicro application knows the intended destination address for the data and the frequency with which the images should be sent. These values could be set using some form of configuration utility, but the transmission of uncompressed video data represents a significant loading on the network, and it would be very unfortunate if the PICmicro continued sending images when the PC was unable to receive them. The easiest solution is for the Windows application to request an image using a single UDP datagram and the PICmicro to respond to that address, sending the correct number of blocks (datagrams containing three display lines) to form that image. The request format is

- mode (1 byte),
- block number (1 byte), and
- block count (1 byte).

The mode byte allows the PC to halt or run the image capture, and the block number and block count allow the PC to request a few blocks at a time. If the block count is zero, the whole image is sent.

```
typedef struct              // Data format for video request (my Tx)
{
    BYTE vmode;                 // Video mode
    BYTE blocknum;              // Starting block number
    BYTE nblocks;               // Number of video blocks,
} VIDREQ;
#define VMODE_RUN    0          // Run video capture
#define VMODE_HALT   1          // Halt video capture
```

This allows the Windows application to vary the rate at which it requests images, depending on the speed of display of the PC and the permissible loading on the PICmicro and the network.

Microcontroller Software

The microcontroller software uses the same UDP techniques as previously described. The main design issue is the efficient transfer of data from field memory to the Ethernet controller, which can be achieved using remarkably straightforward code.

```c
/* Copy the given byte count from capture RAM into the NIC RAM */
void capnic_data(WORD n)
{
    outnic(RBCR0, (BYTE)n);          // Byte count
    outnic(RBCR1, n>>8);
    outnic(CMDR, 0x12);              // Start, DMA remote write
    NIC_ADDR = DATAPORT;
    CAP_OE = 1;                      // Enable RAM O/P
    while (n--)
    {
        CAP_RE = 1;                  // Read from RAM
        CAP_RE = 0;
        NIC_IOW_ = 0;                // Write to NIC
        NIC_IOW_ = 1;
    }
    CAP_OE = 0;                      // Disable RAM O/P
}
```

The rest of the network interface code uses the same UDP transfer techniques as discussed earlier in the chapter. A significant amount of code deals with hardware-specific issues, such as the detection of video-synchronization pulses, but this is way beyond the scope of this book, so I will not discuss it further.

Client Software

The source-code package also includes a Windows utility to display the network images. To get smooth image updates, a double-buffering technique is used, and the image quality is remarkably good for such a low-cost system; Figure 14.6 (page 464) shows an image obtained in the lab using a low-cost (U.S.$60) videoconferencing camera.

A simple Linux utility fetches the data a few blocks at a time and stores it as an uncompressed GIF file. This makes the transfer process much more controllable and reliable by the inclusion of a retry strategy if a block is missed.

This isn't a Windows or Linux programming book, but many of you might want to interface microcontroller systems to hosts such as these, so it is worth highlighting the key network-programming techniques I have used.

Figure 14.6   Sample video image.

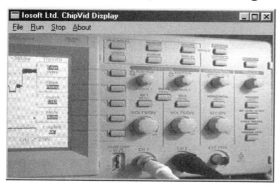

Windows Client

The design of the Windows client software is dominated by the event-driven nature of Windows programming. The client software has to coexist with an arbitrary number of other programs, so it must not hog the CPU. There is no main() software loop. The bulk of time is spent within the operating system itself, and control is returned to the client program only on specific *events*.

When run in *asynchronous* mode, the *Winsock* network interface is capable of generating such events. These events, and timeout events from a free-running timer, provide the main triggers for the client software to take action. The main functionality is relatively simple.

- If a timeout, send a request for the whole video frame.
- If a UDP block is received, update the video image in memory.
- If the network is idle, update the screen display from the memory image.

The use of a video buffer in memory (double buffering) is important to avoid an irritating flicker as each network block arrives.

All network transfers are performed though a *socket*, which is created as follows (error handling has been removed for clarity).

```
SOCKET hsock;
int newsize = . . .          // Max UDP datagram size
char hostname[MAXNAMELEN+1]; // IP addr string of host
SOCKADDR_IN host;            // Host addr

// Create socket
hsock = socket(AF_INET, SOCK_DGRAM, 0);

// Resize socket buffer
setsockopt(hsock, SOL_SOCKET, SO_RCVBUF, (void *)&newsize, sizeof(newsize));
```

```
// Set async. mode
WSAAsyncSelect(hsock, hwnd, WSA_ASYNC, FD_READ);


// Get host IP address
host.sin_addr.s_addr = inet_addr((LPSTR)hostname);
cdp->host.sin_family = PF_INET;
cdp->host.sin_port = htons(HOSTPORT);
```

Using `WSAAsyncSelect()` means that the message handler will get a `WSA_ASYNC` message every time a datagram arrives. This isn't the only way of handling incoming datagrams, but it does mean that a single message handler can drive all the state machine operations.

```
typedef struct                    // Data format for video request (my Tx)
{
    BYTE vmode;                       // Video mode
    BYTE blocknum;                    // Starting block number
    BYTE nblocks;                     // Number of video blocks,
} VIDREQ;

// Window message handler
LRESULT CALLBACK WndProc (HWND hwnd, UINT message, . . . )
{
    . . .
    switch (message)
    {
    case WM_CREATE:                          // New window
        . . .

    case WM_TIMER:                           // Timeout, send new request
        . . .
            ret = sendto(hsock, (void *)&vidreq, sizeof(vidreq), 0,
                        (LPSOCKADDR)&host, sizeof(SOCKADDR));
        . . .
    case WSA_ASYNC:                          // Network message received
        . . .
        len = recvfrom(hsock, blk, MAX_BLKLEN, 0,
                    (struct sockaddr *)&host, &addrsize);

        . . .

    case WM_COMMAND:                         // User interface commands
        . . .
```

```
    case WM_PAINT:                          // Repaint the image

        . . .

    case WM_DESTROY:                        // Destroy the window

        . . .

    }

        . . .

}
```

Because UDP is connectionless, you don't have to open the socket in the usual sense (i.e., open a connection between it and the corresponding remote socket). Use the `sendto()` and `recvfrom()` calls ad hoc to communicate with any given host as required.

Linux Client

As with Windows, it is important that the bulk of time is spent within the operating system, and the client code is called only when necessary (when a datagram is received, or on timeout). It is possible to use an event-driven programming model with Linux, but I wanted to find the most economical way of fetching and storing an image, with a minimum of operating-system complexity. In the end, I used the conventional *blocking* network calls (i.e., calls that don't return unless there is some data), but I used the *alarm* capability to generate a timeout in case nothing was received.

First, a socket is created and then "connected" to the remote system as an alternative to `sendto()` and `recvfrom()` (the error handling has been removed for clarity).

```
#define HOSTPORT    1502            // UDP port number on host

int sock;                          // Socket handle
struct sockaddr_in host;           // Host addr
char *hostname;                    // Host name

// Create socket
sock = socket(AF_INET, SOCK_DGRAM, 0);

// Get host address
memset(&host, 0, sizeof(host));
host.sin_family = AF_INET;
host.sin_port = htons(HOSTPORT);
host.sin_addr.s_addr = inet_addr(hostname);

// Connect to host
connect(sock, (struct sockaddr *)&host, sizeof(host));
```

I have established a supposed connection to the remote host. Because I'm using UDP, this is purely an operating-system fiction, and no network transmissions have taken place, but it does allow the use of read() and write().

```
#define RESP_TIMEOUT    2           // Response timeout in seconds
int rlen;                           // Rx length in bytes
typedef struct                      // Data format for video request (my Tx)
{
    BYTE vmode;                     // Video mode
    BYTE blocknum;                  // Starting block number
    BYTE nblocks;                   // Number of video blocks,
} VIDREQ;
char rdata[MAX_BLKLEN];             // Rx buffer
. . .
write(sock, &req, sizeof(req));         // Send network request
alarmcall(0);                           // Prepare timeout
rlen = read(sock, rdata, sizeof(rdata));// Get network response
if (rlen > 0)                           // If video data..
{
    . . .
}
. . .
/* Alarm signal handler; call with 0 signal value to set up alarm */
void alarmcall(int signum)
{
    signal(SIGALRM, alarmcall);
    if (!signum)
        alarm(RESP_TIMEOUT);
}
```

Normally, read() waits forever until a datagram arrives, but alarm() terminates it after two seconds, so an error recovery (retry) strategy can be employed without having to invoke other operating-system processes.

The complete source code for the Windows and Linux clients is available from the Iosoft Ltd. See www.iosoft.co.uk for more information.

Summary

UDP is particularly useful on microcontrollers with limited resources, and a lot can be done with it. I've described an application that fetches the time from a network server that allows all the clients to be locked into a common time standard, making their hardware simpler because a real-time clock chip and corresponding user interface are no longer required.

To demonstrate the speed of UDP communications, a video-server application has been outlined. Actual video data rates up to 2Mbits per second have been measured, which is a remarkable achievement for a low-cost microcontroller such as the Microchip PIC18.

Although I concentrate on embedded systems in this book, it is often necessary to interface to larger systems, so some code fragments have been presented to illustrate the kind of techniques that might be employed on Windows and Linux systems.

Source Files

The time client uses the following files:

p16web.c	Main program
p16_drv.c	Low-level driver
p16_eth.c	Ethernet interface
p16_ip.c	IP
p16_lcd.c	LCD interface
p16_time.c	Time client
p16_udp.c	UDP interface
p16_usr.c	User interface and configuration

The following compiler-specific files are also required:

ccs_p16.h	CCS-specific PIC16 and PIC18 definitions
ht_p16.h	Hitech-specific PIC16 and PIC18 definitions
ht_utils.c	Hitech-specific utility functions

The following are default settings:

Time server	10.1.1.1
Netmask:	255.0.0.0
Router	10.1.1.100

To change these values, edit the definitions at the top of the main program file.

Because the CCS compiler doesn't have a linker, all the source files are included in the main program file, so only that file needs to be submitted to the compiler.

Definitions at the top of the main program select the functionality required so that the appropriate files are included. The following configuration was used for the time client testing:

```
/* TCP protocols: enabled if non-zero */
#define INCLUDE_HTTP 0      // Enable HTTP Web server
#define INCLUDE_SMTP 0      // Enable SMTP email client
#define INCLUDE_POP3 0      // Enable POP3 email client
```

```
/* UDP protocols: enabled if non-zero */
#define INCLUDE_DHCP 0      // Enable DHCP auto-configuration
#define INCLUDE_VID  0      // Video demo: 1 to enable, 2 if old video h/w
#define INCLUDE_TIME 1      // Time client: polls server routinely


/* Low-level drivers: enabled if non-zero */
#define INCLUDE_ETH   1     // Ethernet (not PPP) driver
#define INCLUDE_DIAL  0     // Dial out on modem (needs PPP)
#define INCLUDE_LCD   1     // Set non-zero to include LCD driver
#define INCLUDE_CFG   1     // Set non-zero to include IP & serial num config
```

The video server uses the following additional file:

> p16cap.c   Video interface

One definition must be changed in the main program:

```
#define INCLUDE_VID  1      // Video demo: 1 to enable, 2 if old video h/w
```

Chapter 15

Dynamic Host Configuration Protocol: DHCP

Overview

So far, the microcontroller TCP/IP software has relied on its IP address being stored in the on-chip nonvolatile memory. This means that each system must be individually configured with the correct IP address, which can be highly inconvenient in large installations. What is needed is an autoconfiguration capability, which would obtain an IP address (and also the gateway address and subnet mask) automatically from a central server.

The standard way of achieving this is through the Dynamic Host Configuration Protocol (DHCP). It is still necessary to equip each network node with a unique, six-byte Ethernet (MAC) address; otherwise, the nodes would be indistinguishable from each other. However, armed with this MAC address, a node can interrogate a DHCP server to obtain its IP address and other optional parameters.

DHCP Methodology

Although DHCP uses UDP as the basis for all its communications, it does make extensive use of broadcasts, in a fashion similar to ARP. Each network is assumed to have at least one DHCP server, which has a database containing the configuration information for the whole network.

When the client requests an IP address, it is not assigned permanently to that node but is issued in the form of a *lease*, which will expire after a specific time and have to be renewed. This allows a large number of nodes to share a smaller number of addresses, so long as the total number of nodes online doesn't exceed the limit at any one time. The servers are intelligent in the way they allocate leases. If a node is reset and requests a new lease, it will, if possible, be assigned the same IP address as before.

Sequence of Operations

The sequence of operations at startup is as follows (Figure 15.1):

Discover   Client broadcasts a DHCP discovery request, indicating the configuration information it needs.

Offer   One or more servers send a response, offering an IP address.

Request   Having selected one of the offers, the client sends a request to a specific server to use that address.

Acknowledge   The server acknowledges the request and updates its database.

The client must renew its lease within the allotted time.

Request   The client requests a renewal from the server that issued the lease.

Acknowledge   The server acknowledges the request.

Several DHCP servers on the network might share a common node database. If the original DHCP server isn't responding, the client must *rebind* to use the same IP address with a different server.

Request   The client broadcasts a request for renewal.

Acknowledge   A new server acknowledges the request.

If the client cannot rebind the old lease to a new server, it falls back to using the same discovery mechanism as at startup to obtain a new lease from a new server.

Timing

A single DHCP server can receive requests from hundreds of nodes, so the timing of those requests can be important. Imagine that the nodes are all powered up simultaneously at a specific time each morning and that they all have the same timeout value. At power-up, and after each timeout, the server will receive a flood of DHCP requests, whereas, at other times, it will be relatively idle.

To counteract this problem, DHCP clients are supposed to introduce some randomness into their timings. Instead of issuing requests immediately on boot-up, they should wait a random time up to 10 seconds.

Figure 15.1   DHCP client–server transaction.

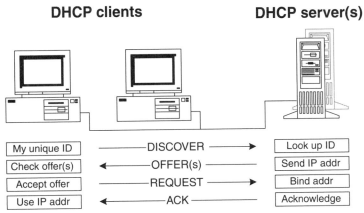

It is quite difficult to generate randomness on a microcontroller without a real-time clock. A boot time, the only parameter that can differentiate any two nodes, is their MAC address, so this is randomized to generate a delay value.

The randomizer uses a simple polynomial to generate a single bit and then loops to expand that into a byte value.

```
/* Randomize lower 8 bits of 16-bit number */
WORD rand_byte(WORD val)
{
    BYTE n;

    for (n=0; n<8; n++)
        val = rand_bit(val);
    return(val);
}


/* Randomize l.s. bit of 16-bit number */
WORD rand_bit(WORD val)
{
    BYTE n=0;

    if (val & 0x8000) n = !n;
    if (val & 0x4000) n = !n;
    if (val & 0x1000) n = !n;
    if (val & 0x0008) n = !n;
    return(val+val+n);

}
```

Message Format

DHCP was designed for backward compatibility with an older protocol, BOOTP, so it uses a rather strange message format (Figure 15.2). A large percentage of the UDP data is redundant, and the main area of interest is at the end of a large amount of padding.

When contacting a DHCP server, most of the fields are zero, apart from the following:

Opcode	1 for request, 2 for reply
Address type	1 for Ethernet hardware
Address length	6 for Ethernet hardware
Transaction ID	A random number chosen by the client to associate a message with a response
Your IP address	The client IP address
Client hardware address	The MAC address of the client

Options

The DHCP options are used to convey details about the DHCP request to the server. Each option consists of:

Option code	A one-byte value to identify the option
Option length	A one-byte value giving the length of option data to follow
Option data	Up to 255 bytes of data

Common DHCP options are

53	Message type	This is compulsory because it indicates the type of DHCP message: discover, offer, request, decline, ACK, NAK, or release
12	Host name	Confusingly, this is the name that the client uses to identify itself to the server so an appropriate IP address can be allocated (see the "Client Name" section below)
61	Client identifier	This is generally the MAC address of the client, which will be used for indexing the DHCP server's database
54	Server identifier	A server identification that helps the client make the correct choice during discovery
1	Subnet mask	The subnet mask value provided by the server for use by the client
3	Routers	The IP addresses of any routers
6	Domain name servers	The IP addresses of any domain name servers
51	IP address lease time	The time (in seconds) of the lease being offered by the server

When analyzing a DHCP offer from the server, the client takes its new IP address from the *your IP address* field and combines it with the lease time and subnet mask given in the options field. If it obtains responses from several servers, it can choose which it prefers according to its own criteria; for example, it can be programmed to reject any offers that don't come from a few trusted hosts. It then sends a DHCP request to the server it has accepted. It need not send any messages to the servers it declines. Once it receives an ACK from the selected server, the client can start using the new IP address, although it may also choose to do some sanity checks first (e.g., seeing if there is any response to an ARP request to that address).

Figure 15.2   DHCP message format.

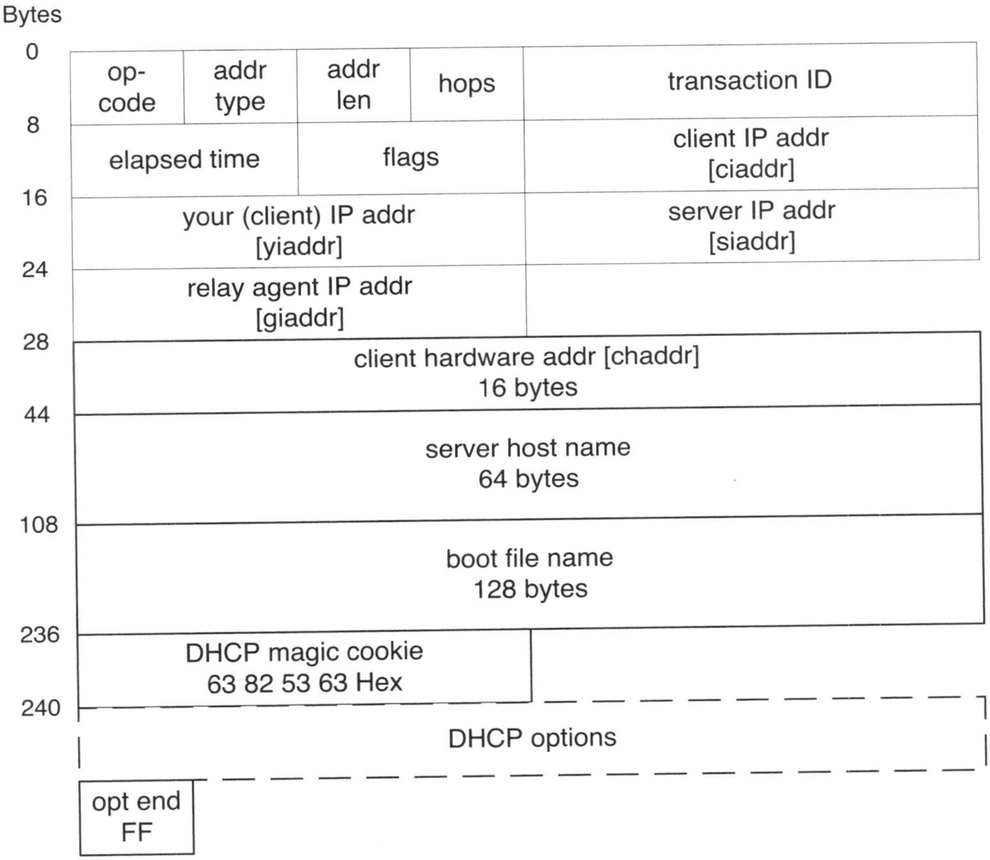

Bytes

op-code	addr type	addr len	hops	transaction ID
elapsed time		flags		client IP addr [ciaddr]
your (client) IP addr [yiaddr]				server IP addr [siaddr]
relay agent IP addr [giaddr]				
client hardware addr [chaddr] 16 bytes				
server host name 64 bytes				
boot file name 128 bytes				
DHCP magic cookie 63 82 53 63 Hex				
DHCP options				
opt end FF				

0
8
16
24
28
44
108
236
240

Client Name

One of the DHCP options mentioned above is a client name, and you might wonder why it is necessary for the client to provide a name to the server. Isn't the whole point of DHCP that the client only needs a unique MAC address in order to boot? The answer is that the client name isn't essential to the boot process but can be useful in resolving the underlying problem of changing (*dynamic*) IP addresses.

First, imagine that you have a node that boots using DHCP and then uses its assigned IP address to contact a server of some sort, say, an email server. It really doesn't matter which IP address the node has been given because the mail server will accept communications from any node within its domain.

Figure 15.3 DHCP client name example.

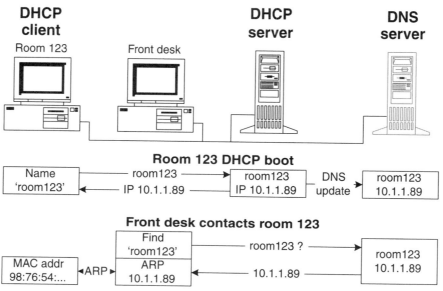

Now imagine that the same node also has a Web-server capability, and you want to contact the node to browse its Web pages. You need to know the node's IP address to initiate an HTTP session, but you don't have that information. Only the DHCP server knows which node has which IP address, and even if you could interrogate the DHCP server, it'd give you only the MAC address to IP mappings, which may not be a lot of use.

The simple way around the problem is to force the DHCP server to assign specific IP addresses to known MAC addresses, but this is an inflexible solution: if a node is faulty and has to be replaced, its MAC address will probably change, and the DHCP database will have to be updated manually. A far more flexible solution is to use a Domain Name System (DNS).

So far I have avoided any mention of DNS because it is beyond the scope of the book, but suffice it to say that it is the global database system that converts a network name such as www.iosoft.co.uk into an IP address such as 212.87.81.113. DNS systems needn't be used only for Internet accesses, they can also be useful for small stand-alone networks and can be used to solve the problem addressed here. If the DNS server is linked to the DHCP server (a process called *auto update*), then the client name supplied as part of the DHCP discover message can end up as an entry in the local DNS database.

Figure 15.3 shows an example of this, assuming that a hotel has been equipped with DHCP-booted systems in each of its rooms, and the client name has been set to the word room followed by the room number. As each system boots, the DNS server database is updated to reflect the client name (as a DNS name) and IP address. If the front-desk PC needs to contact a particular room, it makes a request to the DNS server for, say, room123, and that server responds with the appropriate IP address, The front-desk PC then does an ARP request as usual to resolve that IP address into a MAC address and uses the MAC address for further communications.

DNS and dynamic updates are quite complex and have a reputation for being difficult to configure. However, there is a lot of helpful information on the Web, and PC packages are starting to emerge that are relatively easy to set up. I used the excellent Simple DNS Plus from JH software (www.jhsoft.com) to demonstrate the method described above.

Sample Transaction

The following DHCP transaction is of a Windows PC (named Phoenix) booting from a Simple DNS server. The null entries and Microsoft-specific requests have been removed for clarity.

All the messages use broadcast addresses, and the (unknown) client IP address is all zeros.

Discovery

```
UDP - User Datagram Protocol
   Source Port:            68  Bootstrap (BOOTP Client)
   Destination Port:       67  Bootstrap Protocol Server
   Length:                308
   Checksum:           0xA318
BootP - Bootstrap Protocol
   Operation:               1  Boot Request
   Hardware Address Type:   1  Ethernet (10Mb)
   Hardware Address Length: 6  bytes
   Transaction ID:          4262503227
   Seconds Since Boot Start: 15269
   Client Hardware Address:  00:80:C8:8E:9B:D6
DHCP - Dynamic Host Configuration Protocol
   DHCP Magic Cookie:        0x63825363
   Message Type- DHCP Option
      Option Code:       53  Message Type
      Option Length:      1
      Message Type:       1  Discover
   Client Identifier- DHCP Option
      Option Code:       61  Client Identifier
      Option Length:      7
      Hardware Type:      1
      Hardware Address:   00:80:C8:8E:9B:D6
   Host Name Address- DHCP Option
      Option Code:       12  Host Name
      Option Length:      7
      String:            phoenix
   DHCP Option End
      Option Code:      255  End
```

Offer

```
UDP - User Datagram Protocol
   Source Port:            67  Bootstrap (BOOTP Server)
   Destination Port:       68  Bootstrap Protocol Client
   Length:                 301
   Checksum:               0xA892
BootP - Bootstrap Protocol
   Operation:              2  Boot Reply
   Hardware Address Type:  1  Ethernet (10Mb)
   Hardware Address Length: 6  bytes
   Transaction ID:         4262503227
   IP Address Known By Client:  0.0.0.0  IP Address Not Known By Client
   Client IP Addr Given By Srvr: 10.1.1.210
   Client Hardware Address:  00:80:C8:8E:9B:D6
DHCP - Dynamic Host Configuration Protocol
   DHCP Magic Cookie:      0x63825363
   Message Type- DHCP Option
      Option Code:         53  Message Type
      Option Length:       1
      Message Type:        2  Offer
   Client Identifier- DHCP Option
      Option Code:         61  Client Identifier
      Option Length:       7
      Hardware Type:       1
      Hardware Address:    00:80:C8:8E:9B:D6
   Subnet Mask- DHCP Option
      Option Code:         1  Subnet Mask
      Option Length:       4
      Address:             255.0.0.0
   Server Identifier- DHCP Option
      Option Code:         54  Server Identifier
      Option Length:       4
      Address:             10.1.1.201
   IP Address Lease Time- DHCP Option
      Option Code:         51  IP Address Lease Time
      Option Length:       4
      Value:               7200
   Routers- DHCP Option
      Option Code:         3  Routers
      Option Length:       4
      Address:             10.1.1.100  Router
   Domain Name- DHCP Option
      Option Code:         15  Domain Name
      Option Length:       8
      String:              LOCALNET
   Domain Name Servers- DHCP Option
      Option Code:         6  Domain Name Servers
      Option Length:       4
      Address:             0.0.0.0
   DHCP Option End
      Option Code:         255  End
```

Request

```
UDP - User Datagram Protocol
   Source Port:              68  Bootstrap (BOOTP Client)
   Destination Port:         67  Bootstrap Protocol Server
   Length:                   308
   Checksum:                 0x1979
BootP - Bootstrap Protocol
   Operation:                1   Boot Request
   Hardware Address Type:    1   Ethernet (10Mb)
   Hardware Address Length:  6   bytes
   Transaction ID:           4262503227
   Seconds Since Boot Start: 15269
   Client Hardware Address:  00:80:C8:8E:9B:D6
DHCP - Dynamic Host Configuration Protocol
   DHCP Magic Cookie:        0x63825363
   Message Type- DHCP Option
      Option Code:           53  Message Type
      Option Length:         1
      Message Type:          3   Request
   Client Identifier- DHCP Option
      Option Code:           61  Client Identifier
      Option Length:         7
      Hardware Type:         1
      Hardware Address:      00:80:C8:8E:9B:D6
   Requested IP Address- DHCP Option
      Option Code:           50  Requested IP Address
      Option Length:         4
      Address:               10.1.1.210
   Server Identifier- DHCP Option
      Option Code:           54  Server Identifier
      Option Length:         4
      Address:               10.1.1.201
   Host Name Address- DHCP Option
      Option Code:           12  Host Name
      Option Length:         7
      String:                phoenix
   DHCP Option End
      Option Code:           255 End
```

Acknowledgment

```
UDP - User Datagram Protocol
   Source Port:              67  Bootstrap (BOOTP Server)
   Destination Port:         68  Bootstrap Protocol Client
   Length:                   301
   Checksum:                 0xA592
BootP - Bootstrap Protocol
   Operation:                2   Boot Reply
   Hardware Address Type:    1   Ethernet (10Mb)
   Hardware Address Length:  6   bytes
```

```
   Transaction ID:                 4262503227
   Flags:                          0x0000
   Client IP Addr Given By Srvr: 10.1.1.210
   Client Hardware Address:        00:80:C8:8E:9B:D6
DHCP - Dynamic Host Configuration Protocol
   DHCP Magic Cookie:              0x63825363
   Message Type- DHCP Option
      Option Code:        53  Message Type
      Option Length:      1
      Message Type:       5  ACK
   Client Identifier- DHCP Option
      Option Code:        61  Client Identifier
      Option Length:      7
      Hardware Type:      1
      Hardware Address:   00:80:C8:8E:9B:D6
   Subnet Mask- DHCP Option
      Option Code:        1  Subnet Mask
      Option Length:      4
      Address:            255.0.0.0
   Server Identifier- DHCP Option
      Option Code:        54  Server Identifier
      Option Length:      4
      Address:            10.1.1.201
   IP Address Lease Time- DHCP Option
      Option Code:        51  IP Address Lease Time
      Option Length:      4
      Value:              7200
   Routers- DHCP Option
      Option Code:        3  Routers
      Option Length:      4
      Address:            10.1.1.100  Router
   Domain Name- DHCP Option
      Option Code:        15  Domain Name
      Option Length:      8
      String:             LOCALNET
   Domain Name Servers- DHCP Option
      Option Code:        6  Domain Name Servers
      Option Length:      4
      Address:            0.0.0.0
   DHCP Option End
      Option Code:        255  End
```

Once this transaction is complete, the DHCP diagnostic window shows the new node name (qualified by the domain name), its IP and MAC addresses, and the time its lease expires.

```
phoenix.localnet    10.1.1.210    00:80:C8:8E:9B:D6    2002-01-03 16:48:00
```

When developing DHCP software, it is useful to have access to a local DHCP server with a diagnostic display such as this, in order to ensure the boot transactions are working correctly.

DHCP Implementation

State Machine

The DHCP state machine is implemented as a single function, which checks for the various timeout conditions and takes appropriate action.

The time constants have been chosen to provide a reasonably speedy response for demonstration purposes. They would need to be lengthened if a large number of nodes are sharing the one DHCP server.

```
#define DHCP_DISCTIME (SECTICKS * 5) // Delay between DHCP discovery requests
#define DHCP_MAXDISCTICKS (SECTICKS*120) // Upper limit on discovery timeout
#define LEASE_MINSECS 60            // Minimum lease renewal time (sec)
#define LEASE_MAXSECS 43200         // Maximum lease renewal time (sec)
```

The timing is quite complex because of the need to avoid DHCP server overload, in the event that a large number of nodes all boot at once (e.g., are all powered up at the same time). A certain amount of "fuzziness" is introduced through a random number derived from the node serial number, and the retry time is doubled for every timeout.

```
BYTE dhcp_state=0;              // Current DHCP state
WORD dhcp_delticks;            // Current delay tick count
WORD dhcp_secticks;           // ..and seconds tick counter
WORD dhcp_deltickval;         // Current delay value (in ticks)
WORD dhcp_delbaseval;         // Base value for delay (in ticks)
WORD dhcp_secs;               // Seconds counter

/* Do a DHCP request, return 0 if no DHCP address available yet */
SEPARATED BOOL check_dhcp(void)
{
    WORD sernum;

    if (dhcp_state == DHCP_INIT)
    {
        myip.l = 0;
        sernum = swapw(*(WORD *)&myeth[4]);
        dhcp_secs = 0;
        timeout(&dhcp_delticks, 0);
        timeout(&dhcp_secticks, 0);
        dhcp_delbaseval = dhcp_deltickval =
            SECTICKS + (rand_byte(sernum) & 0x1f);
        xid.w[0] = sernum;
        xid.w[1] = dhcp_delbaseval;
```

```
            dhcp_state = DHCP_DISCOVER;
    }
    else if (dhcp_state==DHCP_DISCOVER || dhcp_state==DHCP_REQUEST ||
            dhcp_state==DHCP_RENEWING)
    {
        if (timeout(&dhcp_delticks, dhcp_deltickval))
        {
            checkflag = checkhi = checklo = 0;
            dhcp_tx(dhcp_state);
            if (dhcp_deltickval < DHCP_MAXDISCTICKS/2)
                dhcp_deltickval += dhcp_deltickval;
        }
    }
    if (dhcp_state==DHCP_BOUND || dhcp_state==DHCP_RENEWING)
    {
        if (timeout(&dhcp_secticks, SECTICKS))
        {
            dhcp_secs++;
            if (dhcp_secs == renew_secs)
            {
                dhcp_deltickval = dhcp_delbaseval;
                dhcp_state = DHCP_RENEWING;
            }
            else if (dhcp_secs >= lease_secs)
                dhcp_state = DHCP_INIT;
        }
        return(1);
    }
    return(0);
}
```

DHCP Transmission

The DHCP format is unusual in that it contains a large amount of unused (null) data to retain backward compatibility with the older BOOTP protocol. It is fortunate that the buffering scheme allows full-size Ethernet messages to be created using small amounts of RAM because the PIC16 microcontrollers have insufficient RAM to store the DHCP request as it is constructed.

Only two options are included in a discovery message: the client ID (MAC address) and the confusingly titled host name (client name). The DHCP request has two additional options: the server ID (IP address) and the new client IP address.

```
/* Send a DHCP message */
SEPARATED void dhcp_tx(BYTE state)
{
    BYTE n;
    WORD sernum;

    print_lcd = print_serial = FALSE;
    print_net = TRUE;
    host.eth.pcol = PCOL_IP;
    ipcol = PUDP;
    locport = DHCPCLIENT_PORT;
    remport = DHCPSERVER_PORT;
    setpos_txin(UDPIPHDR_LEN);
    put_byte(1);                            // DHCP request
    put_byte(1);                            // Ethernet h/w, 6-byte addr
    put_byte(MACLEN);
    put_byte(0);                            // Zero hop count
    put_lword(&xid);
    put_nulls(4);
    put_lword(&myip);
    put_nulls(12);
    put_data(myeth, MACLEN);                // My hardware addr
    while (txin < 236+UDPIPHDR_LEN)
        put_byte(0);                        // Rest of hostname & bootfile name
    put_byte(99);                           // DHCP magic cookie
    put_byte(130);
    put_byte(83);
    put_byte(99);

    put_byte(53);                           // DHCP message type
    put_byte(1);
    put_byte(state & DHCP_TYPE_MASK);
    put_byte(61);                           // Client ID
    put_byte(7);
    put_byte(1);                                // ..addr type 1 (Ethernet)
    put_data(myeth, MACLEN);                    // ..my MAC address
```

```
        put_byte(12);                          // Host name
        print_net = 0;                             // ..get serial num string length
        sernum = swapw(*(WORD *)&myeth[4]);
        n = disp_decword(sernum);
        print_net = 1;
        put_byte(n+1);                         // ..put out length+1
        put_byte('P');                         // ..put out 'P' prefix
        disp_decword(sernum);                  // ..put out serial number
        if (state == DHCP_REQUEST)
        {                                      // If request..
            put_byte(54);                      // ..send server ID (IP addr)
            put_byte(4);
            put_lword(&dhcp_hostip);
            put_byte(50);                      // .. and requested IP addr
            put_byte(4);
            put_lword(&dhcp_newip);
        }
        put_byte(255);                         // End of DHCP options
        if (state == DHCP_RENEWING)
        {                                          // If renewing, use unicast addr
            memcpy(host.eth.srce, dhcp_hosteth, MACLEN);
            remip.l = dhcp_hostip.l;
        }
        else
        {                                          // ..otherwise use broadcast
            memset(host.eth.srce, 0xff, MACLEN);
            memset(&remip, 0xff, sizeof(remip));
        }
        udp_xmit();
}

/* Put the given number of nulls into the Tx buffer */
void put_nulls(BYTE n)
{
    while (n--)
        put_byte(0);
}
```

DHCP Reception

The handler for incoming DHCP messages is called from the UDP handler.

```
if (locport == ECHOPORT)              // Echo: return copy of data
{
    . . .
}
else if (locport == DAYPORT)          // Daytime: return string
{
    . . .
}
else if (locport == DHCPCLIENT_PORT)  // DHCP client
    dhcp_handler();
```

The important items that have to be extracted from the message are

- the new IP address;
- the message type: discover, offer, request, decline, ACK, NAK, or release;
- the lease time;
- the subnet mask; and
- the router address.

It is possible, that for each discovery request, more than one response can be received (from more than one DHCP server). Although the current implementation always takes the first offer, it might be necessary to include code that filters out unwanted responses (e.g., that accepts responses only from a few, known servers).

```
/* Handle an incoming DHCP datagram */
void dhcp_handler(void)
{
    BYTE code, len, type=0;
    LWORD time, newip;

    rx_checkoff = 1;
    if (match_byte(2) &&                      // Boot reply?
        skip_byte() && match_byte(6) &&       // 6-byte MAC address?
        skip_data(13) &&                      // Skip to..
        get_lword(&newip) &&                  // ..my new IP address
        skip_data(8) &&
        match_data(myeth, MACLEN) &&
        skip_data(10 + 64 + 128))
```

```
{
    print_lcd = print_net = FALSE;
    print_serial = TRUE;
    if (match_byte(99) && match_byte(130) &&
        match_byte(83) && match_byte(99))
    {
        while (get_byte(&code) && code!=255 && get_byte(&len))
        {
            if (code==53 && len==1)      // Message type?
                get_byte(&type);
            else if (code==51 && len==4)// Lease time?
                get_lword(&time);
            else if (code==1 && len==4 && type==DHCP_ACK)
                get_lword(&netmask);     // Subnet mask?
            else if (code==3 && len>=4 && type==DHCP_ACK)
                get_lword(&router_ip);  // Router IP addr?
            else
                skip_data(len);
        }
        if (dhcp_state==DHCP_DISCOVER && type==DHCP_OFFER)
        {
            memcpy(dhcp_hosteth, host.eth.srce, MACLEN);
            dhcp_hostip.l = remip.l;
            dhcp_newip.l = newip.l;
            dhcp_deltickval = dhcp_delbaseval;
            dhcp_state = DHCP_REQUEST;
        }
        else if ((dhcp_state==DHCP_REQUEST || dhcp_state==DHCP_RENEWING) &&
                 type==DHCP_ACK)
        {
            lease_secs = LEASE_MAXSECS;
            if (time.l < LEASE_MAXSECS)
                lease_secs = time.w[0];
            if (lease_secs < LEASE_MINSECS)
                lease_secs = LEASE_MINSECS;
            renew_secs = lease_secs >> 1;
            dhcp_secs = 0;
            myip.l = dhcp_newip.l;
            disp_myip();
```

```
                    put_ser("\r\n");
                    dhcp_state = DHCP_BOUND;
                    net_ok = 1;
              }
          }
       }
   }
```

When the DHCP server is configured, a default lease time will have been specified. This implementation accepts very short lease times (one minute), because it's useful when testing the lease-renewal logic, but lease times are generally much longer — a few hours at least.

Summary

DHCP is useful for booting a large number of systems without having to configure each individually, although it does raise some important issues.

- Each node must have a unique MAC address on the network.
- A MAC address is traditionally associated with a given set of network hardware. If the hardware changes, the MAC address changes.
- Each node must have a unique identity (which may just be the MAC address) to identify itself to the DHCP server.
- If a node is to be contacted by others, they must be aware of its unique identity and have a method of deriving an IP and MAC address from it.

These aren't just DHCP issues but are the consequences of the shift in addressing philosophy from fixed IP addresses to a dynamic-allocation scheme. For clients, this change has little or no effect, but network servers with dynamically assigned IP addresses do require significant extra complication, such as auto-update DNS.

Source Files

The DHCP client uses the following files:

p16web.c	Main program
p16_dhcp.c	DHCP
p16_drv.c	Low-level driver
p16_eth.c	Ethernet interface
p16_ip.c	IP
p16_lcd.c	LCD interface
p16_udp.c	UDP interface
p16_usr.c	User interface and configuration

The following compiler-specific files are also required:

`ccs_p16.h`	CCS-specific PIC16 and PIC18 definitions
`ht_p16.h`	Hitech-specific PIC16 and PIC18 definitions
`ht_utils.c`	Hitech-specific utility functions

Because the CCS compiler doesn't have a linker, all the source files are included in the main program file, so only that file needs to be submitted to the compiler.

Definitions at the top of the main program select what functionality is required, so the appropriate files are included. The following configuration was used for initial DHCP client testing:

```
/* TCP protocols: enabled if non-zero */
#define INCLUDE_HTTP 0        // Enable HTTP Web server
#define INCLUDE_SMTP 0        // Enable SMTP email client
#define INCLUDE_POP3 0        // Enable POP3 email client


/* UDP protocols: enabled if non-zero */
#define INCLUDE_DHCP 1        // Enable DHCP auto-configuration
#define INCLUDE_VID  0        // Video demo: 1 to enable, 2 if old video h/w
#define INCLUDE_TIME 1        // Time client: polls server routinely


/* Low-level drivers: enabled if non-zero */
#define INCLUDE_ETH  1        // Ethernet (not PPP) driver
#define INCLUDE_DIAL 0        // Dial out on modem (needs PPP)
#define INCLUDE_LCD  1        // Set non-zero to include LCD driver
#define INCLUDE_CFG  1        // Set non-zero to include IP & serial num config
```

Chapter 16

TCP Clients, SMTP, and POP3 Email

Overview

Although I have already discussed one microcontroller TCP implementation, it was restricted to server use only. It could not be used for client software, such as an SMTP (Simple Mail Transfer Protocol) client to send emails or a POP3 (Post Office Protocol 3) client to fetch emails from a server.

The fundamental problem with running TCP/IP on a microcontroller with minimal RAM is that, in order to make a reliable network connection, TCP must be able to retransmit any data that goes astray. How can TCP retransmit a 1Kb message when it has only 300 bytes of RAM to store that message in? The previous TCP stack circumvented the problem in a manner that severely restricted its versatility, so that it could be used only for small server applications.

This chapter takes the TCP stack one step further and makes it much more versatile. By way of demonstration, an SMTP client is created that can send a complete email on demand. I have also included a POP3 client that can scan a mailbox and extract specific information from the messages in it (e.g., the sender's address and the subject line).

The new TCP stack enables a wide range of TCP applications on the microcontroller while keeping the RAM usage as small as possible.

TCP Client Techniques

The previous TCP implementation minimized RAM usage in all possible areas. Instead of creating a TCP message using a structure in RAM, it is created as needed in a small (42-byte) transmit buffer. When this buffer is full, it is emptied into the Network Interface Controller (NIC) RAM. An equivalent of a file pointer can be positioned anywhere within the Ethernet frame being created so that the TCP data can be written first; then the pointer is repositioned to fill in the TCP and IP headers before the whole RAM buffer is flushed and the NIC frame is transmitted.

Unfortunately, all this effort can't get around the central problem: What happens if the transmitted message is lost in transit and has to be re-sent? There are various ways of working around this problem.

- Keep all messages very short so a copy can be retained in local RAM.
- If the data came from a read-only device, such as a Web page in ROM, then just refetch the data from the same place; it won't have changed in the interim.
- Freeze all data acquisition for the duration of the transfer so that, if the outgoing message has to be recreated from scratch, the data it was derived from won't have changed.

All these techniques place restrictions on the client code, such that it is a lot harder to write and debug. It is important that the client code can take a "write and forget" attitude toward the outgoing data. Any attempt by the TCP stack to dictate the actions of the client (e.g., "go back 1KB and recreate the data you sent") makes for exceedingly complex client code.

Secondary NIC Transmit Buffer

There is only one solution to this problem: if sufficient RAM has to be found to buffer one or more outgoing messages and if the microcontroller doesn't have enough RAM, then you must use the RAM in the NIC. It has an amazing 8KB, compared with 360 bytes in the PIC16 microcontroller. A quick solution that doesn't involve extensive modifications to the existing code is to create a secondary transmit buffer within the NIC RAM. This buffer can be used for TCP transmissions. If a message is lost in transit, the buffer contents are simply resent. While waiting for an acknowledgment, the node can receive other network requests (e.g., an ARP request), in which case the primary buffer must be used for any responses; otherwise, the TCP data would be overwritten. That is why two buffers are needed.

Of course, this approach has its limitations.

- For each extra buffer, only one TCP segment can be sent before an acknowledgment is required.
- For each extra buffer, only one TCP connection can be active at any one time.
- The maximum size of the Ethernet frames may have to be restricted, depending on the amount of NIC RAM used for transmit buffers.

These restrictions aren't as problematic as they may appear:

- In many protocols (such as SMTP), the node always must wait for a response before issuing the next command.
- The microcontroller has very limited resources, so it probably can't handle several simultaneous requests.

- So long as the frames are larger than the 576-byte minimum IP limit, they should be adequate for most applications.

Allocation of NIC RAM

It is worth reviewing the current NIC RAM usage to establish what would be available for a secondary transmit buffer (Figure 16.1).

Figure 16.1   NIC RAM allocation.

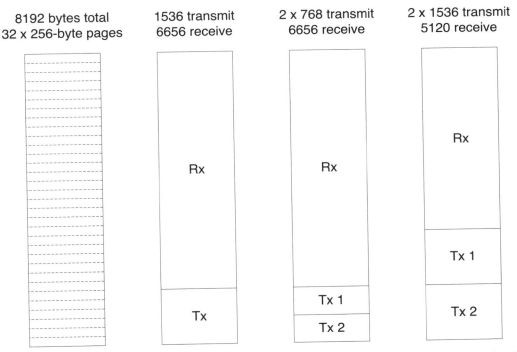

The 8Kb of NIC RAM is divided into 256-byte pages. Normally, the transmit buffer is dimensioned to accommodate the maximum Ethernet frame size (1.5KB), and the rest is used for the receive buffer. A second full-size transmit buffer can be accommodated if the receive buffer is reduced to 5,120 bytes; alternatively, two smaller 768-byte transmit buffers could be used without altering the size of the receive buffer.

The choice of buffer size is influenced by the following criteria:

- All buffers must be on the 256-byte page boundaries.
- The minimum size for any IP datagram is 576 bytes.
- Administrative network traffic tends to be in frames of a few hundred bytes or less.
- A microcontroller node wouldn't normally need to receive large amounts of data because there is nowhere to store it.
- It is easy to throttle the flow of incoming data by reducing the TCP window size.

For the current implementation, two transmit buffers of 1.5KB are used.

Modifications to Ethernet Driver

To support a secondary NIC buffer, minor modifications to the Ethernet driver are needed. First, the page-allocation definitions have to be changed. The transmit-buffer space is below the receive space, so the latter has to be shifted up.

```
/* NIC RAM definitions */
#define RAMPAGES 0x20          /* Total number of 256-byte RAM pages */
#define TXSTART  0x40          /* Tx buffer start page */
#define TXPAGES  6             /* Pages for one Tx buffer */
#define RXSTART  (TXSTART+(TXPAGES*NET_TXBUFFERS))  /* Rx buffer start page */
#define RXSTOP   (TXSTART+RAMPAGES-1)               /* Rx buffer end page */
```

When the transmit buffer is initialized, a parameter specifies which of the NIC buffers is to be used, and this value is stored for later use.

```
BYTE txbuffnum;               // Number of Tx buffer, if using more than 1

/* Initialize the transmit buffer
** If more than one Tx frame, select which */
void init_txbuff(BYTE buffnum)
{
    txbuffnum = buffnum;
    txin = net_txlen = txbuffin = 0;
    checkflag = checkhi = checklo = 0;
    setnic_tx(0);
}
```

The function that repositions the NIC transmit pointer needs to be modified to reference one of the two transmit buffers. It is assumed that only two buffers exist, in order to avoid using multiplication.

```
/* Set the 'remote DMA' address in the NIC Tx packet buffer */
void setnic_tx(WORD addr)
{                                       /* Add on Tx buffer offset */
    addr += (TXSTART << 8) + ETHERHDR_LEN;
    if (txbuffnum)                      /* ..additional offset if 2nd buffer */
        addr += (TXPAGES << 8);
    outnic(ISR, 0x40);                  /* Clear remote DMA interrupt flag */
    outnic(RSAR0, (BYTE)addr);          /* Remote DMA addr */
    outnic(RSAR1, (BYTE)(addr>>8));
    nic_tx_transfer = 1;
}
```

When transmitting, the frame length is saved for use in case the frame needs to be retransmitted. To test the new error handling, a facility has been introduced to deliberately drop transmit frames. If the DROP_TX equate is nonzero, then it determines the ratio of frames to be dropped, so a value of 3 means that one in three frames isn't transmitted.

```c
#define DROP_TX 0          /* If non-zero, drop 1 in n Tx packets for debug */

WORD txbuff_lens[NET_TXBUFFERS];          // Length of last buffer transmission

/* Transmit the Ethernet frame */
void transmit(void)
{
    WORD dlen;
#if DROP_TX
    static BYTE dropcount=0;
#endif
    dlen = net_txlen;
    save_txbuff();
    setnic_tx(-ETHERHDR_LEN);
    txbuffin = 0;
    put_data(host.eth.srce, MACLEN);      /* Destination addr */
    put_data(myeth, MACLEN);              /* Source addr */
    put_byte((BYTE)(host.eth.pcol>>8));   /* Protocol */
    put_byte((BYTE)host.eth.pcol);
    dlen += ETHERHDR_LEN;                 /* Bump up length for MAC header */
    putnic_txbuff();
    if (dlen < MINFRAME)
        dlen = MINFRAME;                  /* Constrain length */
    txbuff_lens[txbuffnum] = dlen;
    outnic(TPSR, txbuffnum ? (TXSTART+TXPAGES) : TXSTART);
    outnic(TBCR0, (BYTE)dlen);            /* Set Tx length regs */
    outnic(TBCR1, (BYTE)(dlen >> 8));
#if DROP_TX
    if (++dropcount == DROP_TX)
    {
        dropcount = 0;
        return;
    }
#endif
    outnic(CMDR, 0x24);                   /* Transmit the packet */
}
```

To retransmit a frame, specify only the buffer number; the length was stored when the frame was originally transmitted.

```
/* Retransmit the Ethernet frame */
void retransmit(BYTE buffnum)
{
    WORD dlen;

    outnic(TPSR, buffnum ? (TXSTART+TXPAGES) : TXSTART);
    dlen = txbuff_lens[buffnum];
    outnic(TBCR0, (BYTE)dlen);              /* Set Tx length regs */
    outnic(TBCR1, (BYTE)(dlen >> 8));
    outnic(CMDR, 0x24);                     /* Transmit the packet */
}
```

TCP Client Implementation

Socket Definitions

It's no surprise that a state machine is needed to control the client TCP stack. For economy, the ARP cycle can be incorporated as well, and some of the TCP closing states can be merged.

```
#define CLIENT_INIT      0       // Starting idle state
#define CLIENT_ARPS      1       // ARP sent, awaiting response
#define CLIENT_SYNS      2       // SYN sent, awaiting SYN ACK
#define CLIENT_EST       3       // Established
#define CLIENT_LASTACK   4       // Closed, awaiting final ACK
#define CLIENT_FINWT     5       // FIN sent, awaiting FIN ACK
```

Changes of state are handled by a function, so that the timeout tick count is refreshed, and the user LED states are adjusted to reflect the current state.

```
/* Change client state, refesh timer */
void new_cstate(BYTE news)
{
    timeout(&cticks, 0);
    ctries = CLIENT_TCP_TRIES;
    ctimeout = CLIENT_TCP_TIMEOUT;
    cstate = news;
    flashled1 = (cstate == CLIENT_ARPS);
    flashled2 = (cstate == CLIENT_EST);
}
```

The state variable and all the data specific to a client connection (i.e., the socket data) would normally be collected into a single structure so that an array of these structures can be used to hold the data for several sockets.

Unfortunately, at the time of writing, the compilers seemed to have problems coping with this, especially the LWORD structures nested within the socket data, so, regretfully, a single set of global variables have been used instead. I kept the naming of these variables similar so they can be reintroduced into a structure at a future date.

```
BYTE cstate=0;              // Client TCP state
WORD cticks;                // Tick count for timeout
WORD ctimeout;              // Timeout value
BYTE ctries;                // Number of retries left
LWORD ctseq;                // Transmit sequence number
LWORD ctack;                // Transmit acknowledgment number
LWORD chostip;              // Remote host IP address
BYTE chosteth[MACLEN];      // Remote host MAC address
```

Startup

If a client application wants to start up the TCP stack (i.e., establish a connection with a remote host), it calls start_client(). The remote port number is specified as an argument, but a global variable is used for the remote IP address, to economize on code space.

```
/* Start up the TCP client */
BOOL start_client(WORD hport)
{
    if (cstate == CLIENT_INIT)
    {
        init_ip(&chostip, MAILSERVER_ADDR);
        remport = hport;
        USERLED1 = USERLEDON;
        USERLED2 = USERLEDOFF;
        arp_host(&chostip);
        timeout(&cticks, 0);
        ctries = CLIENT_ARP_TRIES;
        new_cstate(CLIENT_ARPS);
        return(1);
    }
    return(0);
}
```

Only one socket is supported at present, so a check is made to ensure the socket is idle before starting the ARP cycle to locate the remote host. The ARP function uses the subnet mask to determine whether the remote node can be contacted directly or whether all communications must be passed through a router, and accordingly sends an ARP to the server or to the router. This procedure has already been explored when describing the UDP clients, so I will not discuss it further.

Receive Handler

The handler for received TCP segments has to be modified to redirect the client traffic to a specific handler. The original server code is retained to provide HTTP support.

```
/* Receive a TCP 'segment' */
SEPARATED BOOL tcp_recv(void)
{
    BYTE hlen, rflags, tflags=TACK;
    WORD addr;

    checkflag = 0;                                    // Clear checksum
    checkhi = checklo = 0;
    if (get_word(&remport) && get_word(&locport) && // Source & dest ports
        get_lword(&rseq) && get_lword(&rack) &&       // Seq & ack numbers
        get_byte(&hlen) && get_byte(&rflags) &&       // Header len & flags
        skip_data(6) && iplen>=TCPIPHDR_LEN &&        // Window, csum, urgent ptr
        locip.l==myip.l)                              // Addr match
    {
        check_word(iplen - IPHDR_LEN);                // Check pseudoheader
        check_lword(&myip);
        check_lword(&remip);
        check_byte(0);
        check_byte(PTCP);                             // Skip IP options
        while (hlen>TCPHDR_LEN*4 && iplen>0 && skip_lword())
        {
            iplen -= 4;
            hlen -= 16;
        }
        rpdlen = iplen - TCPIPHDR_LEN;
        addr = rxout;                                 // Check the Rx data
        skip_data(rpdlen);
        setpos_rxout(addr);
        if (checkhi==0xff && checklo==0xff)
        {
            init_txbuff(0);                           // Prepare Tx buffer
            setpos_txin(TCPIPHDR_LEN);
```

```
            d_checkhi = d_checklo = 0;
            print_lcd = print_serial = FALSE;
            print_net = TRUE;

            if (locport == cport)                   // Client port number?
            {
                tcp_client_recv(rflags);            // Call client handler
                rflags = tflags = 0;
            }
            . . . TCP server code is here . . .
            return(1);
        }
    }
    return(0);
}
```

State Machine

Once `start_client()` starts the state machine and issues the first ARP, the main trigger for any state transitions comes from incoming packet reception, so the bulk of the state machine can be included in the receive handler.

```
/* Receive a response to the TCP client */
SEPARATED void tcp_client_recv(BYTE rflags)
{
    BYTE tflags;

    rflags &= (TFIN+TSYN+TRST+TACK);       // Mask out PUSH & URG flags
    init_txbuff(1);                         // Set up Tx buffer..
    setpos_txin(TCPIPHDR_LEN);              // ..assuming default TCP+IP hdr
```

The transmit buffer pointer has been positioned after a notional TCP/IP header so that any put_() calls will write out TCP payload data.

The incoming TCP segment must be checked to ensure it comes from the current transaction. Once the connection has been established, only packets that match the current sequence (SEQ) and acknowledgment (ACK) values and that come from the correct host can be accepted.

```
    if (cstate<=CLIENT_SYNS || ((ctack.l==rseq.l) && (ctseq.l==rack.l) &&
        remip.l==chostip.l))
    {                                       // Check current TCP state
        switch (cstate)
        {
```

The action to be taken depends on the current state of the TCP client. If it is currently closed or sending ARPs, then any TCP segments are to be rejected.

```
case CLIENT_INIT:                          // Closed or sending ARPs?
case CLIENT_ARPS:
    client_rst_xmit();
    break;
```

If a SYN has been sent, then SYN ACK should have been received, and the incoming ACK value must match the outgoing SEQ value plus one (for the SYN). An ACK is sent to indicate the connection is established. This is also a good opportunity to clear down the client state machine (in this case, the email state machine) before entering the Established state.

If a SYN ACK isn't received, the packet is probably a stray from a past connection, so a RESET must be sent to kill off that old connection. Note the use of a different buffer (zero) for sending this. At all stages of the TCP negotiation, the last transmission may need to be resent, so I don't want to overwrite it with an unconnected TCP reset.

```
case CLIENT_SYNS:                          // SYN sent and SYN ACK value OK?
    if (rflags==TSYN+TACK && rack.l==CLIENT_SEQ_START+1)
    {
        rpdlen++;                          // Send ACK
        client_xmit(TACK);
        mstate = MAIL_INIT;                // ..init mail state machine
        new_cstate(CLIENT_EST);            // ..and go established
        USERLED2 = USERLEDON;
    }
    else
    {
                                           // SYN sent and not SYN ACK?
        init_txbuff(0);
        client_xmit(TRST+TACK);            // Send RESET
    }
    break;
```

The connection is established, so data can be exchanged. It is important that an application doesn't tie up the connection for too long, so the retry timer is only refreshed if data is received. If no data is received, the connection will eventually time out and reset. This isn't standard practice in TCP, where connections can remain open for hours without any data exchange, but it is essential in this application if the meager resources aren't going to be permanently tied up by a host that has crashed.

The client data handler returns the TCP flags it wants to send. If none are set, no TCP transmission is needed. If there is data to transmit, ACK will be set, and if the connection is to be closed, FIN will be set as well, and a state change is made.

If a FIN flag was received, then the remote server wants to close the connection, and this client accepts that by sending FIN ACK and changing state. Half-Close (the node continues to send data after receiving a FIN) is not included because very few applications require it.

```
            case CLIENT_EST:                    // Established
                if (rpdlen)
                    new_cstate(CLIENT_EST);     // Refresh timer
                if (remport == SMTP_PORT)        // SMTP client data?
#if INCLUDE_SMTP
                    tflags = smtp_client_data(rflags);
#else
                    client_rst_xmit();
#endif
                if (remport == POP3_PORT)        // POP3 client data?
#if INCLUDE_POP3
                    tflags = pop3_client_data(rflags);
#else
                    client_rst_xmit();
#endif
                if (rflags & TFIN)               // If remote close..
                {
                    tflags = TFIN+TACK;          // ..send FIN ACK
                    rpdlen++;
                    new_cstate(CLIENT_LASTACK);
                }
                else if (tflags & TFIN)          // If local close..
                    new_cstate(CLIENT_FINWT);    // ..wait for FIN ACK
                if (tflags)
                    client_xmit(tflags);         // If anything to send, do so
                break;
```

If a FIN has been received and FIN ACK has been sent, then it is necessary only to wait for a final ACK before going idle.

```
            case CLIENT_LASTACK:
                if (rflags == TACK)              // FIN ACK sent, waiting for ACK
                    new_cstate(CLIENT_INIT);     // ..if received, go idle
                break;
```

If a FIN has been sent, then a FIN ACK is required, which is a FIN flag plus an acknowledgment of the transmitted FIN. In a simultaneous close scenario, it is possible to receive a FIN but no acknowledgment of the outgoing FIN. In this case, the sequence-number checking at the head of this function will reject the TCP segment, and this stack will, after a timeout, resend its FIN to prompt the remote into sending a FIN ACK.

If the remote is half-Closed, it will continue to send data before sending a FIN, so this data is acknowledged and discarded.

```
      case CLIENT_FINWT:
          if (rflags & TFIN)                // FIN sent, waiting for FIN ACK
          {
              rpdlen++;
              client_xmit(TACK);            // ..if received, ACK and go idle
              new_cstate(CLIENT_INIT);
          }
          else if (rpdlen)                  // ..if only data, ACK it
              client_xmit(TACK);
          break;
      }
   }
}
```

Timeouts

A polling function checks whether the server has been ARPed and whether there are any timeouts in the TCP state machine.

If an ARP has been sent and a response received (or the server was already ARPed by a previous client), then the TCP stack can send a SYN to the server. A new client port is chosen from the available range of 1024 to 4999 to prevent a stray message from a previous negotiation from disrupting the new connection. The choice of these ports is a bit old-fashioned; modern systems generally use ports 49152 to 65535 instead.

Once the TCP stack is active, any timeout causes the last Ethernet frame to be resent, and the timeout value is doubled (1, 2, 4 seconds, etc.) until the maximum retry count is exceeded, in which case a TCP reset is sent.

```
#define CLIENT_ARP_TIMEOUT (2*SECTICKS)
#define CLIENT_ARP_TRIES    4
#define CLIENT_TCP_TIMEOUT (1*SECTICKS)
#define CLIENT_TCP_TRIES    5


#define MIN_CLIENT_PORT     1024
#define MAX_CLIENT_PORT     5000


/* Check the state of the TCP client */
void check_client(void)
```

```
{
    if (cstate == CLIENT_ARPS)                  // If sending ARPs
    {
        if (host_arped(&chostip))               // ..and response received
        {
            memcpy(chosteth, remeth, MACLEN);
            init_txbuff(1);                     // Choose next client port num
            if (++cport<MIN_CLIENT_PORT || cport>=MAX_CLIENT_PORT)
                cport = MIN_CLIENT_PORT;
            locport = cport;
            rseq.l = 0;
            rack.l = CLIENT_SEQ_START;           // Set starting SEQ number
            client_xmit(TSYN);                   // Send SYN
            new_cstate(CLIENT_SYNS);
        }
        else if (timeout(&cticks, CLIENT_ARP_TIMEOUT))
        {
            if (ctries--)                        // If ARP timeout..
                arp_host(&chostip);              // ..resend..
            else                                 // ..or stop if no more retries
            {
                stop_client();
                USERLED1 = USERLEDON;
            }
        }
    }
    else if (cstate>CLIENT_ARPS && timeout(&cticks, ctimeout))
    {
        if (ctries--)                            // If TCP timeout..
        {
            ctimeout += ctimeout;                // ..resend last packet
            retransmit(1);
        }
        else                                     // ..or stop if no more retries
        {
            if (cstate > CLIENT_SYNS)            // ..sending RESET if connected
            {
                init_txbuff(0);
                client_xmit(TRST+TACK);
            }
```

```
                stop_client();
                USERLED2 = USERLEDON;
            }
        }
    }
```

Transmit

By default, the transmit function assumes that all the incoming data is to be acknowledged, so the ACK value is the received SEQ value plus the data length (which has been adjusted for any SYN or FIN flags). The transmitted SEQ value does not reflect the amount of data sent (it indicates the *start* of the transmitted data block), but the saved value is adjusted for the transmitted data plus flags because it is used to validate any responses.

```
/* Transmit on the given socket */
void client_xmit(BYTE flags)
{
    WORD tpdlen=0;

    save_txbuff();
    add_lword(&rseq, rpdlen);           // My ACK = remote SEQ + datalen
    ctack.l = rseq.l;                   // Save my ACK & SEQ values
    ctseq.l = rack.l;
    if (net_txlen > TCPIPHDR_LEN)       // Get my TCP Tx data length
        tpdlen = net_txlen - TCPIPHDR_LEN;
    if (flags & (TSYN+TFIN))            // SYN/FIN flags count as 1 byte
        tpdlen++;
    add_lword(&ctseq, tpdlen);          // My next seq = SEQ + datalen
    tcp_xmit(flags);                    // Transmit the packet
}

/* Send a reset response */
void client_rst_xmit(void)
{
    init_txbuff(0);
    client_xmit(TRST+TACK);
}
```

SMTP Email Client

A good demonstration of the new TCP client software is to use it to send and receive emails using SMTP and POP3.

The distinction between the two is important: SMTP defines the protocol when your system has an email to send and contacts a mailserver to send it. Once the SMTP transfer is complete, your system can forget about the email. The mailserver will attempt to forward it to the correct recipient (or recipients) without any further intervention. In contrast, POP3 is used to scan a mailbox on a mailserver to see if any messages are available; it is essentially a read-only protocol, while SMTP (in normal usage) is a write-only protocol.

SMTP Transaction

If you intend to work with SMTP, I strongly recommend that you read the specification RFC (Request For Comment) 821. The Internet RFC specifications are written in an easy-to-read format and are readily available online from a wide variety of sources. Just enter "SMTP" or "rfc 821" into a search engine. There is little point in my reproducing large chunks of the specification here, so I will concentrate on the practical aspects of implementing the protocol.

The easiest way to get an appreciation of SMTP is to initiate a transaction and monitor the results. Here is a Microsoft Exchange client sending mail to an Internet Service Provider (ISP). The message is going from jpb@xyz.co.uk to jeremy@uvw.co.uk via the ISP abc.com.

The client initiates a TCP connection to port 25 on the server, and then the conversation proceeds as follows. All lines are terminated with a carriage return and line feed (<CR><LF>).

Once the TCP connection is established, the server introduces itself.

```
220 mta07-svc.abc.com ESMTP server... ready Mon, 7 Jan 2002 18:43:43 +0000
```

The line begins with a three-digit number, which allows the client to check the response without looking any further. A leading 2 indicates the server is happy, 3 is used for the data transfers (see later), and a value of 4 or 5 would indicate an error condition. The client then logs on using its Windows system name, which happens to be chips.

```
HELO chips
```

The server responds with an acknowledgment.

```
250 mta07-svc.abc.com
```

The client then indicates sender and recipient of the email, which is acknowledged.

```
MAIL FROM: <jpb@xyz.co.uk>
250 Sender <jpb@xyz.co.uk> Ok
RCPT TO: <jeremy@uvw.co.uk>
250 Recipient <jeremy@uvw.co.uk> Ok
```

The client initiates the transfer of email data, and the server acknowledges this.

```
DATA
354 Ok Send data ending with <CRLF>.<CRLF>
```

The email data is then sent down the TCP connection. The sender and recipient addresses have to be repeated, so they appear in the final text.

```
From: <jpb@xyz.co.uk>
To: <jeremy@uvw.co.uk>
```

The subject field will also appear in the final email.

```
Subject: Test message
```

As with an HTTP request from a Web client, there is copious additional information.

```
Date: Mon, 7 Jan 2002 18:43:53 -0000
Message-ID: <PFBBLMCGMLKNOKAJABMEEEAICCAA.jpb@xyz.co.uk>
MIME-Version: 1.0
Content-Type: text/plain;
        charset="iso-8859-1"
Content-Transfer-Encoding: 7bit
X-Priority: 3 (Normal)
X-MSMail-Priority: Normal
X-Mailer: Microsoft Outlook IMO, Build 9.0.2416 (9.0.2911.0)
Importance: Normal
X-MimeOLE: Produced By Microsoft MimeOLE V5.00.2314.1300
```

The header is terminated with a single blank line, and the email text begins.

```
This is a test

Jeremy Bentham
Xyz Ltd.
```

To terminate the data transfer, a single full stop appears at the start of an otherwise empty line. The server acknowledges receipt of the message, and the transaction is complete.

```
.
250 Message received: 20020107184344.PUJL327.mta07-svc.abc.com@chips
```

The client closes the SMTP session, and the server acknowledges.

```
QUIT
221 mta07-svc.abc.com ESMTP server closing connection
```

When the message is finally received (for example, by another Microsoft Outlook client), all the headers are stripped off for display.

```
From: jpb@xyz.co.uk
To: jeremy@uvw.co.uk
Subject: Test message

This is a test

Jeremy Bentham
Xyz Ltd.
```

By selecting message options, the headers on the received message can be viewed.

```
Return-Path: <jpb@xyz.co.uk>
Received: from punt-1.mail.def.net by mailstore for jeremy@uvw.co.uk
         id 1010429025:10:02865:1; Mon, 07 Jan 2002 18:43:45 GMT
Received: from mta07-svc.abc.com ([62.253.162.47]) by punt-1.mail.def.net
          id aa1121191; 7 Jan 2002 18:43 GMT
Received: from chips ([213.107.104.1]) by mta07-svc.abc.com
         (InterMail vM.4.01.03.23 201-229-121-123-20010418) with SMTP
         id <20020107184344.PUJL327.mta07-svc.abc.com@chips>
         for <jeremy@uvw.co.uk>; Mon, 7 Jan 2002 18:43:44 +0000
From: <jpb@xyz.co.uk>
```

```
To: <jeremy@uvw.co.uk>
Subject: Test message
Date: Mon, 7 Jan 2002 18:43:53 -0000
Message-ID: <PFBBLMCGMLKNOKAJABMEEEAICCAA.jpb@xyz.co.uk>
MIME-Version: 1.0
Content-Type: text/plain;
        charset="iso-8859-1"
Content-Transfer-Encoding: 7bit
X-Priority: 3 (Normal)
X-MSMail-Priority: Normal
X-Mailer: Microsoft Outlook IMO, Build 9.0.2416 (9.0.2911.0)
Importance: Normal
X-MimeOLE: Produced By Microsoft MimeOLE V5.00.2314.1300
```

At each hop between servers, the software that performs the transfer (the Mail Transfer Agent, or MTA) has added its own header to the front, but the bulk of the message remains unchanged. This is an important principle of email transmission. Most MTAs, including the widely used Sendmail application, have the ability to reformat the header information in remarkably complex ways, but it is best to avoid this by using a standard sequence of headers because a well-intentioned rewrite by the MTA could cause more problems than it solves.

A real-world example of this problem is an automated data logger that sends emails with the recipient field omitted (i.e., it provides the correct RCPT TO address in the SMTP negotiation but does not include a "To:" field in the SMTP data). The MTA will spot this omission and might ignore it, or it might attempt to fill in a replacement, which it derives from the SMTP transaction.

```
Apparently-To: <test@uvw.co.uk>
```

Alternatively, it might signal its concern at the absence of this information.

```
To: undisclosed-recipients:;
```

Although the emails will probably arrive at their intended destination (because of the SMTP recipient field), the missing information could be a headache for the systems administrator and could hamper your ability to sort incoming emails into the correct mailboxes.

SMTP Implementation

Because of the simple lock-step nature of the commands and responses, this protocol is relatively easy to integrate into the client TCP stack. A state machine is needed to track the sending of the commands.

```
#define MAIL_INIT      0
#define MAIL_HELO      1
#define MAIL_FROM      2
#define MAIL_TO        3
#define MAIL_DATA      4


BYTE mstate=MAIL_INIT;
```

For each command, the response is checked to make sure it has the correct leading character (2, or 3 if starting to send data) and an end-of-line character to ensure the complete response has been received.

```c
/* Handle incoming & outgoing SMTP data while connected
** Return the TCP flags to be transmitted (ACK and/or FIN) */
BYTE smtp_client_data(BYTE rflags)
{
    BYTE tflags=0;
    char c, e;

    NETWORK_OUTPUT;
    rx_checkoff = 1;
    if (get_byte(&c))
    {
        tflags = TACK;
        while (get_byte(&e) && e!='\n')
            ;
        if (e != '\n')
            tflags = 0;
        else if (c == '2')
        {
            if (mstate == MAIL_INIT)
                putstr("HELO chipweb\r\n");
            else if (mstate == MAIL_HELO)
                putstr("MAIL FROM: auto@chipweb\r\n");
            else if (mstate == MAIL_FROM)
                putstr("RCPT TO: test@localhost\r\n");
            else if (mstate == MAIL_TO)
                putstr("DATA\r\n");
            else if (mstate == MAIL_DATA+1)
                putstr("QUIT\r\n");
            else if (mstate == MAIL_DATA+2)
            {
                tflags = TFIN+TACK;
                USERLED1 = USERLEDOFF;
                USERLED2 = USERLEDOFF;
            }
            else
                mstate--;
```

```
            mstate++;
        }
        else if (c == '3')
        {
            if (mstate == MAIL_DATA)
            {
                putstr("From: \"ChipWeb server\" <auto@chipweb>\r\n");
                putstr("To: <test@localhost>\r\n");
                putstr("Subject: ChipWeb test\r\n");
                putstr("Content-type: text/plain\r\n\r\n");
                putstr("Test message\r\n");
                putstr(".\r\n");
                mstate++;
            }
        }
        else
        {
            tflags = TFIN+TACK;
            USERLED1 = USERLEDOFF;
            USERLED2 = USERLEDON;
        }
    }
    return(tflags);
}
```

In this case, the email is sufficiently short that it fits into a single Ethernet frame. If this were not the case, the data would be chopped into convenient chunks for transmission. The state machine would be extended to track the sending of these chunks, ensuring only one was in transit at a time. The TCP acknowledgments would be used for this because SMTP doesn't acknowledge data blocks, only the complete email data.

Initiating Email Transmission

One can imagine various ways that the email transfer could be initiated: for example, when an alarm condition is detected by the system. The microcontroller is capable of sending an email very quickly (under one second, even when using an ISP), so it is important to include some limits to ensure that it doesn't generate a flood of emails when there are a large number of alarm events.

For this demonstration, the push button on the prototype board is used. Every time it is pressed, an email is sent. The two user LEDs flicker briefly to indicate the transaction is successful. If either remains on, the transaction has failed. The first LED indicates the server couldn't be found; the second LED indicates the SMTP transaction failed.

To use the push button, it needs to be debounced, to guard against a single press causing multiple electrical transitions. A convenient place to do this is the timer tick handler. I'll also include code that flashes the user LEDs if a global variable is set.

```
BOOL flashled1, flashled2;      // Flags to enable flashing of user LEDs
BOOL button_iostate;            // Current I/O state of user button
BOOL button_down;               // Debounced state of user button
BOOL button_in;                 // State variable to detect button change


/* Check timer, scan ADCs, toggle LED if timeout */
void scan_io(void)
{
    restart_wdt();              // Kick watchdog
    if (geticks())             // Get tick count
    {                           // Debounce user push button
        if (USER_BUTTON == button_iostate)
            button_down = !button_iostate;
        else
            button_iostate = !button_iostate;
        if (tickcount & 1)
        {
            if (flashled1)      // Fast flash user LEDs
                USERLED1 = !USERLED1;
            if (flashled2)
                USERLED2 = !USERLED2;
        }
    }
    get_adcvals();              // Read ADC values
    if (timeout(&ledticks, LEDTIME))
        SYSLED = !SYSLED;       // Toggle system LED
}
```

The debounce logic checks that the button is in the same state for two clock ticks (2 × 50ms) before registering the new state. In the main program loop, the TCP client is polled and the button is checked to see whether it was just pressed; if so, the client TCP state machine is started. No further action is required because the SMTP state machine is driven from the TCP client polling.

```
#define SMTP_PORT 25

. . .
check_client();
if (button_down != button_in)
```

```
{
    button_in = button_down;
    if (button_down)
        start_client(SMTP_PORT);
}
. . .
```

POP3 Email Client

POP3 is used to receive emails. It allows you to list your emails on a mailserver, selecting which are to be downloaded or deleted. As with SMTP, you are encouraged to read the standard (RFC 1939). I will concentrate on practical implementation issues, rather than describing the protocol in detail.

POP3 Transaction

The following analysis shows the previous SMTP test email being retrieved from a POP3 server. As before, all lines are terminated with <CR><LF>, and the client commands are shown in bold.

As soon as the client makes a TCP connection to port 110, the server responds.

```
+OK POP3 server ready
```

Instead of the SMTP numeric response codes, POP3 uses a leading + or - to indicate success or failure.

The client must now log in with a name and password, so the server knows which set of emails (*maildrop*) is to be accessed. The mapping between the emails in the server's database and those displayed in a specific maildrop is highly dependent on the configuration of the email system. In the case of an ISP, I might log on as xyz if that is my company name, then receive emails for jeremy@xyz.co.uk, fred@xyz.co.uk, and so on. If the mailserver is on a local network, then I might log in as jeremy and only receive emails for jeremy@xyz.co.uk. If POP3 doesn't give you access to the emails you want, then this is an email server-configuration issue and has nothing to do with the POP3 protocol itself.

```
USER xyz
+OK
PASS secret
+OK
```

Having been authorized to access the server, the STAT command gives an on-line summary of the maildrop status.

```
STAT
+OK 1 1115
```

The format of this response is tightly defined. There should be only one space after the OK indication, then the total number of messages in the maildrop, then another space and the total size of the maildrop in bytes.

Each email is assigned a message number, starting at 1, for the duration of the POP3 transaction. Issuing the RETR command with a message number of 1 retrieves the first message.

```
RETR 1
+OK
Return-Path: <jpb@xyz.co.uk>
Received: from punt-1.mail.def.net by mailstore for jeremy@uvw.co.uk
         id 1010429025:10:02865:1; Mon, 07 Jan 2002 18:43:45 GMT
Received: from mta07-svc.abc.com ([...]) by punt-1.mail.def.net
          id aa1121191; 7 Jan 2002 18:43 GMT
Received: from chips ([...]) by mta07-svc.abc.com
         (InterMail vM.4.01.03.23 201-229-121-123-20010418) with SMTP
         id <20020107184344.PUJL327.mta07-svc.abc.com@chips>
         for <jeremy@uvw.co.uk>; Mon, 7 Jan 2002 18:43:44 +0000
From: <jpb@xyz.co.uk>
To: <jeremy@uvw.co.uk>
Subject: Test message
Date: Mon, 7 Jan 2002 18:43:53 -0000
Message-ID: <PFBBLMCGMLKNOKAJABMEEEAICCAA.jpb@xyz.co.uk>
MIME-Version: 1.0
Content-Type: text/plain;
         charset="iso-8859-1"
Content-Transfer-Encoding: 7bit
X-Priority: 3 (Normal)
X-MSMail-Priority: Normal
X-Mailer: Microsoft Outlook IMO, Build 9.0.2416 (9.0.2911.0)
Importance: Normal
X-MimeOLE: Produced By Microsoft MimeOLE V5.00.2314.1300

This is a test

Jeremy Bentham
Xyz Ltd.
.
```

As with the SMTP transfer, the whole message, including headers, is sent down the TCP connection, the end being indicated by a single full stop at the start of a blank line.

The email can be left on the server or marked for deletion if the client has received it successfully.

```
DELE 1
+OK
```

If the STAT command indicated more than one email was present, the RETR, DELE, or both commands would be repeated for the rest.

When the POP3 session is terminated any messages marked for deletion will be deleted.

```
QUIT
+OK
```

POP3 Line Buffering

I'll be honest: when I first considered a POP3 implementation for this book, I was sorely tempted to take the easy way out and just get the software to contact a mailserver and dump the entire content of all the emails out on the serial port. The much more difficult alternative is to display selected fields only (e.g., the sender and subject line) so they can be validated. This would be a lot more useful for embedded systems. The subject line could carry coded instructions for the system, in which case the sender's address must be checked to ensure these instructions come from the correct source. So why is this selective display so difficult?

- Support for string manipulation on the PICmicro is poor because string constants have to be stored in code space and then copied into data space before use.
- The emails can be arbitrarily large. It would be unacceptable for the system to malfunction because an overly large (multi-megabyte) email had been sent, perhaps maliciously.
- There is no correlation between the TCP data blocks received and the text lines contained in the email header. One text line might straddle the boundary between two TCP blocks.

The last issue brings into focus the implicit lock-step arrangement used in SMTP. For that protocol, it was quite acceptable to assume that each incoming TCP data block (*segment*) contained a single text line and that no line ever crossed the boundary between blocks. For POP3, it is unlikely that the subject or sender's address will straddle two segments, because most servers normally use an Ethernet block of maximum size, but it is still quite possible; for example, the server might run low on resources and send shorter segments, or the mail header might grow very large after passing through many MTAs.

To accommodate these chopped-up lines, it is necessary to keep a line buffer that can reassemble them. To keep memory usage low, you can restrict the maximum line length and truncate the line if it is any longer.

```
#define LBUFFLEN    80
BYTE lbuff[LBUFFLEN];
BYTE lbuff_in=0, lbuff_len=0;

/* Get a line of text into the line buffer, eliminate CR and LF characters
** If there is already a partial line in the buffer, append the new data
** If line is too long for buffer, truncate it */
BOOL get_lbuff(void)
{
    char c=0;

    while (get_byte(&c) && c!='\r')
    {
        if (lbuff_in<LBUFFLEN-1 && c!='\n')
        {
            lbuff[lbuff_in++] = c;
        }
    }
    if (c == '\r')
    {
        lbuff[lbuff_in] = 0;
```

```
        lbuff_len = lbuff_in;
        lbuff_in = 0;
        return(1);
    }
    return(0);
}
```

Having got a complete line in the buffer, it is possible to perform string comparisons, although these are slightly more awkward than I would wish, because all constants have to be copied into RAM before use.

```
char from_str[]="From:";
char subj_str[]="Subject:";


. . .

if (!strncmp(lbuff, from_str, sizeof(from_str)-1) ||
    !strncmp(lbuff, subj_str, sizeof(subj_str)-1))
{
    . . .
}
```

POP3 Implementation

The POP3 state machine is very similar to that used for SMTP.

```
#define POP_INIT    0
#define POP_USER    1
#define POP_PASS    2
#define POP_STAT    3
#define POP_RETR    4
#define POP_DATA    5
#define POP_QUIT    6

BYTE mstate=0;
```

The client code is structured around these states and the actions to be taken.

```
/* Handle incoming & outgoing SMTP data while connected
** Return the TCP flags to be transmitted (ACK and/or FIN) */
BYTE pop3_client_data(BYTE rflags)
{
    BYTE tflags=0;
    static BYTE count;
```

```
NET_OUTPUT;
rx_checkoff = 1;
if (rxleft())
    tflags = TACK;
if (get_lbuff())
{
    if (lbuff[0]!='+' && mstate!=POP_DATA)
    {
        putstr("QUIT\r\n");
        mstate = POP_QUIT;
    }
    else if (mstate == POP_RETR)
        mstate = POP_DATA;
    else if (mstate == POP_DATA)
    {
        do
        {
            if (lbuff[0]=='.' && lbuff_len==1)
            {
                put_ser("\r\n");
                PRINTF2("RETR %u\r\n", ++count);
                mstate = POP_RETR;
            }
            if (!strncmp(lbuff, from_str, sizeof(from_str)-1) ||
                !strncmp(lbuff, subj_str, sizeof(subj_str)-1))
            {
                ser_puts(lbuff);
                put_ser("\r\n");
            }
        } while (get_lbuff());
    }
    else if (mstate == POP_INIT)
    {
        putstr("USER test\r\n");
        mstate++;
    }
    else if (mstate == POP_USER)
    {
        putstr("PASS secret\r\n");
        mstate++;
```

```
        }
        else if (mstate == POP_PASS)
        {
            putstr("STAT\r\n");
            mstate++;
        }
        else if (mstate == POP_STAT)
        {
            count = atoi(&lbuff[4]);
            SERIAL_OUTPUT;
            PRINTF2("\r\n%u messages\r\n\r\n", count);
            NETWORK_OUTPUT;
            putstr("RETR 1\r\n");
            count = 1;
            mstate++;
        }
        else if (mstate == POP_QUIT)
        {
                tflags = TFIN+TACK;
                USERLED1 = USERLEDOFF;
                USERLED2 = USERLEDOFF;
        }
    }
    return(tflags);
}
```

Initiating Email Poll

Because the POP3 client makes use of the 9600-baud serial link for its display, it is initiated using a key press on the serial terminal. As with SMTP, a polling function in the main loop keeps the TCP client and POP3 state machine alive.

```
#define POP3_PORT 110
. . .
check_client();
if (kbhit())
{
    getch();
    start_client(POP3_PORT);
}
. . .
```

Pressing the space bar on the terminal causes the mailserver to be scanned for messages. If the server is set up to return the SMTP messages, a display similar to the following is obtained.

```
3 messages

From: "ChipWeb server" <auto@chipweb>
Subject: ChipWeb test

From: "ChipWeb server" <auto@chipweb>
Subject: ChipWeb test

From: "ChipWeb server" <auto@chipweb>
Subject: ChipWeb test
```

Summary

It's been a long haul, from creating a miniature Web server that ran over a SLIP link to the email clients running over Ethernet. Implementing a TCP/IP network node using a minimal amount of RAM has certainly been a challenging task. Although the 368 bytes in a PIC16F877 does preclude its use for some of the protocols I've described, the overwhelming 1.5Kb in the PIC18xxx series does provide an excellent platform for general-purpose networking.

In an age where a few hundred megabytes of RAM costs a few tens of dollars, it might seem strange discussing systems with a few hundred bytes. Why bother, when RAM is so cheap? The answer comes when you look at those hidden systems that drive the world — embedded in central heating controllers, microwave ovens, and automobiles. Small is not only beautiful, it is also more reliable and lower cost. Why use megabytes if a kilobyte can do the job better?

Whatever the future of networking, it's clear that TCP/IP will play an important role. As microcontrollers develop, they could become the most widely used network devices. Will they be running *your* software?

Source Files

The mail clients use the following files:

p16web.c	Main program
p16_drv.c	Low-level driver
p16_eth.c	Ethernet interface
p16_ip.c	IP
p16_lcd.c	LCD interface
p16_mail.c	SMTP and POP3 email clients
p16_tcp.c	TCP

`p16_tcpc.c`	TCP client interface
`p16_usr.c`	User interface and configuration

The following compiler-specific files are also required:

`ccs_p16.h`	CCS-specific PIC16 and PIC18 definitions
`ht_p16.h`	Hitech-specific PIC16 and PIC18 definitions
`ht_utils.c`	Hitech-specific utility functions

The following are default settings:

Mailserver	`10.1.1.1`
Netmask	`255.0.0.0`
Router	`10.1.1.100`

To change these values, edit the definitions at the top of the main program file.

Because the CCS compiler doesn't have a linker, all the source files are included in the main program file, so only that file needs to be submitted to the compiler.

Definitions at the top of the main program select what functionality is required so that the appropriate files are included. The following configuration was used for the time client testing.

```
/* TCP protocols: enabled if non-zero */
#define INCLUDE_HTTP 0       // Enable HTTP Web server
#define INCLUDE_SMTP 1       // Enable SMTP email client
#define INCLUDE_POP3 1       // Enable POP3 email client

/* UDP protocols: enabled if non-zero */
#define INCLUDE_DHCP 0       // Enable DHCP auto-configuration
#define INCLUDE_VID  0       // Video demo: 1 to enable, 2 if old video h/w
#define INCLUDE_TIME 0       // Time client: polls server routinely

/* Low-level drivers: enabled if non-zero */
#define INCLUDE_ETH  1       // Ethernet (not PPP) driver
#define INCLUDE_DIAL 0       // Dial out on modem (needs PPP)
#define INCLUDE_LCD  1       // Set non-zero to include LCD driver
#define INCLUDE_CFG  1       // Set non-zero to include IP & serial num config
```

Appendix A

Configuration Notes

Network Configuration

If you intend to experiment with any of the utility programs, you will need to set up a network and alter the default configuration file to suit. The configuration file is in ASCII text, which can be edited with any plain-text editor, such as Windows Notepad. Each line begins with a keyword, followed by optional arguments, delimited by spaces. Comment lines begin with any nonalphanumeric character, such as #. The types of network configurations are direct-driven network card, serial link, and packet driver.

Direct-Drive Network Card

The software has the capability to drive one or more network cards directly, and this is the preferred option for simplicity and ease of debugging and because it is supported by all the compilers. The network card must be a Novell NE2000–compatible, or 3COM 3C509 Etherlink III card. You will need to know the base I/O address: I use a value of 280h for the NE2000 and 300h for the 3C509. The configuration file entries are one of the following lines:

```
net ether ne 0x280
net ether 3c 0x300
```

It is important to check that the network card isn't already in use by the operating system. If you're using Windows, access the System settings in the Control Panel and check the Network-adapters category in the Device Manager. If the card is plug-and-play, the operating system might insist on installing it, whether you want to or not. If so, check the Control Panel settings for Network to make sure there are no protocol drivers attempting to use the card, or return to Network adapters in the Device Manager option of the System control panel, bring up the adapter's properties, and select the Device Usage checkbox that will "Disable in this hardware profile."

My ideal development system has two network cards, one of which is used by Windows and the other direct-driven by my software. This allows me to run a standard browser and my Web server all on the one machine.

Serial Link

The software supports a point-to-point serial link between two PCs as an alternative to a "proper" network. An alternative configuration file, `slip.cfg`, contains the following parameters:

```
net slip pc com2:38400,n,8,1
```

The baud rate of 38,400 represents a reasonable compromise between speed and simplicity of the driver software; only the COM port number may need to be changed. If the software is running under Win32, it automatically uses the Windows serial device driver. If it's running under DOS, it drives the COM port hardware directly. The SLIP drivers are not currently available when using the DJGPP compiler.

The simplest serial configuration is where both systems are running the TCPLEAN software, and it is strongly recommended that you first test the serial link in this way. A more complex configuration has one system running a standard Windows serial network interface, because there are a large number of settings that must be correct. Refer to "Windows SLIP Configuration" on page 520 in this appendix for more information.

When interconnecting two PC serial ports, a crossover, or null modem, cable is needed. This is obtainable from computer retailers.

Packet Driver

If you want to use an unsupported network card and you are using either of the Borland compilers, you can employ a "packet driver" interface between the card and my software. The packet driver (often known as a Crynwr driver) is supplied by the network card manufacturer and must be loaded before my software is run. The TCPLEAN configuration file should have the following entry:

```
net ether pktd
```

When starting, the TCPLEAN software will check for the presence of a packet driver; if several are present, it will choose the first one. Packet drivers offer the greatest flexibility and highest performance, but they are more complicated than the other options.

Addressing

Each computer on the network must have a unique address. Addressing is discussed in detail in Chapter 3, but for now, change the address field on the configuration file to be a unique value starting with 10.1.1.11 for Ethernet connections or 172.16.1.11 for serial connections; for example, the first Ethernet system would have in its configuration file

```
ip 10.1.1.11
```

and the next would have

```
ip 10.1.1.12
```

and so on.

Testing the Network

Once you have made the necessary configuration-file changes, it is worth checking the network. On the first system, enter the following lines

```
c:\>cd tcplean
c:\tcplean>ping
```

and it should report its IP address and that it is entering server mode.

```
PING v0.17
IP 10.1.1.11 mask 255.0.0.0 Ethernet 00:c0:26:b0:0a:93
Server mode - ESC or ctrl-C to exit
```

On the second system, enter

```
c:\>cd tcplean
c:\tcplean>ping 10.1.1.11
```

and if all is well, you should see a response every second until the escape key is pressed.

```
IP 10.1.1.12 mask 255.0.0.0 Ethernet 00:c0:26:b0:0b:56
Resolving 10.1.1.1 - ESC or ctrl-C to exit
ARP OK
Reply from 10.1.1.1 seq=1 len=32 OK
Reply from 10.1.1.1 seq=2 len=32 OK
Reply from 10.1.1.1 seq=3 len=32 OK

ICMP echo: 3 sent, 3 received, 0 errors
```

If the remote unit isn't responding, check for problems in the following places:

IP addresses.   Make sure that the two Ping utilities show the correct IP addresses, similar to those above.

Ethernet addresses.   If the Ethernet addresses are all zeros or all ones, the card settings do not match the configuration file, or there is a hardware clash.

Network indicators. The Ethernet card and/or hub usually has a light-emitting diode (LED) indicator that should flash every second when the Ping utility sends its message. If a collision indicator is illuminated, the network cabling is probably not installed correctly.

Operating system conflicts. The network card may be in use by the operating system.

Serial conflicts. Other software installations may be competing for the same serial port. If you have used a serial application in a DOS box, it will monopolize that serial device until the DOS box is closed, even though the application may no longer be running. Conversely, a DOS application may be denied use of a serial port, without an error message, if the port happens to be in use by a Windows application.

If problems persist, it is worth rebooting the systems into DOS without any network drivers loaded. You can then be sure the hardware is functional before investigating any Windows configuration problems.

Windows SLIP Configuration

If you are using a serial link and want to use the standard Windows TCP/IP utilities, you need to configure a Windows SLIP connection. A detailed discussion of Windows networking is outside the scope of this book, but the following steps should help.

1. In Control Panel, select Modems
2. Click Add and then No to the suggestion that Windows search for the hardware
3. Select Modem (to be added) and that you want to select from a list
4. Select Standard Modem Types - Standard 28800 bps modem
5. Select COM port for the SLIP connection; the modem should then be installed
6. Select the Dial-up networking folder in Accessories (or Accessories - Communications), and Make new connection
7. Enter a suitable name such as "SLIP connection"
8. Select the Standard 28800 bps modem; click Configure and check that the maximum speed is 38400 baud
9. Enter any digit for the phone number; click Finish
10. Edit the new connection by right-clicking it in the Dial-up networking folder and selecting Properties
11. Set the server type as SLIP - Unix Connection; click TCP/IP Settings ...
12. Select Specify an IP address; set to 172.16.1.2; deselect Use IP header compression; select Use default gateway on remote network; hit OK.
13. Select Scripting and enter `c:\tcplean\null.scp`; hit OK.

The file `null.scp` is a do-nothing script file to disable any login procedures.

```
;
; This dummy script file
; to establish a slip connection with a host.
;
proc main

endproc
```

For the connection to work with standard Windows TCP/IP software, the remote system must support modem emulation and have an IP address such as 172.16.1.1, which is in the same domain as the local IP address you entered in step 12 above.

To use this connection, double-click it in the Dial-up networking folder. If prompted, enter a dummy password; then click Connect. If all is well, the connection should be established to the remote system within a few seconds, and then all the standard Windows TCP/IP software (such as Web browsers) can be used simply by entering the remote IP address. To cancel the connection, right-click it in the Dial-up networking folder, and click Disconnect.

Appendix B

Resources

Publications

TCP/IP Illustrated Volume 1: The Protocols
W. Richard Stevens, Addison-Wesley, 1994
A comprehensive description of the TCP/IP protocols.

TCP/IP
Sidnie Feit, McGraw-Hill, 1998
Another good book on TCP/IP.

Webmaster in a Nutshell
Stephen Spainbour and Valerie Quercia, O'Reilly, 1996
An excellent quick-reference book on HTML and HTTP.

Microchip Technical Library CD-ROM
Microchip Technology Inc., http://www.microchip.com
Complete set of data sheets and application notes for PICmicro® microcontrollers and other devices.

LM75 Digital Temperature Sensor
Data sheet, National Semiconductor, http://www.national.com

PCF8583 Clock Calendar with 256 x 8-Bit Static RAM
Data sheet, Philips Semiconductors, `http://www-us.semiconductors.philips.com`

M24256 256Kbit Serial I2C Bus EEPROM
Data sheet, ST Microelectronics, `http://www.st.com`

Etherlink III … Adapter Drivers Technical Reference
3COM Corporation, `http://www.3com.com`

RFCs (requests for comment)
`http://www.faqs.org/rfcs`
The standardization documents for TCP/IP and Internet protocols.

RTL8019AS Ethernet Controller
Data sheet, Realtek Semiconductor Corp.
http://www.realtek.com.tw

Hardware

Elan104-NC single-board PC-compatible with PC104 and network interfaces
Arcom Control Systems, `http://www.arcomcontrols.com`
Single-board computer with DOS in ROM: an excellent test bed for the software in this book.
It also can be used with NETMON for tracing serial or Ethernet transactions.

PWEB board with signal conditioning
Io Ltd., Cambridge, U.K., `http://www.ioltd.co.uk`
Industrial PICmicro board, software-compatible with the author's design.

RICE17A PICmicro emulator, PIC-ICD PICmicro in-circuit debugger
Advanced Transdata Corp. Dallas, TX, `http://www.adv-transdata.com`
Excellent emulator that is fully compatible with the PCM compiler.

EMP20 Device Programmer
Needham's Electronics, `http://www.needhams.com`
Good parallel port device programmer. It can program the PICmicro microcontroller and
E^2ROM using personality modules.

RICE17A emulator and **EMP20 programmer**
Smart Communications, Borehamwood, U.K. (U.K. distributor)
`http://www.smartcom.co.uk`

LogicFlex single-board computer
JK Microsystems Inc., `http://www.jkmicro.com`
Excellent DOS-compatible SBC; development kit has a fully licensed version of Borland C++
v4.52.

PICmicro devices and development boards
Crownhill Associates, Ely, U.K., `http://www.crownhill.co.uk`
Supplier of compatible devices and development kits

Software

Ethereal protocol analyzer
Free network protocol analyzer for Linux and Windows, http://www.ethereal.com

PCM compiler for the PICmicro
Custom Computer Services, http://www.ccsinfo.com
Excellent low-cost compiler for the PICmicro, with support for on-chip peripherals.

PICC compiler for the PICmicro
Hitech Software, http://www.htsoft.com
ANSI-standard compiler for the PICmicro

DJGPP compiler v2.02 with RHIDE v1.4
Delorie Software, http://www.delorie.com
RHIDE is an excellent front end to the DJGPP development system and the GNU compiler and is very similar to the Borland IDE.

Borland C++ v3.1 and v4.52
Borland Inprise, http://www.borland.com
Older 16-bit versions are no longer available from Borland, but they are supplied with some C++ tutorial books. Borland C++ v4.52 is available from JK Microsystems Inc. See above.

Visual C++ v6
Microsoft Inc., http://www.microsoft.com
The standard 32-bit development system.

Visual SlickEdit
MicroEdge Inc., http://www.slickedit.com
A very good cross-platform text editor.

Updates and commercial licenses for the TCP/IP Lean software
Iosoft Ltd., Cambridge, U.K., http://www.iosoft.co.uk
The author's company!

DNS and DHCP server
JH Software, http://www.jhsoft.com
A Windows DNS and DHCP server, very useful for experimentation

Appendix C

Software on the CD-ROM

The enclosed CD-ROM contains complete source code for you, as purchaser of the book, to experiment with. However, the author retains full copyright to the software, and it may be distributed only in conjunction with the book. For example, you may not post any of the source code on the Internet or misrepresent its authorship by extracting fragments or altering the copyright notices. It is experimental software, and the author offers no warranties as to its fitness for purpose. Use it at your own risk or not at all.

The software on the CD-ROM may have been updated since publication. See the Iosoft Ltd. Web site (`http://www.iosoft.co.uk`) for bug reports and details of any upgrades.

If you want to sell anything that contains any part of the software, a license is required to cover commercial incorporation. Aside from the purchase of one book for each member of staff using the software, there are no additional development-licensing costs. The volume incorporation licensing charges are kept low to encourage commercial usage. See the Iosoft Ltd. Web site (`www.iosoft.co.uk`) or contact `license@iosoft.co.uk` for more information.

When experimenting with the software in this book, you are strongly advised to create a "scratch" network, completely isolated from any others. Relatively minor additions or changes to the software can result in very significant volumes of traffic, which may disrupt other devices on the network.

The following statement applies to all the software on the CD-ROM:

ARPSCAN

Utility	IP address scanner
Usage	`arpscan [options] [start_IP_address]`
	Enters server mode if no IP address given
Options	`-c name` Configuration filename (default `tcplean.cfg`)
	`-n count` Number of addresses to scan (default 20)
Example	`arpscan -n 10 10.1.1.1`
Keys	Ctrl-C or Esc to exit
Config	`net` to identify network type
	`ip` to identify IP address
Modes	Defaults to server mode (ARP responder) unless IP address given
Notes	Issues ARP requests for the given address range; displays responses
	In server mode, just responds to ARP requests

DATAGRAM

Utility	General-purpose UDP interface
Usage	`datagram [options] [IP_address [port [data]]]`
	Enters server mode if no IP address given
Options	`-b` Binary mode (hex data display)
	`-c name` Configuration filename (default `tcplean.cfg`)
	`-i name` Data input filename
	`-u` UDP datagram display mode
	`-v` Verbose display mode
	`-x` Hex packet display
Example	`datagram -v 10.1.1.1 echo "hello there"`
Keys	Ctrl-C or Esc to exit
Config	`net` to identify network type
	`ip` to identify IP address
	`mask` optional subnet mask
	`gate` optional gateway IP address
Modes	Defaults to server mode (echo and daytime) if no IP address
Notes	The port can be a number or one of the following:
	`echo`
	`daytime`
	`time`
	`snmp`

NETMON

Utility	Simple network monitor; displays all network traffic
Usage	`netmon [options]`
Options	`-c name` Configuration filename (default `tcplean.cfg`)
	`-t` Display TCP segments
	`-v` Verbose packet display
	`-x` Hexadecimal display of raw data
Example	`netmon -c slip.cfg`
Keys	Ctrl-C or Esc to exit
Config	`net` to identify network type (multiple entries allowed)
Notes	Can be overloaded by moderate network traffic

PICmicro® Software

There are two sets of PICmicro software on the CD: the original PWEB project from Chapters 9–11 and the newer ChipWeb in Chapters 12–16. Refer to the individual chapters for more information or to Appendix D for a discussion of PICmicro-specific issues.

PING

Utility	Emulation of standard Ping utility, with server capabilities	
Usage	`ping [options] [IP_address]`	
	Enters server mode if no IP address given	
Options	`-c name`	Configuration filename (default `tcplean.cfg`)
	`-f`	Flood mode
	`-l xxx`	Length of ICMP data in bytes (default 32)
	`-v`	Verbose display mode
	`-w xxx`	Waiting time in milliseconds (default 1000)
Example	`ping -c slip -v 172.16.1.1`	
Keys	Ctrl-C or Esc to stop pings	
Config	`net`	to identify network type
	`ip`	to identify IP address
	`mask`	optional subnet mask
	`gate`	optional gateway IP address
Modes	Defaults to server mode (Ping responder) unless IP address given	

ROUTER

Utility	Simple router for two or more networks	
Usage	`router [options]`	
Options	`-c name`	Configuration filename (default `router.cfg`)
	`-v`	Verbose display mode
Example	`router -v`	
Keys	Ctrl-C or Esc to exit	
Config	At least two networks must be defined with	
	`net`	to identify network type
	`ip`	to identify IP address
	`mask`	optional subnet mask
	`gate`	optional gateway IP address
Notes	The default config file is `router.cfg`, not `tcplean.cfg`	
	Networks can be a mixture of Ethernet and SLIP	

SCRATCHP

Utility	Test bed for a nonstandard protocol	
Usage	`scratchp [configfile]`	
	Reads `tcplean.cfg` from default directory if no file specified	
Options	None	
Example	`scratchp test.cfg`	
Interface	Single keypress with user prompts	
	[I]	Identify remote node
	[O]	Open connection to remote node
	[Q]	Quit
	When connected	
	[D]	Directory of remote
	[E]	Echo data test
	[G]	Get file from remote
	[P]	Put file into remote
Config	`net`	to identify network type
	`ident`	to identification string for node
Modes	Defaults to server mode unless otherwise directed	

TELNET

Utility	Simple emulation of standard telnet, with server capability	
Usage	`telnet [options] [IP_address [port]]`	
	Enters server mode (Echo, Daytime, and HTTP) if no IP address given	
Options	`-c name`	Configuration filename (default `tcplean.cfg`)
	`-s`	State display mode
	`-t`	TCP segment display mode
	`-v`	Verbose display mode
	`-x`	Hex packet display
Example	`telnet -v -c slip 172.16.1.1 daytime`	
Keys	Ctrl-C or Esc to exit	
	Keystrokes sent on network when connected to telnet port	
Config	`net`	to identify network type
	`ip`	to identify IP address
	`mask`	optional subnet mask
	`gate`	optional gateway IP address
Modes	Defaults to server mode (Echo, Daytime, HTTP) if no IP address	
Notes	The port may be a number or one of the following:	

```
echo
daytime
http
```

WEBROM

Utility	Creates ROM image of all the files in a directory
Usage	`webrom rom_file directory_name`
Example	`webrom test.rom c:\temp\romdocs`
Notes	Adds HTTP header for following extensions:

```
.htm
.egi
.txt
.gif
.xbm
```

WEBSERVE

Utility	Web server with file directory capability
Usage	`webserve [options] [directory]`
	Uses directory `.\webdocs` if none given
Options	`-c name` Configuration filename (default `tcplean.cfg`)
	`-s` State display mode
	`-t` TCP segment display mode
	`-v` Verbose display mode
	`-w` Web (HTTP) display mode
	`-x` Hex packet display
Example	`webserve c:\temp\webdocs`
Keys	Ctrl-C or Esc to exit
Config	`net` to identify network type
	`ip` to identify IP address
	`mask` optional subnet mask
	`gate` optional gateway IP address

WEB_EGI

Utility	Web server with embedded gateway interface (EGI)
Usage	`web_egi [options] [directory]`
	Uses directory `.\webdocs` if none given
Options	`-c name` Configuration filename (default `tcplean.cfg`)
	`-s` State display mode
	`-t` TCP segment display mode
	`-v` Verbose display mode
	`-w` Web (HTTP) display mode
	`-x` Hex packet display
Example	`web_egi -v -w c:\temp\webdocs`
Keys	Ctrl-C or Esc to exit
Config	`net` to identify network type
	`ip` to identify IP address
	`mask` optional subnet mask
	`gate` optional gateway IP address

Appendix D

PICmicro®-Specific Issues

Compiler Support

When I started PICmicro development, I happened to use the Custom Computer Services (CCS) PIC16 compiler, and the early software could be used only with this. As development progressed, I added compatibility with the Hitech PIC16 compiler. Through the judicial use of macros and conditional compilation, the same set of source files could be used for both compilers without any modifications.

As the software developed, I needed extra RAM and ROM for some protocols (or some combinations of protocols), so I adapted the code to be compatible with the PIC18 families. With yet more macros and conditionals, all four compiler/processor combinations could be supported from one set of source code.

Well, that was the theory. In practice, some problems remain with the different approach taken by the two PIC16 compilers, and also with the newness of the PIC18 development tools.

PIC16 Compilers

Aside from the cosmetic issues (such as the format of I/O pin definitions), there are some fundamental differences between the CCS and Hitech PIC16 compilers. Note that these restrictions apply to the current compiler versions at the time of writing. They may have been fixed by the time you read this.

- **Linking**   The CCS compiler has no linker, so all source files have to be included in the main file — a minor inconvenience.

- **Variable sizes**   The early CCS compilers had strange variable sizes (eight-bit integers and 16-bit longs) and no native support for 32-bit arithmetic. The version 3 compilers fixed these problems, although my software still bears the scars of this unfortunate legacy.

- **Dead code elimination**   The Hitech compiler doesn't eliminate unused code, so source files have to be pruned to remove unused functions.

- **RAM allocation**   If 16-bit data pointers are enabled, the CCS compiler will automatically position the data areas within the RAM space and insert the appropriate bank-switching code. The Hitech compiler will autoallocate variables only within the lower two RAM banks. For the rest, specific memory bank prefixes must be used, taking care not to include any interbank pointer references because they generate obscure error messages.

- **Stack usage**   The CCS compiler can merge a function into the calling code, saving one level on the eight-level hardware stack. This process happens automatically if the function is called only once, or it can be controlled manually through the use of the `inline` and `separate` directives.

- **Print output**   Quite a lot of formatted printing is sent to the serial port, LCD, and network, so it is important that this code is compact. CCSs provide a strange `printf()` variant that outputs the characters to a user-defined function, whereas with Hitech, you have to provide your own version of `putch()`. Having looked at the relative code penalties, a system of flags was created, so the print output can be redirected to any or all of the destinations.

- **Serial support**   The Hitech compiler doesn't provide built-in functions for RS232 and I^{2}C I/O, so these were created in a separate file.

- **Case insensitivity**   By default, the CCS compiler is insensitive to case. This feature (?) can be disabled in the latest compilers, but the software does not yet take advantage of this.

The RAM-allocation issue caused major problems. The TCP/IP stack used almost all the available RAM, and every time a new variable was needed, I would struggle to find room for it under the Hitech PIC16 compiler, to the extent that a significant amount of PIC16 code is only CCS compatible.

The stack usage also has a profound effect on the software structure. Under the CCS compiler, I used to write relatively large numbers of small functions (they're easier to document) and rely on the compiler "inlining" them to keep stack usage down. After switching to the Hitech compiler, I had to merge the code manually, resulting in larger functions.

PIC16 Definitions

Some rather strange macro definitions have been included for compatibility reasons, so I explain their purposes here.

Variable Sizes

To accommodate the different definitions of INT and LONG, new definitions have to be created.

- BOOL is one bit
- BYTE is unsigned eight bits
- WORD is unsigned 16 bits
- LWORD is a union of bytes, words, and a long

The LWORD is a hangover from the earlier CCS compilers that didn't support 32-bit arithmetic.

```
typedef union                  // Longword (not native for old CCS compiler)
{
    BYTE b[4];
    WORD w[2];
    unsigned INT32 l;
} LWORD_;
```

RAM Banks

For Hitech, explicit RAM-bank overrides have to be used on many variables.

```
#define BANK1 bank1
#define BANK2 bank2
#define BANK3 bank3
#define LOCATE(x, y)

. . .
BANK2 WORD concount;
```

For the CCS compiler, a single statement enables automatic RAM-bank handling, so the bank overrides do nothing.

```
#device *=16
#define BANK1
#define BANK2
#define BANK3
```

Very occasionally, the CCS allocation scheme fails to make optimum use of the memory space, and a large variable has to be located manually.

```
#define LOCATE(var, addr) #LOCATE var=addr

. . .
BYTE txbuff[TXBUFFLEN];    // Tx buffer
LOCATE(txbuff, 0x190)
```

Stack Usage

If a function is called only once, the CCS compiler will insert the code in-line. Sometimes, the overall code size is reduced if a function is forced to be separate, rather than in-line.

```
#define SEPARATED #separate
. . .
SEPARATED void tcp_client_recv(BYTE rflags);
```

I/O Definitions

Byte port definitions.

```
CCS:     #byte    PORTD = 8
Hitech:  already defined
```

Bit port definitions.

```
DEFBIT_2(PORTA, USERLED1)          // User LED 1, 0 when on
..translates to..
CCS:     #BIT USERLED1=PORTA.2
..or..
HITECH: static volatile bit USERLED1 @ (unsigned)&PORTA*8 + 3;
```

In case you wondered, the bit number can't be a macro parameter because of a problem with the CCS preprocessor macro substitution.

PIC18 Compilers

Changing the PIC16 code to the PIC18 is remarkably easy. The main change involves the I/O address map.

```
#define PORTA_ADDR 5           // I/O port A address for PIC16
..or..
#define PORTA_ADDR 0xF80       // I/O port A address for PIC18
```

The fuse settings are completely different.

```
__CONFIG(1, OSCSEN & HSPLL & UNPROTECT);
__CONFIG(2, BOREN & BORV42 & WDTDIS);
__CONFIG(3, 0);
__CONFIG(4, STVREN);
```

Configuration Fuse Settings

Before burning a program into a PICmicro device, it is important to check that the fuse settings have been propagated correctly to the device programmer. At the time of writing, the PIC18 software tools were relatively new, so it has proved impossible to make a thorough test of PIC18 functionality in time for publication of this book.

I use the following fuse settings for the PIC16:

Oscillator	HS
Watchdog	Disable
Powerup timer	Enable
Brown-out detect	Enable
Code protect	Off
Low-volt program	Disable
Code protect EE	Off
EECON write	Enable

The PIC18 settings differ slightly:

Oscillator	Hs
OSC enable	Disable
Watchdog	Disable
Code protect	Off
Powerup timer	Enable
Brown-out detect	Enable
CCP2 mux	I/O mux with RC1
STVREN	Cause reset

Incorrect fuse settings can cause a complete malfunction (if the PICmicro doesn't run) or subtle problems because I/O pins were reassigned for other purposes. If a program works using an emulator but fails when programmed into a device, the configuration settings are probably at fault.

The watchdog capability is very useful in real-world applications but has been disabled for test purposes. Before enabling it, check that any time-consuming processes include a regular call to refresh the watchdog timer.

For software updates and application notes, consult the Iosoft Ltd. Web site, www.iosoft.co.uk.

Function Index

Structure Index

Index

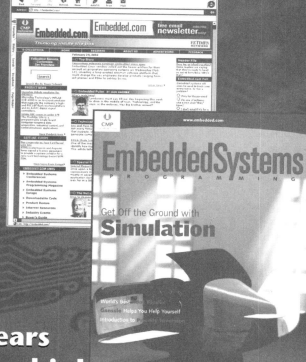

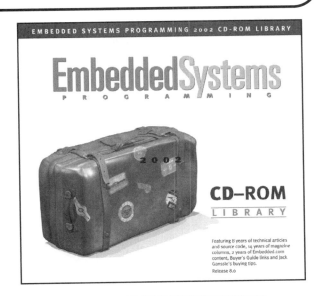

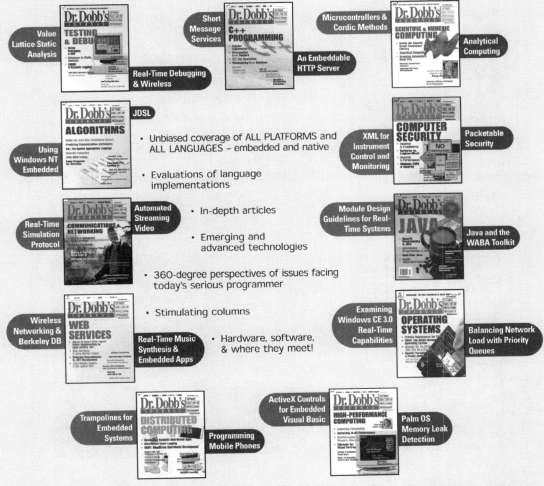

Designing with FPGAs & CPLDs

by Bob Zeidman

Choose the right programmable logic devices with this guide to the technologies and internal architectures. Master the important design, verification, synthesis, and testing issues as well as the different EDA tools available. Includes a Universal Design Methodology that facilitates the optimal allocation of resources and manpower. 220pp, ISBN 1-57820-112-8, $44.95

Microcontroller Projects Using the Basic Stamp

Second Edition

by Al Williams

Harness the power of the new BS2P Basic Stamp! Beginners find clear explanations of the principles needed to design hardware and program the Basic Stamp. Seasoned designers discover a fast and easy means of prototyping microcontrollers. Includes additional advanced projects, a Basic Stamp emulator, and a tool to program PIC chips. CD-ROM included, 456pp, ISBN 1-57820-101-2, $44.95

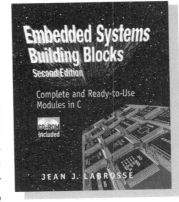

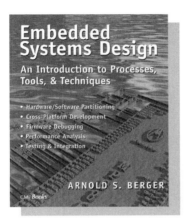

What's on the CD-ROM?

The CD-ROM that accompanies **TCP/IP Lean: Web Servers for Embedded Systems, Second edition,** contains complete source code to everything in this book so that you, as purchaser of the book, can experiment. However, the author retains full copyright to the software, and it may be distributed only in conjunction with the book.

CD-ROM TCPLEAN Subdirectories

BC31	Borland C v3.1 files
BC45	Borland C v4.5 files
DJGPP	DJGPP (GNU compiler plus RHIDE) files
INFO	Application notes
PCM	C and project files for Chapters 9–11
ROMDOCS	Web pages for the PWEB server
SOURCE	Source code for all PC projects
VC6	Visual C v6 files
WEBDOCS	Web pages for the WEBSERVE and WEB_EGI utilities

CD-ROM CHIPWEB Subdirectories

P16WEB	Source files for Chapters 12–16
PCM	Original source files for Chapter 12

The Development Environment

The following four PC compilers are supported.

Borland C++ v3.1.   An excellent DOS-hosted compiler with an integrated-development environment.

Borland (Inprise) C++ v4.52.   Windows-hosted compiler, which seems to be the latest version that can generate executable files for DOS.

Microsoft Visual C++ v6.   Windows-hosted compiler that can generate Win32 console applications.

DJGPP v2.02 with RHIDE v1.4.   Part of the GNU project, this is a remarkably good clone of the Borland 3.1 development environment, which runs in a 32-bit extended DOS environment and can be downloaded free of charge.

The Borland compilers, though ostensibly obsolete, may be found on the CD-ROM of some C programming tutorial books or purchased from the supplier mentioned in Appendix B. Please note that not all compilers support all network interfaces — see the book for details.

For PICmicro® development, the CCS and Hitech compilers are supported, for both the PIC16xxx and PIC18xxx families (see Appendix D for more information).

For more information about installation, running and rebuilding the utilities, the packet driver, the PICmicro Web server, licensing, and so on, see the Readme.txt file on the CD-ROM and the Preface. Software updates are available from the Iosoft Ltd. Web site, www.iosoft.co.uk

---